Nuclear Back-end and Transmutation Technology for Waste Disposal

Ken Nakajima
Editor

Nuclear Back-end and Transmutation Technology for Waste Disposal

Beyond the Fukushima Accident

Editor
Ken Nakajima
Research Reactor Institute
Kyoto University
Sennan-gun, Osaka
Japan

ISBN 978-4-431-55110-2   ISBN 978-4-431-55111-9 (eBook)
DOI 10.1007/978-4-431-55111-9
Springer Tokyo Heidelberg New York Dordrecht London

Library of Congress Control Number: 2014949895

© The Editor(s) (if applicable) and the Author(s) 2015. The book is published with open access at SpringerLink.com
Open Access. This book is distributed under the terms of the Creative Commons Attribution Noncommercial License which permits any noncommercial use, distribution, and reproduction in any medium, provided the original author(s) and source are credited.
All commercial rights are reserved by the Publisher, whether the whole or part of the material is concerned, specifically the rights of translation, reprinting, re-use of illustrations, recitation, broadcasting, reproduction on microfilms or in any other way, and storage in data banks. Duplication of this publication or parts thereof is permitted only under the provisions of the Copyright Law of the Publisher's location, in its current version, and permission for commercial use must always be obtained from Springer. Permissions for commercial use may be obtained through RightsLink at the Copyright Clearance Center. Violations are liable to prosecution under the respective Copyright Law.
The use of general descriptive names, registered names, trademarks, service marks, etc. in this publication does not imply, even in the absence of a specific statement, that such names are exempt from the relevant protective laws and regulations and therefore free for general use.
While the advice and information in this book are believed to be true and accurate at the date of publication, neither the authors nor the editors nor the publisher can accept any legal responsibility for any errors or omissions that may be made. The publisher makes no warranty, express or implied, with respect to the material contained herein.

Printed on acid-free paper

Springer is part of Springer Science+Business Media (www.springer.com)

Foreword

On March 11, 2011, a massive earthquake and the resultant tsunami struck the Tohoku area of Japan, causing serious damage to the Fukushima Daiichi Nuclear Power Plant (NPP) and the release of a significant quantity of radionuclides into the surrounding environment. This accident underlined the necessity of establishing more comprehensive scientific research for promoting safety in nuclear technology. In this situation, the Kyoto University Research Reactor Institute (KURRI) established a new research program called the "KUR Research Program for Scientific Basis of Nuclear Safety" in 2012.

Nuclear safety study includes not only the prevention of nuclear accidents but also the safety measures after the accident from a wider point of view ensuring the safety of residents. A long time is needed for the improvement of the situation, but the social needs for the reinforcement of nuclear safety increases rapidly. The advancement of disaster prevention technology for natural disasters such as earthquakes and tsunamis, the reinforcement of measures to counter the effects of accidents, and the reinforcement of the safety management of spent fuels and radioactive wastes are demanded, not to mention the reinforcement of nuclear reactor safety. Also required are the underlying mechanism investigation and accurate assessment for the effect of radiation on the human body and life. As with all such premises, detailed inspection and analysis of the accident are indispensable.

In the Research Program for the Scientific Basis of Nuclear Safety, an annual series of international symposia was planned along with specific research activities. The first in the series of symposia, entitled "The International Symposium on Environmental Monitoring and Dose Estimation of Residents after Accident of TEPCO's Fukushima Daiichi Nuclear Power Stations", was held on December 14, 2012, concerning the radiological effects of the accident on the public, and covering a wide range of monitoring and dose assessment activities after the accident. Although the proceedings of the symposium had been published, a more comprehensive and conclusive book was published with open access at the requests of many people including residents near the accident site.

Following the first one, the second annual symposium in this series was held on November 28, 2013, dealing with nuclear back-end issues and the role of nuclear transmutation technology after the accident at TEPCO's Fukushima Daiichi NPP. The accident has called upon us to focus our attention on the large amount of spent nuclear fuels stored in NPPs as well as on the impacts of the accident. In fact, public anxiety regarding the treatment and disposal of high-level radioactive wastes which require long-term control is now growing, while the government policy on the back-end of the nuclear fuel cycle is unpredictable in the aftermath of the accident. The issues are not simply technical, they are critically important not only for dealing with the accident but also for pursuing nuclear energy production in the world.

This publication summarizes the current status of the back-end issues and of research and development on nuclear transmutation technology for radioactive waste management. It is expected to contribute to better understanding and further discussion of the issues.

On behalf of KURRI, I wish to thank all the contributors to this book as well as the reviewers. KURRI hopes that this publication will promote further progress in nuclear safety research and will contribute to the reduction of public anxiety after the accident.

Kyoto University Research Reactor Institute   Hirotake Moriyama
Kyoto, Japan

Preface

The accident at the Fukushima Daiichi Nuclear Power Plant (NPP), which occurred on March 11, 2011, has caused us to focus our attention on a large amount of spent nuclear fuels stored in NPPs. In addition, public anxiety regarding the treatment and disposal of high-level radioactive wastes that require long-term control is growing. The Japanese policy on the back-end of the nuclear fuel cycle is still unpredictable in the aftermath of the accident; moreover, these back-end issues are inevitable as long as nuclear energy is used. Therefore, research and development for enhancing the safety of various processes involved in nuclear energy production is being actively pursued worldwide. In particular, the nuclear transmutation technology—employed for reducing the toxicity of highly radioactive wastes—has been drawing significant attention.

In the KUR Research Program for Scientific Basis of Nuclear Safety, the Kyoto University Research Reactor Institute organized an international symposium "Nuclear back-end issues and the role of nuclear transmutation technology after the accident of TEPCO's Fukushima Daiichi Nuclear Power Stations" in November 28th, 2013. Under such circumstances in which nuclear back-end issues and the role of nuclear transmutation technology after the accident at TEPCO's Fukushima Daiichi NPP is gaining greater concern, this timely publication highlights the following topics: (1) Development of accelerator-driven systems (ADS), which is a brand-new reactor concept for transmutation of highly radioactive wastes; (2) Nuclear reactor systems from the point of view of the nuclear fuel cycle. How to reduce nuclear wastes or how to treat them including the debris from TEPCO's Fukushima nuclear power stations is discussed; and (3) Environmental radioactivity, radioactive waste treatment, and geological disposal policy.

State-of-the-art technologies for overall back-end issues of the nuclear fuel cycle as well as the technologies of transmutation are presented here. The chapter authors are actively involved in the development of ADSs and transmutation-related technologies. The future of the back-end issues is very uncertain after the accident in Fukushima Daiichi NPP, and this book provides an opportunity for readers to consider the future direction of those issues.

Sennan-gun, Osaka, Japan   Ken Nakajima

Cooperators

Jun-ichi Katakura	Nagaoka University of Technology, Nagaoka, Japan
Tomohiko Iwasaki	Tohoku University, Sendai, Japan
Akio Yamamoto	Nagoya University, Nagoya, Japan
Kengo Hashimoto	Kinki University, Osaka, Japan
Ken Kurosaki	Osaka University, Osaka, Japan
Kazufumi Tsujimoto	Japan Atomic Energy Agency, Ibaraki, Japan
Kazuya Idemitsu	Kyusyu University, Fukuoka, Japan
Yoshihiro Ishi	Research Reactor Institute, Kyoto University, Osaka, Japan
Yasushi Saito	Research Reactor Institute, Kyoto University, Osaka, Japan
Cheol Ho Pyeon	Research Reactor Institute, Kyoto University, Osaka, Japan
Toshiyuki Fujii	Research Reactor Institute, Kyoto University, Osaka, Japan
Toshihiro Yamamoto	Research Reactor Institute, Kyoto University, Osaka, Japan
Joonhong Anh	University of California, Berkeley, CA, USA
Hamid Aït Abderrahim	Studiecentrum voor Kernenergie—Centre d'étude de l'Energie Nucléaire, Mol, Belgium
Thierry Dujardin	Organisation for Economic Co-operation and Development/Nuclear Energy Agency, Paris France

Contents

Part I Basic Research for Nuclear Transmutation and Disposal: Physical and Chemical Studies Relevant to Nuclear Transmutation and Disposal Such as Measurement or Evaluation of Nuclear Cross-Section Data

1   Nuclear Transmutation of Long-Lived Nuclides with Laser Compton Scattering: Quantitative Analysis by Theoretical Approach.. 3
    Shizuka Takai and Kouichi Hagino

2   Recent Progress in Research and Development in Neutron Resonance Densitometry (NRD) for Quantification of Nuclear Materials in Particle-Like Debris........................ 13
    M. Koizumi, F. Kitatani, H. Tsuchiya, H. Harada, J. Takamine, M. Kureta, H. Iimura, M. Seya, B. Becker, S. Kopecky, W. Mondelaers, and P. Schillebeeckx

3   Development of Nondestructive Assay to Fuel Debris of Fukushima Daiichi NPP (1): Experimental Validation for the Application of a Self-Indication Method.............. 21
    Jun-ichi Hori, Tadafumi Sano, Yoshiyuki Takahashi, Hironobu Unesaki, and Ken Nakajima

4   Development of Nondestructive Assay of Fuel Debris of Fukushima Daiichi NPP (2): Numerical Validation for the Application of a Self-Indication Method.............. 31
    Tadafumi Sano, Jun-ichi Hori, Yoshiyuki Takahashi, Hironobu Unesaki, and Ken Nakajima

5   Precise Measurements of Neutron Capture Cross Sections for LLFPs and MAs............................ 39
    S. Nakamura, A. Kimura, M. Ohta, T. Fujii, S. Fukutani, K. Furutaka, S. Goko, H. Harada, K. Hirose, J. Hori,

M. Igashira, T. Kamiyama, T. Katabuchi, T. Kin, K. Kino,
F. Kitatani, Y. Kiyanagi, M. Koizumi, M. Mizumoto, M. Oshima,
K. Takamiya, Y. Toh, and H. Yamana

6 Development of the Method to Assay Barely Measurable
   Elements in Spent Nuclear Fuel and Application
   to BWR 9×9 Fuel.................................... 47
   Kenya Suyama, Gunzo Uchiyama, Hiroyuki Fukaya,
   Miki Umeda, Toru Yamamoto, and Motomu Suzuki

Part II Development of ADS Technologies: Current Status of Accelerator-Driven System Development

7 Contribution of the European Commission to a European
   Strategy for HLW Management Through Partitioning
   & Transmutation: Presentation of MYRRHA and Its Role
   in the European P&T Strategy............................. 59
   Hamid Aït Abderrahim

8 Design of J-PARC Transmutation Experimental Facility........ 73
   Toshinobu Sasa

9 Accelerator-Driven System (ADS) Study in Kyoto University
   Research Reactor Institute (KURRI)........................ 81
   Cheol Ho Pyeon

Part III Mechanical and Material Technologies for ADS: Development of Mechanical Engineering or Material Engineering-Related Technologies for ADS and Other Advanced Reactor Systems

10 Heat Transfer Study for ADS Solid Target: Surface
   Wettability and Its Effect on a Boiling Heat Transfer........... 95
   Daisuke Ito, Kazuki Hase, and Yasushi Saito

11 Experimental Study of Flow Structure and Turbulent
   Characteristics in Lead–Bismuth Two-Phase Flow.............. 107
   Gen Ariyoshi, Daisuke Ito, and Yasushi Saito

Part IV Basic Research on Reactor Physics of ADS: Basic Theoretical Studies for Reactor Physics in ADS

12 Theory of Power Spectral Density and Feynman-Alpha
   Method in Accelerator-Driven System and Their
   Higher-Order Mode Effects................................ 119
   Toshihiro Yamamoto

13	Study on Neutron Spectrum of Pulsed Neutron Reactor .... Takanori Kitada, Thanh Mai Vu, and Noboru Dobuchi	129

Part V Next-Generation Reactor Systems: Development of New Reactor Concepts of LWR or FBR for the Next-Generation Nuclear Fuel Cycle

14	Application of the Resource-Renewable Boiling Water Reactor for TRU Management and Long-Term Energy Supply .... Tetsushi Hino, Masaya Ohtsuka, Renzo Takeda, Junichi Miwa, and Kumiaki Moriya	141
15	Development of Uranium-Free TRU Metallic Fuel Fast Reactor Core .... Kyoko Ishii, Mitsuaki Yamaoka, Yasuyuki Moriki, Takashi Oomori, Yasushi Tsuboi, Kazuo Arie, and Masatoshi Kawashima	155
16	Enhancement of Transmutation of Minor Actinides by Hydride Target .... Kenji Konashi and Tsugio Yokoyama	169
17	Method Development for Calculating Minor Actinide Transmutation in a Fast Reactor .... Toshikazu Takeda, Koji Fujimura, and Ryota Yamada	179
18	Overview of European Experience with Thorium Fuels .... Didier Haas, M. Hugon, and M. Verwerft	197

Part VI Reactor Physics Studies for Post-Fukushima Accident Nuclear Energy: Studies from the Reactor Physics Aspect for Back-End Issues Such as Treatment of Debris from the Fukushima Accident

19	Transmutation Scenarios after Closing Nuclear Power Plants .... Kenji Nishihara, Kazufumi Tsujimoto, and Hiroyuki Oigawa	207
20	Sensitivity Analyses of Initial Compositions and Cross Sections for Activation Products of In-Core Structure Materials .... Kento Yamamoto, Keisuke Okumura, Kensuke Kojima, and Tsutomu Okamoto	233
21	Options of Principles of Fuel Debris Criticality Control in Fukushima Daiichi Reactors .... Kotaro Tonoike, Hiroki Sono, Miki Umeda, Yuichi Yamane, Teruhiko Kugo, and Kenya Suyama	251

22 Modification of the STACY Critical Facility for Experimental
    Study on Fuel Debris Criticality Control 261
    Hiroki Sono, Kotaro Tonoike, Kazuhiko Izawa,
    Takashi Kida, Fuyumi Kobayashi, Masato Sumiya,
    Hiroyuki Fukaya, Miki Umeda, Kazuhiko Ogawa,
    and Yoshinori Miyoshi

Part VII Nuclear Fuel Cycle Policy and Technologies: National Policy,
          Current Status, Future Prospects and Public Acceptance
          of the Nuclear Fuel Cycle Including Geological Disposal

23 Expectation for Nuclear Transmutation 271
    Akito Arima

24 Issues of HLW Disposal in Japan 279
    Kenji Yamaji

25 Considering the Geological Disposal Program of High-Level
    Radioactive Waste Through Classroom Debate 289
    Akemi Yoshida

Part VIII Environmental Radioactivity: Development of Radioactivity
           Measurement Methods and Activity of Radionuclides in the
           Environment Monitored After the Accidents at TEPCO's
           Nuclear Power Stations

26 Environmental Transfer of Carbon-14 in Japanese
    Paddy Fields 303
    Nobuyoshi Ishii, Shinichi Ogiyama, Shinji Sakurai,
    Keiko Tagami, and Shigeo Uchida

27 Development of a Rapid Analytical Method for
    ^{129}I in the Contaminated Water and Tree Samples
    at the Fukushima Daiichi Nuclear Power Station 311
    Asako Shimada, Mayumi Ozawa, Yutaka Kameo,
    Takuyo Yasumatsu, Koji Nebashi, Takuya Niiyama,
    Shuhei Seki, Masatoshi Kajio, and Kuniaki Takahashi

Part IX Treatment of Radioactive Waste: Reduction
         of the Radioactivity or Volume of Nuclear Wastes

28 Consideration of Treatment and Disposal of Secondary
    Wastes Generated from Treatment of Contaminated Water 321
    Hiromi Tanabe and Kuniyoshi Hoshino

29 **Volume Reduction of Municipal Solid Wastes Contaminated with Radioactive Cesium by Ferrocyanide Coprecipitation Technique** 329
Yoko Fujikawa, Hiroaki Ozaki, Hiroshi Tsuno,
Pengfei Wei, Aiichiro Fujinaga, Ryouhei Takanami,
Shogo Taniguchi, Shojiro Kimura,
Rabindra Raj Giri, and Paul Lewtas

**Part I
Basic Research for Nuclear Transmutation
and Disposal: Physical and Chemical
Studies Relevant to Nuclear Transmutation
and Disposal Such as Measurement or
Evaluation of Nuclear Cross-Section Data**

Chapter 1
Nuclear Transmutation of Long-Lived Nuclides with Laser Compton Scattering: Quantitative Analysis by Theoretical Approach

Shizuka Takai and Kouichi Hagino

Abstract A photo-neutron (γ, n) reaction with laser Compton scattering γ-rays has been suggested to be effective for the nuclear transmutations of fission products. The photo-neutron reaction occurs via a giant dipole resonance, which has a large cross section and whose properties are smooth functions of mass number. The laser Compton scattering can generate effectively and selectively high-energy photons with a desired energy range. In this chapter, we investigate quantitatively the effectiveness of the transmutation with laser Compton scattering based on the Hauser–Feshbach theory using the TALYS code. We carry out simulations for high-decay heating nuclide ^{137}Cs, in which the cross sections for ^{137}Cs (γ, γ), (γ, n), and $(\gamma, 2n)$ reactions, and the total photonuclear reaction cross sections versus incident photon energy, are calculated. The incident photon energy obtained by laser Compton scattering is also optimized. It is shown that the transmutation with medium-energy photon with a flux of more than 10^{18}/s effectively reduces the radioactivity of the target ^{137}Cs.

Keywords ^{137}Cs • Giant dipole resonance • Laser Compton scattering • Photo-neutron reaction • Radioactive wastes • Transmutation

1.1 Introduction

One of the major problems of the nuclear fuel cycle is the disposal of high-level radioactive waste that contains long-lived nuclides such as ^{129}I and high-decay heating nuclides such as ^{137}Cs. After the severe accident at the Fukushima Daiichi

---

S. Takai (✉)
Nuclear Safety Research Center, Japan Atomic Energy Agency, Tokai-mura, Naka-gun, Ibaraki 319-1115, Japan
e-mail: takai.shizuka@jaea.go.jp

K. Hagino
Department of Physics, Tohoku University, Sendai, Miyagi 980-8578, Japan

Nuclear Power Plant, there is also a problem of ^{137}Cs having been concentrated by treatment of contaminated water. Transmuting such nuclides into short-lived or stable nuclides is one possible way to resolve this problem. Neutron capture reactions have been proposed for transmutations of such fission products. However, the neutron capture cross sections differ significantly from nuclide to nuclide, and this transmutation method is not effective for nuclides with small neutron capture cross sections such as ^{137}Cs.

Recently, photo-neutron (γ, n) reactions with laser Compton scattering γ-rays have been suggested as an alternative method for nuclear transmutations [1, 2]. Figure 1.1 shows a schematic illustration of this transmutation. This transmutation uses γ-rays generated by laser photons backscattered off GeV electrons and photo-nuclear reactions via electric giant dipole resonance (GDR) [3], which has a large cross section for most nuclides. The GDR is a collective excitation of a nucleus that decays mainly by the emission of neutrons, and its total cross section is a smooth function of mass number. Therefore, this method is expected to be effective for transmuting fission products regardless of isotopes.

So far, transmutation with laser Compton scattering for some nuclides has been evaluated only in a simple manner. In this chapter, we investigate more quantitatively the effectiveness of the transmutation with laser Compton scattering, especially for ^{137}Cs.

1.2 Calculation Method

1.2.1 Reaction via Giant Dipole Resonance

Nuclear transmutation with laser Compton scattering uses photonuclear reactions via GDR because the cross section of GDR is quite large and the total cross section is a smooth function of mass number. GDR is a collective excitation of a nucleus involving almost all nucleons, which is interpreted classically as a macroscopic oscillation of a bulk of protons against that of neutrons. The total cross section $\sigma_{\text{GDR}}^{\text{tot}}$, the resonance energy E_R, and the width Γ_R are given by [4]

$$\begin{aligned}
\sigma_{\text{GDR}}^{\text{tot}} &= 60(1+\kappa)\frac{NZ}{A} \text{ mb} \cdot \text{MeV}, \\
E_R &= 31.2\,A^{-1/3} + 20.6\,A^{-1/6}\,\text{MeV}, \\
\Gamma_R &= 0.0026 E_R^{1.91}\,\text{MeV},
\end{aligned} \tag{1.1}$$

where N and Z are the neutron and proton numbers, $A = N+Z$ is the mass number, and κ, which is roughly equal to 0.2 for medium nuclei, is a correction coefficient for the pion exchange.

When a target nucleus is irradiated with a photon and excited to GDR, it often forms a compound nucleus with only a small contribution of a pre-equilibrium reaction [5]. The compound nucleus is an excited state in which the energy brought

Fig. 1.1 Schematic illustration of nuclear transmutation with laser Compton scattering

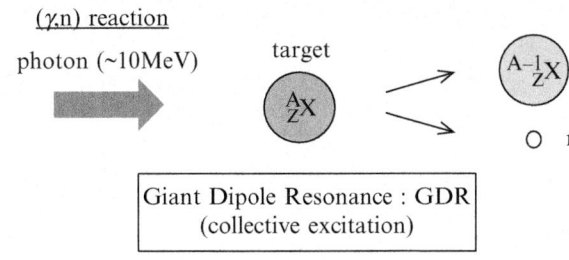

by the incident particle is shared among all degrees of freedom of the nucleus. The reaction cross section from an initial channel α to a final channel β proceeding through a compound nucleus state of spin J can be written by the Hauser–Feshbach formula as

$$\sigma_{\alpha\beta}^{cn} = \frac{\pi}{k_\alpha^2} \sum_J g_J \frac{T_\alpha \langle T_\beta \rangle}{\sum_\gamma \langle T_\gamma \rangle}, \qquad (1.2)$$

where k_α is the wave number in the initial channel, g_J is a statistical factor, T is a transmission coefficient, and $\langle T \rangle$ is the energy average of T. The statistical factor is

$$g_J = \frac{2J+1}{(2i_\alpha + 1)(2I_\alpha + 1)}, \qquad (1.3)$$

where i_α and I_α are the projectile and target spins.

Calculations of reaction cross sections are performed using the nuclear model code TALYS (version 1.4) [6]. The neutron transmission coefficients are calculated via the global optical potential [7]. The gamma-ray transmission coefficients are calculated through the energy-dependent gamma-ray strength function according to Brink [8] and Axel [9]. We employed the level density given by Gilbert and Cameron [10].

Figure 1.2 shows the photonuclear reaction cross sections of ^{137}Cs calculated using the TALYS code. In the incident photon energy $B(n) \leq E_\gamma < B(2n)$, where $B(n)$ and $B(2n)$ are the one- and two-neutron binding energies, respectively, we can see that the (γ, n) reaction mainly occurs. Because the resonance energy of GDR E_R is 15–18 MeV, which is roughly equal to $B(2n)$ for medium nuclides, about half the reactions via GDR are (γ, n) reactions, which occur at $B(n) \leq E_\gamma < B(2n)$.

1.2.2 High-Energy Photons Obtained by Laser Compton Scattering

Laser Compton scattering is a method to obtain high-energy photons by laser photons backscattered off energetic GeV electrons. In the case of head-on collision

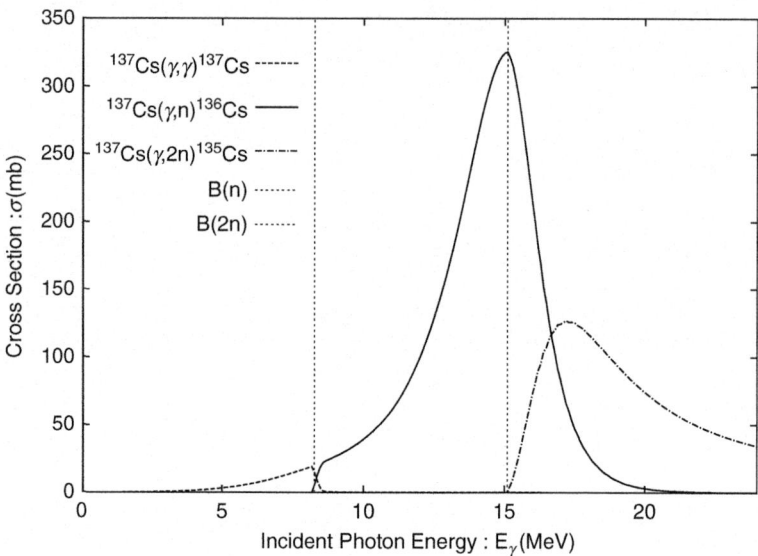

Fig. 1.2 Cross sections for ^{137}Cs (γ,γ), ^{137}Cs (*dashed line*), ^{137}Cs (γ,n), ^{136}Cs (*solid line*), and ^{137}Cs $(\gamma,2n)$. ^{135}Cs (*dash-dot line*) reactions versus incident photon energy: *dotted lines* represent B(n) and B(2n) of ^{137}Cs

between relativistic electrons and laser photons, the energy of scattered photons is given by

$$E_\gamma = \frac{4\gamma^2 E_L}{1 + (\gamma\theta)^2 + 4E_L\gamma/m_e}, \quad (1.4)$$

where $\gamma = E_e/m_e$ is the Lorentz factor of the electron beam with energy E_e, m_e is the rest mass of the electron, E_L is the energy of the laser photon, and θ is the scattering angle. From Eq. (1.4), the energy of the scattered photon is maximum at $\theta = 0$, and it depends on the energy of incident electrons and photons. The minimum energy of the scattered photon can be fixed by controlling θ with collimators.

The scattering cross section of laser Compton scattering is given by the Klein–Nishina formula:

$$\frac{d\sigma}{dE_\gamma} = \frac{\pi r_0^2}{2} m_e^2 E_L E_e^2 \left\{ \frac{m_e^4}{4E_L^2 E_e^2} \left(\frac{E_e}{E_\gamma - E_e}\right)^2 - \frac{m_e^2}{4E_L E_e} \left(\frac{E_e}{E_\gamma - E_e}\right) + \frac{E_e}{E_\gamma - E_e} + \frac{E_\gamma - E_e}{E_e} \right\},$$

$$r_0 = e^2/4\pi m_e,$$

$$N_\gamma = \int dE_\gamma \frac{dN_\gamma}{dE_\gamma} = \int dE_\gamma \frac{d\sigma}{dE_\gamma} \cdot \text{const.}$$

$$(1.5)$$

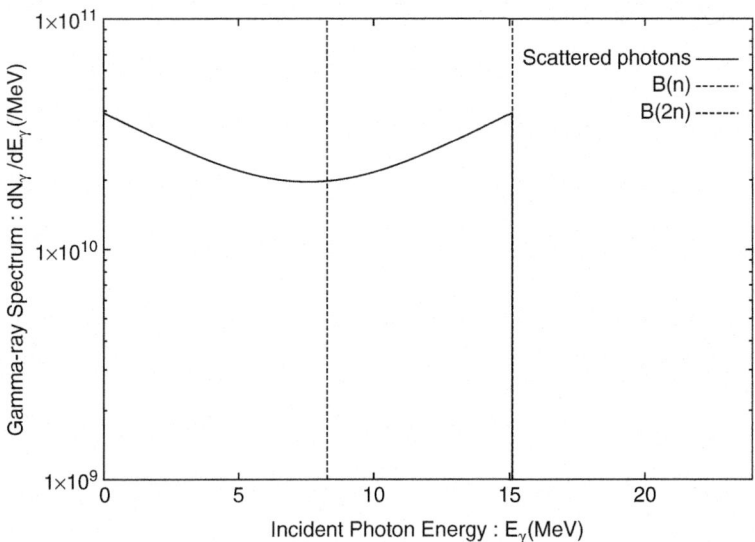

Fig. 1.3 Calculated gamma (γ)-ray spectrum (*solid line*) generated by laser Compton scattering. The maximum energy, 15 MeV, was chosen to be equal to the binding energy $B(2n)$ of ^{137}Cs. The binding energies of $B(n)$ and $B(2n)$ for ^{137}Cs are indicated by *dashed lines*

To have a situation in which (γ, n) reactions occur, the photon beam with energy at $B(n) \leq E_\gamma < B(2n)$ is desired. In case of the free electron laser, we may assume/expect to get the total photon flux $N_\gamma \approx 2 \times 10^{12}$/s/500 mA for $E_e = 1.2$ GeV [1]. From Eq. (1.4), with $E_e = 1.2$ GeV and $E_L = 0.7$ eV, we obtain the maximum photon energy of $E_\gamma = 15$ MeV at $\theta = 0$, which is equal to $B(2n)$ for ^{137}Cs. Figure 1.3 shows the calculated γ-ray spectrum generated by laser Compton scattering using Eq. (1.5), where the total photon flux with energy from 0 to $B(2n)$ is $N_\gamma \approx 2 \times 10^{12}$/s.

From Fig. 1.3, we can see that about half the total scattered photons are in $B(n) \leq E_\gamma < B(2n)$ and contribute to generate the (γ, n) reactions for ^{137}Cs. In contrast, for the Bremsstrahlung that is usually used to generate high-energy photons, the photon intensity decreases rapidly as the photon energy increases, and only a small part of the high-energy tail is available for (γ, n) reactions [11].

1.2.3 Setup of the Calculation for ^{137}Cs

When a target nucleus X is irradiated with a photon beam with energy E_γ, it forms a compound nucleus, which releases one neutron and becomes its isotope X′. The reaction rate of X(γ, n)X′ at time t is given by

$$r_{X\to X'}(t) = \int_{B(n)}^{B(2n)} dE_\gamma \frac{dN_\gamma}{dE_\gamma} \sigma_{X\to X'}(E_\gamma) n_X(t) a, \qquad (1.6)$$

where $\sigma_{X\to X'}(E_\gamma)$ is the reaction cross section of $X(\gamma,n)X'$, $n_X(t)$ is the number of target nucleus per unit area at time t, and a is the attenuation factor of incident photons through a thick target. dN_γ/dE_γ is expressed with Eq. (1.5) and $\sigma_{X\to X'}(E_\gamma)$ is calculated from Eq. (1.2) using the TALYS code.

Figure 1.4 shows a calculation in which the photon beam is generated by the laser Compton scattering of 1.2 GeV electrons and 0.7 eV laser beams. We assume that the cylindrical target of ^{137}Cs of 1 g is irradiated with a photon beam with energy $B(n) \leq E_\gamma < B(2n)$ within a radius $r \approx 0.8$ mm at 2 m from the interaction point. When a target of ^{137}Cs is irradiated with photons and is excited to GDR, the (γ,γ), (γ,n) and $(\gamma,2n)$ reactions mainly occur. We consider ^{137}Cs, ^{136}Cs, ^{135}Cs, and ^{134}Cs as the isotopes generated by the transmutation. The numbers of these isotopes are expressed as

$$n_{137}(t+\Delta t) = n_{137}(t)e^{-\lambda_{137}\Delta t} - r_{137\to 136}(t) - r_{137\to 135}(t), \qquad (1.7)$$

$$n_{136}(t+\Delta t) = n_{136}(t)e^{-\lambda_{136}\Delta t} + r_{137\to 136}(t) - r_{136\to 135}(t) - r_{136\to 134}(t), \qquad (1.8)$$

$$n_{135}(t+\Delta t) = n_{135}(t)e^{-\lambda_{135}\Delta t} + r_{137\to 135}(t) + r_{136\to 135}(t) - r_{135\to 134}(t), \qquad (1.9)$$

$$n_{134}(t+\Delta t) = n_{134}(t)e^{-\lambda_{134}\Delta t} + r_{136\to 134}(t) + r_{135\to 134}(t). \qquad (1.10)$$

One can calculate the number of each isotopes by solving these equations with the Runge–Kutta method.

1.3 Results and Discussion

1.3.1 Nuclear Transmutation of ^{137}Cs with Laser Compton Scattering

Figure 1.5 shows the dependence of the reduction of 1 g ^{137}Cs on the photon flux $N_\gamma = 10^{12}$, 10^{18}, 10^{19}, 10^{20}/s, which is calculated with this setup (Fig. 1.4). The number of ^{137}Cs is effectively reduced with photon flux over 10^{18}/s, that is, the number of ^{137}Cs is reduced by 10 % for 24 h irradiation. Figure 1.6 shows the number of Cs isotopes when 1 g ^{137}Cs is irradiated with photon flux 2×10^{12}/s with the same setup. From this figure, we can see that the reduction rate of ^{137}Cs by the transmutation, which is nearly equal to the generation rate of ^{136}Cs, is two orders of magnitude smaller than the natural decay rate of 1 g ^{137}Cs. Thus, the transmutation of ^{137}Cs is not effective with photon flux 2×10^{12}/s, which is maximum with present accelerator systems.

1 Nuclear Transmutation of Long-Lived Nuclides with Laser Compton Scattering... 9

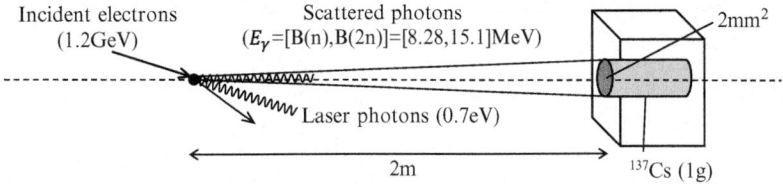

Fig. 1.4 Setup for calculation of transmutation of ^{137}Cs

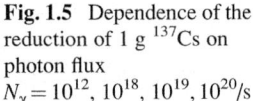

Fig. 1.5 Dependence of the reduction of 1 g ^{137}Cs on photon flux $N_\gamma = 10^{12}, 10^{18}, 10^{19}, 10^{20}$/s

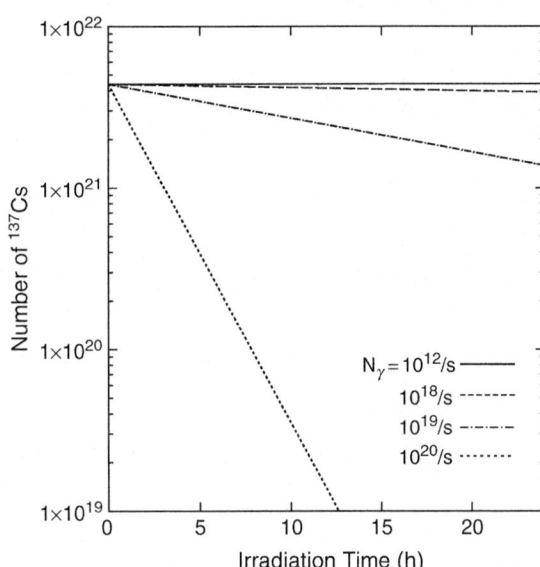

1.3.2 Comparison with Other Nuclides

Table 1.1 summarizes the generation rate of $^{A-1}$X from of 1 g of a fission product ^{A}X with photon flux 2×10^{12}/s also for ^{129}I, ^{135}Cs, and ^{90}Sr in addition to ^{137}Cs. The reduction rate, which is nearly equal to the generation rate of $^{A-1}$X, does not depend significantly on nuclides because the properties of GDR are smooth functions of the mass number. From this table, we can see that the reduction rate for the transmutation of ^{137}Cs can be similar to that of other medium nuclides.

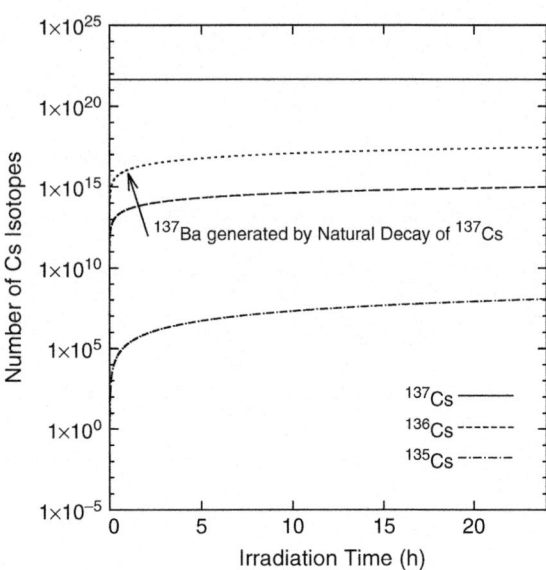

Fig. 1.6 Number of Cs isotopes when 1 g ^{137}Cs is irradiated with photon flux 2×10^{12}/s. *Dotted line* shows the number of ^{137}Ba that are generated by the natural decay of 1 g ^{137}Cs

Table 1.1 Generation rate of $^{A-1}$X from 1 g of fission product ^{A}X with photon flux 2×10^{12}/s

Target (^{A}X)	[B (n), B (2n)] (MeV)	E_{GDR} (MeV)	σ_{GDR}^{tot} (b·MeV)	N ($^{A-1}$X) (/s)
^{129}I	[8.83, 15.7]	15.3	2.25	1.24×10^{10}
^{135}Cs	[8.76, 15.7]	15.2	2.31	1.31×10^{10}
^{137}Cs	[8.28, 15.1]	15.1	2.37	1.19×10^{10}
^{90}Sr	[7.81, 14.2]	16.7	1.58	4.71×10^{9}

1.4 Conclusion

In this work, the effectiveness of transmutation with laser Compton scattering for reducing fission products was quantitatively investigated. The transmutation of ^{137}Cs is effective with photon flux greater than 10^{18}/s, which results in 10 % reduction for 24 h irradiation. However, transmutation with photon flux 2×10^{12}/s, which is achievable with present maximum accelerator systems, is not effective, and the reduction rate is approximately two orders of magnitude less than the natural decay rate.

Nuclear transmutation with laser Compton scattering can transmute selectively a medium mass nuclide ^{A}X into $^{A-1}$X, and its reduction rate is independent of isotopes. Because the transmutation with laser Compton scattering can almost exclusively generate desired nuclides, this method will be useful for the generation of isotopes for medicine [1].

Open Access This chapter is distributed under the terms of the Creative Commons Attribution Noncommercial License, which permits any noncommercial use, distribution, and reproduction in any medium, provided the original author(s) and source are credited.

References

1. Ejiri H, Shima T, Miyamoto S, Horikawa K, Kitagawa Y, Asano Y, Date S, Ohashi Y (2011) J Phys Soc Jpn 80:094202
2. Li D, Imasaki K, Horikawa K, Miyamoto S, Amano S, Mochizuki T (2009) J Nucl Sci Technol 46:8
3. Bohr A, Mottelson B (1969, 1975) Nuclear structure, vols I and II. Benjamin, New York
4. Capote R, Herman M, Oblozinsky P, Young PG, Goriely S, Belgya T, Ignatyuk AV, Koning AJ, Hilaire S, Plujko V, Avrigeanu M, Bersillon O, Chadwick MB, Fukahori T, Kailas S, Kopecky J, Maslov VM, Reffo G, Sin M, Soukhovitskii E, Talou P, Yinlu H, Zhigang G (2009) Nucl Data Sheets 110:3107
5. Maeda K, Shibata T, Ejiri H (1983) Phys Rev C 28:635
6. Koning AJ, Hilaire S, Duijvestijn MC (2008) TALYS-1.4. In: Bersillon O, Gunsing F, Bauge E, Jacqmin R, Leray S (eds) Proceedings of the International Conference on Nuclear Data for Science and Technology, vol 211, Nice, April 22–27 2007. EDP Sciences, . See also http://www.talys.eu/documentation/
7. Koning AJ, Delaroche JP (2003) Nucl Phys A 713:231
8. Brink DM (1967) Nucl Phys 4:215
9. Axel P (1962) Phys Rev 126:671
10. Gilbert A, Cameron AGW (1965) Can J Phys 43:1446
11. Bunatiana GG, Nikolenko VG, Popov AB (2010) arXiv 1012.5002v1 [nucl-exp, Dec 2010]. http://arxiv.org/pdf/1012.5002.pdf

Chapter 2
Recent Progress in Research and Development in Neutron Resonance Densitometry (NRD) for Quantification of Nuclear Materials in Particle-Like Debris

M. Koizumi, F. Kitatani, H. Tsuchiya, H. Harada, J. Takamine, M. Kureta,
H. Iimura, M. Seya, B. Becker, S. Kopecky, W. Mondelaers,
and P. Schillebeeckx

Abstract To quantify special nuclear materials (SNM) in particle-like debris, a technique named neutron resonance densitometry (NRD) has been proposed. This method is a combination of neutron resonance transmission analysis (NRTA) and neutron resonance capture analysis (NRCA) or prompt gamma-ray analysis (PGA). In NRTA, neutron transmission rate is measured as a function of neutron energy with a short flight path time-of-flight (TOF) system. Characteristic neutron transmission dips of Pu and U isotopes are used for their quantification. Materials in the samples (H, B, Cl, Fe, etc.) are measured by the NRCA/PGA method. For the NRD measurements, a compact TOF facility is designed. The statistical uncertainties of the obtained quantities of the SNMs in a sample are estimated. A high-energy-resolution and high-S/N γ-ray spectrometer is under development for NRCA/PGA. Experimental studies of systematic uncertainties concerning the sample properties, such as thickness and uniformity, are in progress at the TOF facility GELINA of European Commission (EC), Joint Research Centre (JRC), Institute for Reference Materials and Measurements (IRMM).

Keywords Capture • Fukushima • GELINA • Neutron resonance densitometry
• NRD • Nuclear security • Severe accident • Transmission

M. Koizumi (✉) • F. Kitatani • H. Tsuchiya • H. Harada • J. Takamine • M. Kureta • H. Iimura
Nuclear Science and Engineering Center, Japan Atomic Energy Agency, Tokai-mura,
Naka-gun, Ibaraki 319-1195, Japan
e-mail: koizumi.mitsuo@jaea.go.jp

M. Seya
Integrated Support Center for Nuclear Nonproliferation and Nuclear Security, Japan Atomic
Energy Agency, Tokai-mura, Naka-gun, Ibaraki 319-1118, Japan

B. Becker • S. Kopecky • W. Mondelaers • P. Schillebeeckx
European Commission, Joint Research Centre, Institute for Reference Materials
and Measurements, Retieseweg 111, 2440, Geel, Belgium

2.1 Introduction

Quantifying nuclear materials (NM) in the debris of melted fuel (MF) formed in a severe accident is considered to be difficult because of their variety of size, shape, unknown composition, and strong radioactivity. Although techniques of nondestructive assay (NDA) are indispensable for the evaluation of NM in debris, quantification methods have not been established so far [1]. In the cases of TMI-2 or Chernobyl-4, accounting for the NM was based on some estimations.

We have proposed a technique called neutron resonance densitometry (NRD) [2, 3] to quantify NM in particle-like debris that is assumed to be produced in the rapid cooling processes of a severe accident [4]. Small pieces are also produced when MF are cut or broken down to be taken out of the damaged reactors [1].

To examine the NRD method, studies have begun. Some experiments were carried out at the time-of-flight (TOF) facility GELINA [5] of EC-JRC-IRMM under the agreement between JAEA and EURATOM in the field of nuclear materials safeguards research and development.

In this chapter, we briefly describe the concept of NRD, give an overview of the development of NRD, and explain some parts of the recent progress.

2.2 Neutron Resonance Densitometry

2.2.1 The Concept of NRD

Neutron resonance densitometry is a method of a combination of neutron resonance transmission analysis (NRTA) and neutron resonance capture analysis (NRCA) or prompt gamma-ray analysis (PGA). The fundamental principles of NRTA and NRCA are described by Postma and Schillebeeckx [6].

In NRTA, neutron transmission is measured as a function of neutron energy with a TOF technique. Characteristic neutron transmission dips of Pu and U isotopes are observed in the neutron energy in the range of 1–50 eV [7, 8]. Measurements of these transmission spectra can be carried out with a short-flight path TOF system [9, 10].

Although strong γ-ray radiation from MF samples does not interfere with NRTA measurements, reduction of neutron flux caused by nuclei with large total cross section (such as H, B, Cl, Fe) makes accurate NM quantification difficult. Nevertheless, the quantities of these contained nuclei could not be determined by NRTA only, because these nuclei do not resonantly interact with neutrons in this energy range. To identify and to quantity the composing isotopes, the NRCA/PGA method is required. Characteristic prompt γ rays ware utilized. Table 2.1 shows prompt γ-rays emitted from nuclei after neutron capture reaction. Most of these discrete prompt γ-rays have significant intensities. The information obtained by NRCA/PGA enables us to determine the appropriate sample thickness and measurement time. This information also supports NRTA analysis.

Table 2.1 Energies of prominent prompt γ-rays and the first neutron resonances of nuclei

Nucleus	Reaction	Prompt γ rays (KeV)	First resonance (KeV)
^{1}H	^{1}H (n, γ) ^{2}H	2,223	–
^{10}B	^{10}B $(n, \alpha\gamma)$ ^{7}Li	478	170
^{27}Al	^{27}Al (n, γ) ^{28}Al	3,034, 7,724	5.9
^{28}Si	^{28}Si (n, γ) ^{29}Si	3,539, 4,934	31.7
^{53}Cr	^{53}Cr (n, γ) ^{54}Cr	835, 8,885	4.2
^{56}Fe	^{56}Fe (n, γ) ^{57}Fe	7,631, 7,646	1.1
^{59}Co	^{59}Co (n, γ) ^{60}Co	230, 6,877	0.132
^{58}Ni	^{58}Ni (n, γ) ^{59}Ni	465, 8,999	6.9

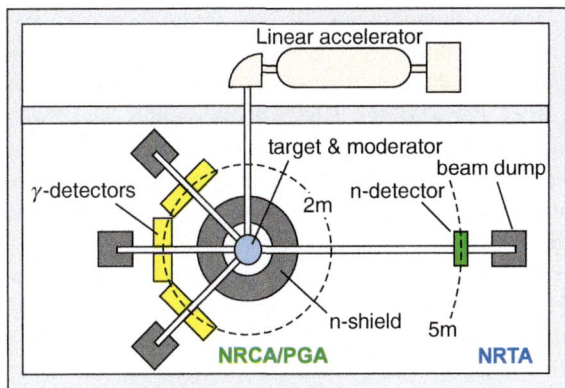

Fig. 2.1 A rough draft of an NRD facility. The neutron flight path length for NRTA is 5 m and that for NRCA/PGA is 2 m

2.2.2 A Rough Draft of an NRD Facility

For practical application, the scale of an NRD facility should be minimized. Figure 2.1 shows a rough draft of the NRD facility. An electron linear accelerator with a power of 1 kW and acceleration voltage of 30 MeV is assumed [11]. High-energy neutrons are generated in the order of 10^{12} n/s by photonuclear reactions following Bremsstrahlung at the electron target. The generated neutrons are slowed down to epithermal energy by collisions in a moderator surrounding the target. Neutrons from the moderator are collimated to supply for NRTA and for NRCA/PGA.

The length of the flight path is important to design a TOF system, because the longer flight path reduces the neutron flux whereas it increases the energy resolution of the system. It may require at least a 5-m flight path to achieve a good enough resolution to resolve resonances of NMs below 50 eV in NRTA [9, 10]. A shorter neutron flight path is feasible for NRCA/PGA because the nuclei in Table 2.1 are identified by the prompt γ-ray energies. We consider that a 2-m flight path is sufficient for NRCA/PGA. The beam line lengths mainly determine the scale of

Table 2.2 Estimated statistical uncertainty of quantities of U and Pu isotopes in a sample

Nucleus	Concentration in a fuel (kg/tHM)	Statistic error (%)
^{238}Pu	0.19	0.85
^{239}Pu	5.25	0.074
^{240}Pu	2.13	0.051
^{241}Pu	1.23	0.23
^{242}Pu	0.48	0.069
^{235}U	14.6	0.049
^{238}U	928	0.010

The measurements are assumed to be carried out for 40 min with a 10^{12} n/s neutron source

the facility. One beam line for NRTA and three beam lines for NRCA/PGA are placed as shown in Fig. 2.1. The sample size for NRTA is assumed to be 10–30 cm in diameter and 1–2 cm in thickness. In comparison, the sample size for NRCA/PGA is smaller; the diameter is 1–2 cm, and the thickness is 1–2 cm. A collimator is placed between the NRCA/PGA sample and the γ-ray detector to reduce the background γ-rays from the sample. Because optimal sample thickness for NRTA strongly depends on the amount of impurities or matrix material, the quantity of the interfering nuclei in debris has to be measured roughly by NRCA/PGA preceding NRTA measurements [12].

The statistical uncertainties of NMs quantified by NRTA were estimated [12]. The size of a MF sample is assumed to be 1 cm in thickness and 30 cm in diameter. The weight of the sample becomes about 4 kg: it consisted of nuclear fuel (64 vol.%), ^{nat}Fe (8 vol.%), ^{nat}B (8 vol.%), and 20 vol.% of vacancy. The composition of the nuclear fuel was taken from Ando and Takano [13] [a fuel of 40 GWd/t burn-up in a boiling water reactor (BWR)]. The measurement was assumed to be carried out for 40 min, in which 20 min was for sample and 20 min for background. Table 2.2 shows the estimated statistical uncertainties of quantified Pu and U isotopes in the sample. The achieved statistical uncertainties are less than 1 %.

With the measurement cycle given here, about 0.15 ton of debris can be handled in a day; this enables us to measure 30 tons of debris in a year (200 working-days are assumed). This amount can be increased with the number of NRTA beam lines.

2.3 Development of a γ-Ray Spectrometer for NRCA/PGA

The γ-ray background from debris is expected to be strong. The strongest radioactive isotope in a MF of the TMI-2 accident was ^{137}Cs, which ranged from 10^6 to 3×10^8 Bq/g [14]. The energies of the prominent γ rays from nuclei listed in Table 2.1 is larger than the 661 keV γ-rays from ^{137}Cs, except for ^{10}B. Accordingly, most of the measurements of the NRCA/PGA will not have interference with the γ-rays from ^{137}Cs. On the other hand, the Compton edge of the 661 keV γ-rays surely overlaps with the 478-keV γ-ray peak originating the ^{10}B$(n, \alpha\gamma)^7$Li reaction.

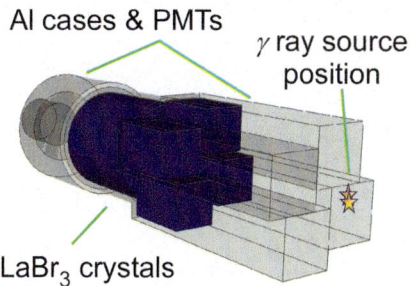

Fig. 2.2 Design of a prototype LaBr$_3$ γ-ray spectrometer, which consists of a cylindrical detector and four square pillar detectors. Collimated γ-rays pass through the square opening of the spectrometer to reach the cylindrical detector

The γ-ray spectrometer used for NRCA/PGA, therefore, requires properties of not only high-energy resolution and fast timing response, but also a high peak-to-Compton ratio.

To satisfy these requirements, a well-type spectrometer made of LaBr$_3$ crystal has been proposed [2, 15]. In a study based on Monte Carlo simulations of a well-type LaBr$_3$ spectrometer [15, 16], the Compton edge was successfully reduced by an order of magnitude. Such reduction enables us to roughly quantify ^{10}B in a sample, even in the presence of high background γ-rays from ^{137}Cs.

A prototype LaBr$_3$ γ-ray spectrometer has been designed (Fig. 2.2). Because of the technical difficulty of producing a crystal with a well, the spectrometer is made of several detectors: a cylindrical detector and four square pillar detectors. The cylindrical crystal is 120 mm in diameter and 127 mm in length; each square pillar crystal is $50.8 \times 50.8 \times 76.2$ mm. An arrangement of the detector pillars opens a square channel of 20×20 mm for the passage of collimated γ-rays from the samples. This spectrometer will be tested soon.

2.4 Experiments for NRD Developments

To evaluate systematic uncertainties in NRD, we have started experimental study at GELINA [5] under the collaboration between JAEA and EC-JRC-IRMM. The items to be studied are as follows: (1) particle size, (2) sample thickness, (3) presence of contaminated materials, (4) sample temperature, and (5) the response of the TOF spectrometer [3, 17]. Some experiments has been performed at GELINA [18–21]. A resonance shape analysis code, REFIT [22], has been adopted for the data analysis.

Experiments on sample thickness were carried out at the 25-m TOF neutron beam line of GELINA. Cu plates with various thicknesses were measured with an NRTA method. Peaks at the 579 eV resonance of ^{63}Cu were analyzed with the REFIT program. The evaluated areal densities are compared with the declared values, which were derived from measurements of the weight and the area of the

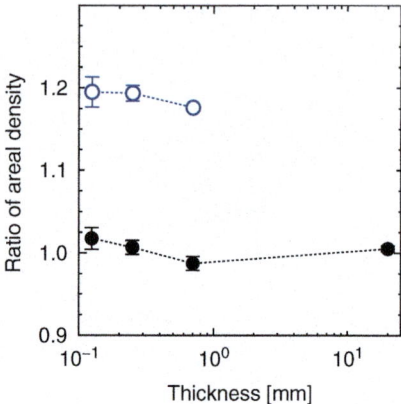

Fig. 2.3 Ratios of evaluated and declared areal densities. The 579 eV transmission peaks of ^{63}Cu were analyzed with REFIT. *Open circles* indicate the results analyzed with the resonance parameter values taken from Mughabghab [23] (#6 in Fig. 2.4), and *closed circles* represent tentatively introduced values (#7 in Fig. 2.4), which reproduce the areal densities of Cu plates better. The lines are guides for the eye. *Note*: We also analyzed the transmission spectrum of a 2-cm-thick copper sample with the parameter #6. The obtained fitted curve, however, did not reproduce the peak shape at all. Thus, the misleading *open circle* data point was removed

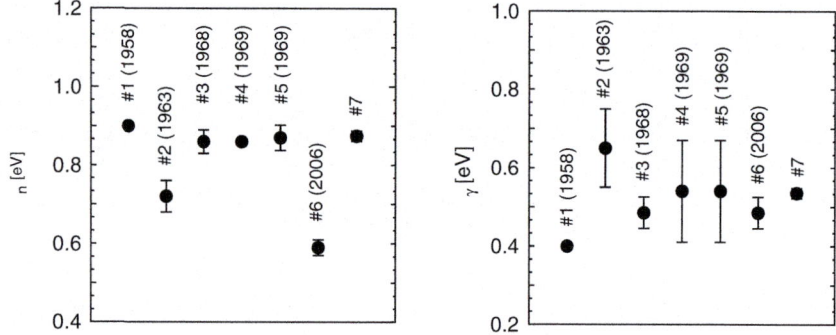

Fig. 2.4 Experimentally obtained 579 eV ^{63}Cu resonance parameters. Each data point is taken from different references (#1 [24], #2 [25], #3 [26], #4 [27], #5 [28], and #6 [23]). The data of #6 were utilized by REFIT originally; the data of #7 are tentatively introduced to reproduce the experimental transmission dips

samples. Figure 2.3 shows the results. The abscissa is the thickness of the Cu plates and the ordinate is the ratio of evaluated and declared areal densities. The open circles are the results analyzed with the resonance parameter values taken from Mughabghab [23] (#6 of Fig. 2.4); the closed circles are the results analyzed with the tentatively introduced values (#7 of Fig. 2.4), which reproduced the areal density of Cu plates better. Figure 2.4 shows measured ^{63}Cu resonance parameters. The 579-eV resonance parameters of ^{63}Cu may require being reevaluated. It should be emphasized that survey of the total cross sections of Pu and U isotopes is quite important to quantify NM.

2.5 Summary

We have proposed NRD for measurements of NM in particle-like debris of MF. The NRD system utilizes a compact neutron TOF system equipped with a neutron detector for NRTA and high-energy-resolution and high-S/N γ-ray detectors for NRCA/PGA. The rough design of a NRD facility is given. The capacity of NM measurements in the facility has been shown. Experiments on systematical uncertainties caused by sample properties, such as sample thickness and uniformity, are in progress under the collaboration between JAEA and EC-JRC-IRMM. The importance of confirmation of nuclear data has been shown in the case of Cu thickness measurements by NRTA.

Acknowledgments The research and development have been carried out under the agreement between JAEA and EURATOM in the field of nuclear materials safeguards research and development and are supported by the Japanese government, the Ministry of Education, Culture, Sports, Science and Technology in Japan (MEXT).

Open Access This chapter is distributed under the terms of the Creative Commons Attribution Noncommercial License, which permits any noncommercial use, distribution, and reproduction in any medium, provided the original author(s) and source are credited.

References

1. Seya M, Harada H, Kitatani F, Koizumi M, Tsuchiya H, Iimura H, Kureta M, Takamine J, Hajima R, Hayakawa T, Shizuma T, Angell C, Bolind AM (2013) In: Proceedings of the INMM 54th annual meeting. The Institute of Nuclear Materials Management (INMM), Chicago, USA
2. Koizumi M, Kitatani F, Harada H, Takamine J, Kureta M, Seya M, Tsuchiya H, Iimura H (2012) In: Proceedings of the 53th INMM annual meeting. The Institute of Nuclear Materials Management (INMM), Chicago, USA
3. Harada H, Kitatani F, Koizumi M, Tsuchiya H, Takamine J, Kureta M, Iimura H, Seya M, Becker B, Kopecky S, Schillebeeckx P (2013) In: Proceedings of the 35th ESARDA. The Institute of Nuclear Materials Management (INMM), Chicago, USA
4. Watari T, Inoue Y, Masuda F (1990) J Atom Energy Soc Jpn 32:12–24 (in Japanese)
5. Mondelaers W, Schillebeeckx P (2006) Notiziario 11:19–25
6. Postma H, Schillebeeckx P (2009) Neutron resonance capture and transmission analysis. In: Meyers RA (ed) Encyclopedia of analytical chemistry. Wiley, New York, pp 1–22
7. Bowman CD, Schrack RA, Behrens JW, Johnson RG (1983) Neutron resonance transmission analysis of reactor spent fuel assemblies. In: Barton JP, von der Hardt P (eds) Neutron radiography. ECSC/EEC/EAEC, Brussels/Belgium/Luxembourg, pp 503–511
8. Behrens JW, Johnson RG, Schrack RA (1984) Nucl Technol 67:162–168
9. Chichester DL, Sterbentz JW (2011) A second look at neutron resonance transmission analysis as a spent fuel NDA technique. INL/C0N-11-20783
10. Sterbentz JW, Chichester DL (2010) Neutron resonance transmission analysis (NRTA): a nondestructive assay technique for the next generation safeguards initiative's plutonium assay challenge. INL/EXT-10-20620
11. Kase T, Harada H (1997) Nucl Sci Eng 126:59–70

12. Kitatani F (2013) 2013 Annual meeting of the Atomic Energy Society of Japan, A55
13. Ando Y, Takano H (1999) Estimation of LWR spent fuel composition. JAERI-Research 99-004
14. Uetsuka H, Nagase F, Suzuki T (1995) Gamma spectrometry of TMI-2 debris (in Japanese). JAERI-Research 95-084
15. Tsuchiya H, Harada H, Koizumi M, Kitatani F, Takamine J, Kureta M, Iimura H (2013) Nucl Instrum Methods Phys Res A 729:338–345
16. Tsuchiya H, Harada H, Koizumi M, Kitatani F, Takamine J, Kureta M, Seya M, Iimura H, Becker B, Kopecky S, Schillebeeckx P (2013) In: Proceedings of the 35th European Safeguards Research & Development Association (ESARDA), Varese, Italy
17. Harada H, Kitatani F, Koizumi M, Tsuchiya H, Takamine J, Kureta M, Iimura H, Seya M, Becker B, Kopecky S, Mondelaers W, Schillebeeckx P (2013) In: Proceedings of the 54th European Safeguards Research & Development Association (ESARDA), Varese, Italy
18. Schillebeeckx P, Abousahl S, Becker B, Borella A, Harada H, Kauwenberghs K, Kitatani F, Koizumi M, Kopecky S, Moens A, Sibbens G, Tsuchiya H (2013) In: Proceedings of the 35th European Safeguards Research & Development Association (ESARDA), Varese, Italy
19. Becker B, Harada H, Kauwenberghs K, Kitatani F, Koizumi M, Kopecky S, Moens A, Schillebeeckx P, Sibbens G, Tsuchiya H (2013) In: Proceedings of the 35th European Safeguards Research & Development Association (ESARDA), Varese, Italy
20. Schillebeeckx P, Becker B, Emiliani F, Kopecky S, Kauwenberghs K, Moens A, Mondelaers W, Sibbens G, Harada H, Kitatani F, Koizumi M, Kureta M, Iimura H, Takamine J, Tsuchiya H, Abousahl S, Moxon M (2013) In: Proceedings of the INMM 54th annual meeting European Safeguards Research & Development Association (ESARDA), Varese, Italy
21. Tsuchiya H, Harada H, Koizumi M, Kitatani F, Takamine J, Kureta M, Iimura H, Becker B, Kopecky S, Kauwenberghs K, Moens A, Mondelaers W, Schillebeeckx P (2013) In: Proceedings of the INMM 54th annual meeting European Safeguards Research & Development Association (ESARDA), Varese, Italy
22. Massimi C, Borella A, Kopecky S, Lampoudis C, Schillebeeckx P, Moxon MC, Vannini G (2011) J Korean Phys Soc 59:1689–1692 (and references therein)
23. Mughabghab SF (2006) Atlas Neutron Resonances. In: 5th ed Resonance Parameters and Thermal Cross Sections. Z=1–100, (2006, Elsevier Science)
24. Cote RE, Bollinger LM, Leblanc JM (1958) Phys Rev 111:288–295
25. Kapchigashev SV, Popov YP (1963) Atomnaya Energiya 15:120–126
26. Weigmann H, Winter J (1968) Z Physik 213:411–419
27. Alves RN, Barros SD, Chevillon PL, Julien J, Morgenstern J, Samour C (1969) Nucl Phys A 134:118–140
28. Julien J, Alves R, Barros SD, Huynh VD, Morgenstern J, Samour C (1969) Nucl Phys A132:129–160

Chapter 3
Development of Nondestructive Assay to Fuel Debris of Fukushima Daiichi NPP (1): Experimental Validation for the Application of a Self-Indication Method

Jun-ichi Hori, Tadafumi Sano, Yoshiyuki Takahashi, Hironobu Unesaki, and Ken Nakajima

Abstract We have proposed a new concept of the "self-indication method" combined with neutron resonance densitometry (NRD) for nondestructive assaying of the distribution of nuclear materials in the fuel debris of Fukushima Daiichi NPP. To verify the method, we performed experiments using a 46 MeV electron linear accelerator at the Kyoto University Research Reactor Institute. First, we measured the area densities of gold foil 10, 20, 30, 40, and 50 μm thick by area analysis at the 4.9 eV resonance region. It was confirmed that the area densities of the target nuclide can be determined by conventional NRD and the self-indication method within 3 % accuracy, respectively. As the next step, we added a silver foil of 50 μm thickness to a gold foil of 10 μm thickness and measured the area density of the gold foil. It was shown that the contribution from the other nuclide (silver foil) can be remarkably suppressed by applying the self-indication method. Finally, we have demonstrated a nondestructive assay of nuclear material using a mixture composed of a natural uranium foil, sealed minor actinide samples of ^{237}Np and ^{243}Am. The results indicated that the self-indication method is useful for assaying a mixture of materials with high activity such as fuel debris.

Keywords Fuel debris • Neutron resonance absorption • Pulsed-neutron source • Self-indication method • TOF

J. Hori (✉) • T. Sano • Y. Takahashi • H. Unesaki • K. Nakajima
Research Reactor Institute, Kyoto University, 2-1010, Asashironishi, Kumatori, Osaka 590-0494, Japan
e-mail: hori@rri.kyoto-u.ac.jp

3.1 Introduction

It is surmised that melted fuel debris is present in the cores at units 1, 2, and 3 of Fukushima Daiichi NPP. Identifying the fuel debris status in the reactors is one of the most important issues for decommissioning. Therefore, we need to determine how to analyze the properties of actual debris collected from those cores in advance of removal work. As the debris contains melted fuel and cladding tube and structure materials heterogeneously in addition to salt content, nondestructive assaying of the distribution of nuclear materials within the debris is absolutely essential for nuclear material accountancy and critical safety.

Neutron resonance densitometry (NRD) [1] with the time-of-flight (TOF) technique based on neutron resonance transmission analysis (NRTA) [2] and neutron resonance capture analysis (NRCA) [3, 4] is a promising way to characterize the debris. However, there are two difficulties in applying those methods to fuel debris. In NRD, many resonances of other nuclides that are contained in the debris may make it difficult to identify and quantify the target nuclide. In NRCA, it is expected that the intense decayed gamma rays from debris result in high background and large dead time of the gamma-ray detector. In this work, we propose a new concept of the "self-indication method" as a complementary assay to overcome those difficulties. In the self-indication method, we set an indicator consisting of target nuclide with a high purity at the beam downstream from a sample. By detecting the reaction products such as neutron capture γ-rays or fission products from the indicator with the TOF method, the transmission neutron can be measured indirectly. The self-indicator is a transmission neutron detector that has high efficiency around the objective neutron resonance energies of the target nuclide, enabling us to quantify effectively the amount of resonance absorption of the target nuclide. Moreover, it is not easily affected by the decayed γ-rays from the debris.

In this work, experimental validation for application of the self-indication method was carried out. A part of the preliminary results is shown in this chapter.

3.2 Experiment

To verify the self-indication method, we have performed three kinds of experiments using a 46 MeV electron linear accelerator (linac) at the Kyoto University Research Reactor Institute. The experimental arrangement is shown in Fig. 3.1. Pulsed neutrons were produced from a water-cooled photo-neutron target assembly, 5 cm in diameter and 6 cm long, which was composed of 12 sheets of tantalum plates with total thickness of 29 mm [5]. This target was set at the center of an octagonal water tank, 30 cm long and 10 cm thick, as a neutron moderator. The linac was operated with a repetition rate of 50 Hz, a pulse width of 100 ns, a peak current of 5 A, and an electron energy of about 30 MeV. We used a flight path in the direction of 135 ° to the linac electron beam. To reduce the gamma flash generated

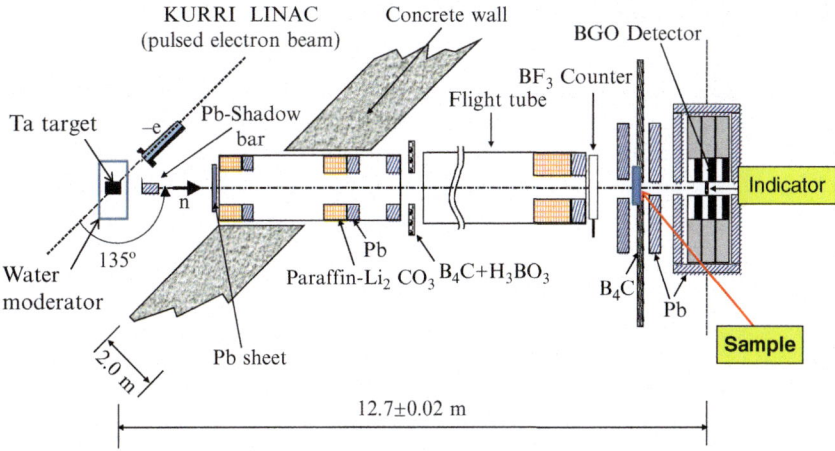

Fig. 3.1 Experimental arrangement for the time-of-flight (TOF) measurement

by the electron burst from the target, a lead block, 7 cm in diameter and 10 cm long, was set in front of the entrance of the flight tube.

First, a gold foil 10, 20, 30, 40, or 50 μm thick was located as a sample at a flight length of 11.0 m, where the neutron beam was well collimated to a diameter of 24 mm. An indicator located at a flight length of 12.7 m was surrounded by a $Bi_4Ge_3O_{12}$ (BGO) assembly, which consists of 12 scintillation bricks each $5 \times 5 \times 7.5$ cm^3 [6]. Prompt-capture γ-rays from the indicator were detected with the BGO assembly in the TOF measurement. A ^{10}B plug 8 mm thick or gold foil 50 μm thick was used as an indicator. Because the former thick indicator can absorb most neutrons with energies below the epi-thermal region, it was equivalent to the conventional NRTA. In the latter case, it was the self-indication measurement. The area densities of the samples with different thickness were estimated by area analysis for the 4.9 eV resonance of ^{197}Au.

As the next step, a 50-μm-thick silver foil was added to a 10-μm-thick gold foil to form a sample and the area density of the gold foil was measured. It is worth noting that silver has a large resonance at 5.2 eV, close to the 4.9 eV resonance of ^{197}Au. The ^{10}B plug 8 mm thick or a gold foil 50 μm thick was also used as an indicator. Moreover, we demonstrated a nondestructive assay for nuclear materials using a mixture composed of a natural uranium plate and sealed minor actinide samples of ^{237}Np and ^{243}Am. The natural uranium plate was 10×20 mm^2 and weighed 5.8 g. The samples of ^{237}Np and ^{243}Am were oxide powder, which was pressed into a pellet 20 mm in diameter and encapsulated in an aluminum disk-shaped container 30 mm in diameter with 0.5-mm-thick walls. The activities of ^{237}Np and ^{243}Am were 26 and 868 MBq, respectively. In the third measurement, the ^{10}B plug 8 mm thick or a natural uranium plate of 10×20 mm^2 and weight 5.8 g was used as an indicator.

3.3 Results and Discussion

The TOF spectra obtained by both methods, NRTA (dotted lines) and the self-indication method (solid lines), around the 4.9 eV resonance of ^{197}Au are shown in Fig. 3.2. Neutron absorption resulting from the 4.9 eV resonance of ^{197}Au can be observed as a dip and a lack of peak for the NRTA and the self-indication method, respectively. Here the net area ratio R, which is defined as the ratio of resonance absorption to the number of incident neutrons, is defined by

$$R = \sum_{i=I_{\min}}^{I_{\max}} (C_{\text{out},i} - C_{\text{in},i}) / \sum_{i=I_{\min}}^{I_{\max}} C_{\text{out},i}, \quad (3.1)$$

where C_i is the net counting rate of the ith channel in the TOF measurement and the subscripts "in" and "out" mean "with" and "without" sample. Background events were estimated by TOF measurement without an indicator and subtracted from the foreground TOF spectrum. The 4.9 eV resonance peak ranges from the $I_{\min}^{\text{th}}$ to $I_{\max}^{\text{th}}$ channel. The net area ratios for the NRTA and the self-indication method can be expressed as follows:

$$R_{\text{NRTA}} = \int_{E_{\min}}^{E_{\max}} \{1 - \exp(-n\sigma_{\text{tot}}(E))\} dE / \int_{E_{\min}}^{E_{\max}} dE \quad (3.2)$$

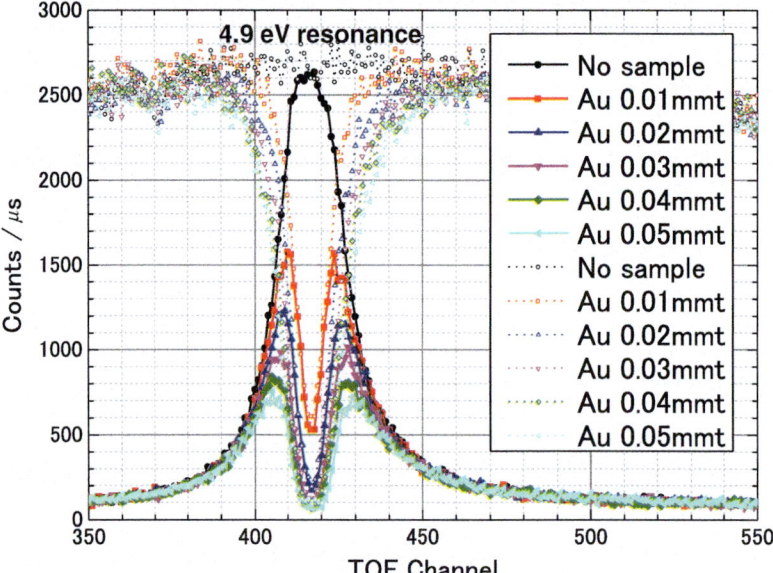

Fig. 3.2 Comparison of TOF spectra obtained by transmission method and self-indication method for 4.9 eV resonance of ^{197}Au

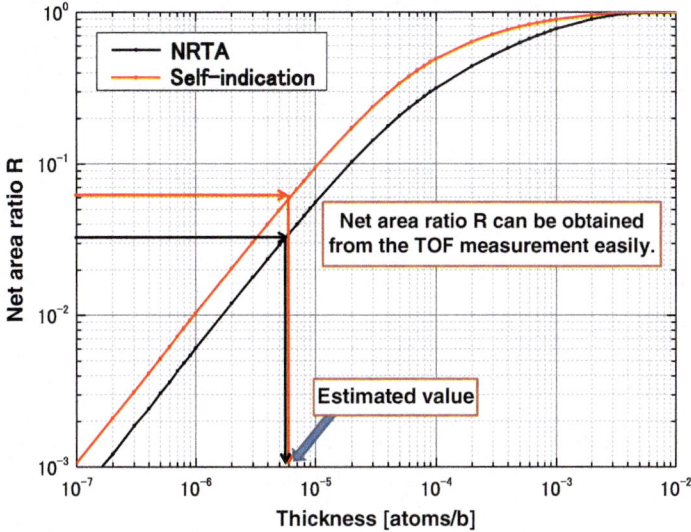

Fig. 3.3 Relationship between net area ratio and sample thickness

$$R_{self} = \int_{E_{min}}^{E_{max}} \{1 - \exp(-n\sigma_{tot}(E))\} n_{ind}\sigma_{cap}(E) dE / \int_{E_{min}}^{E_{max}} n_{ind}\sigma_{cap}(E) dE, \quad (3.3)$$

where n and n_{ind} denote the thickness (the number of target nuclide per unit area) of the sample and indicator, respectively. The quantities σ_{tot} and σ_{cap} represent the energy-dependent neutron total and capture cross sections of the target nuclide, respectively. The integration is performed over the resonance peak region. By applying Eqs. (3.2) and (3.3), the relationships between the net area ratio and the thickness were obtained using point-wise cross-section data of JENDL-4.0 [7], as shown in Fig. 3.3. If $n\sigma_{tot}$ is not large, the net area ratio is proportional to the sample thickness. However, it converges to unity and loses information about thickness as $n\sigma_{tot}$ becomes larger. The thickness of each gold foil sample was derived from the relationship of Fig. 3.3 and the value of R was determined by experiment. Figure 3.4 shows the results of quantitative examination. It was confirmed that the thickness of the target nuclide can be determined by both methods within 3 % accuracy. The accuracy can be improved further by using a smaller resonance ($n\sigma_{tot}$ is not large).

The TOF spectra with silver and gold samples are shown with the NRTA method in Fig. 3.5 and with the self-indication method in Fig. 3.6. In NRTA, the dips of the 4.9 eV resonance of ^{197}Au and the 5.2 eV resonance of ^{109}Ag overlapped around 400 ch. in Fig. 3.5. In the self-indication method, the contribution from impurity was suppressed and a weak 58 eV resonance of ^{197}Au was emphasized around 120 ch. (Fig. 3.6). The TOF spectra for the mixture composed of ^{nat}U, ^{237}Np, and ^{243}Am are shown in Figs. 3.7 and 3.8. Although many resonance dips caused by impurities of ^{237}Np and ^{243}Am were observed (Fig. 3.7), there are no differences

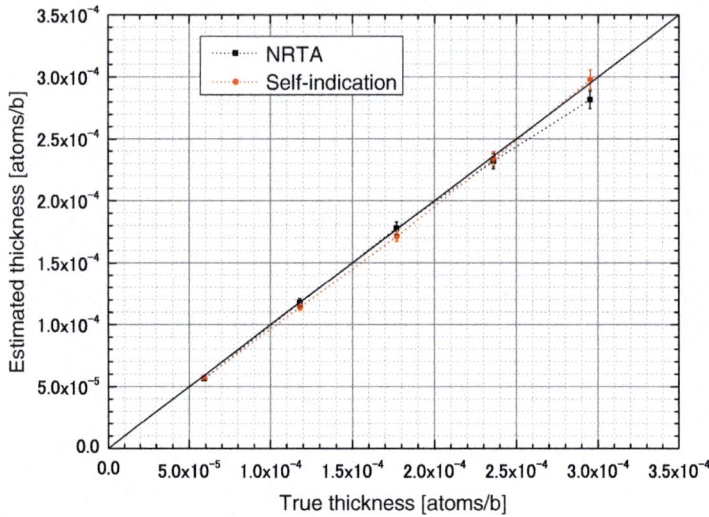

Fig. 3.4 Results of estimated sample thickness for ^{197}Au

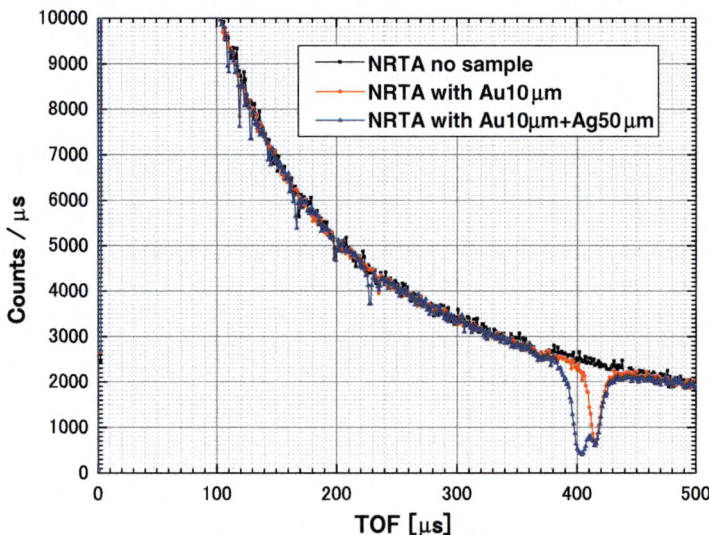

Fig. 3.5 TOF spectra obtained with neutron resonance transmission analysis (*NRTA*) for pure gold and mixture of gold and silver

between TOF spectra with only ^{nat}U (blue line) and the mixture (red line) (Fig. 3.8). This result indicates that the self-indication TOF spectrum was not greatly influenced by nuclide impurity. It was experimentally shown that the contribution from the other nuclide can be remarkably suppressed by applying the self-indication method.

3 Development of Nondestructive Assay to Fuel Debris of Fukushima Daiichi...

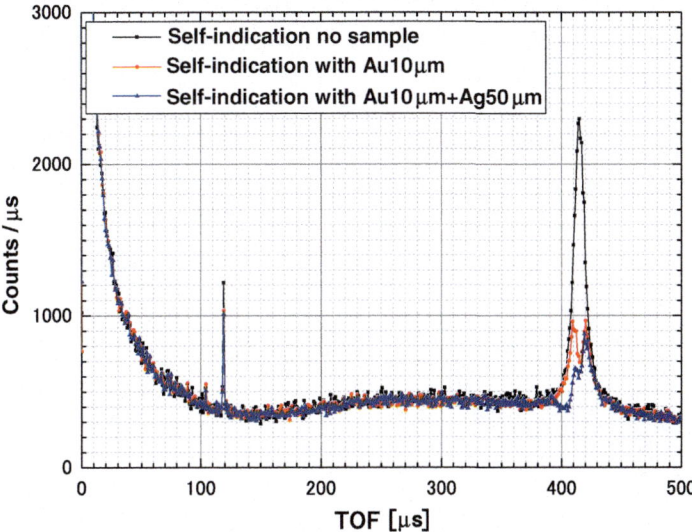

Fig. 3.6 TOF spectra obtained with the self-indication method for pure gold and mixture of gold and silver

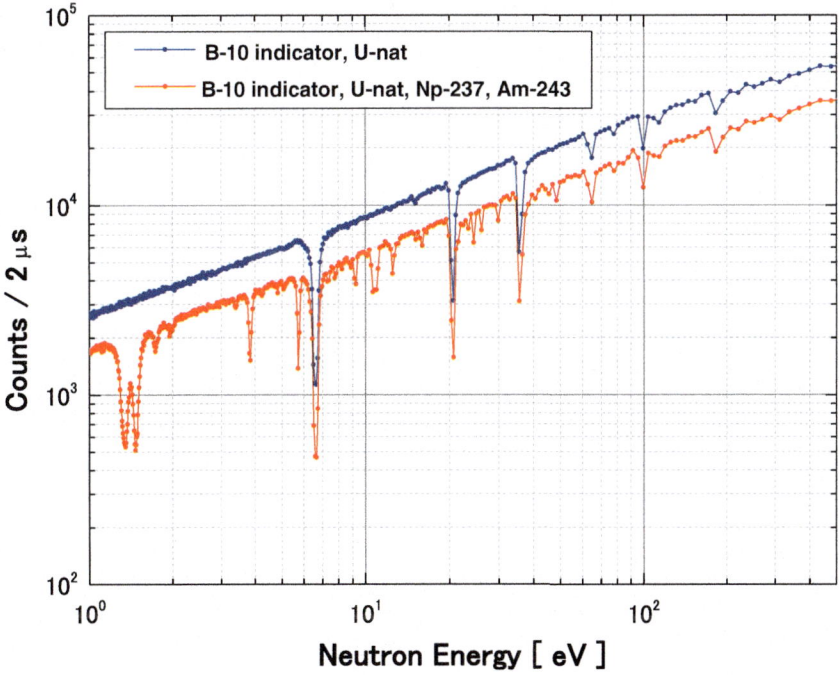

Fig. 3.7 TOF spectra obtained with NRTA for ^{nat}U and mixture of ^{nat}U, ^{237}Np, and ^{243}Am

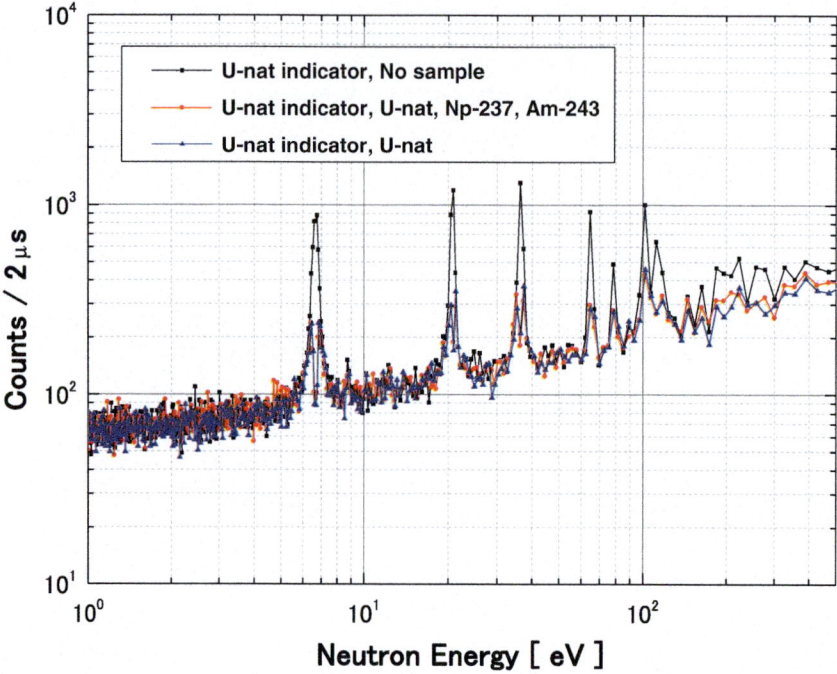

Fig. 3.8 TOF spectra obtained with the self-indication method for ^{nat}U and mixture of ^{nat}U, ^{237}Np, and ^{243}Am

3.4 Summary

In this work, we proposed a new concept of the "self-indication method" as a complementary nondestructive assay for the fuel debris of Fukushima Daiichi NPP. We carried out experimental validation for application of the self-indication method. It was confirmed that the area density (thickness of the target nuclide) can be determined within 3 % accuracy by simple area analysis without a resonance fitting process. Moreover, it was experimentally shown that the contribution from the other nuclide can be remarkably suppressed by applying the self-indication method. The self-indication method combined with the TOF technique will be a useful tool for nondestructive assaying of the distribution of nuclear material in the melted fuel debris, which contains many impurities and has high activities.

Acknowledgments This work was supported by JSPS KAKENHI Grant Number 24760714.

Open Access This chapter is distributed under the terms of the Creative Commons Attribution Noncommercial License, which permits any noncommercial use, distribution, and reproduction in any medium, provided the original author(s) and source are credited.

References

1. Postma H et al (2001) Radioanal Nucl Chem 248:115–120
2. Behrens JW, Johnson RG, Schrack RA (1984) Nucl Technol 67:162–168
3. Kiyanagi Y et al (2005) Measurement of eV-region pulse shapes of neutrons from KENS thermal neutron source by a neutron resonance absorption method. J Nucl Sci Technol 42(3):263–266
4. Pietropaolo A et al (2010) A neutron resonance capture analysis experimental station at the ISIS spallation source. Appl Spectrosc 64(9):1068–1071
5. Kobayashi K et al (1987) KURRI-Linac as a neutron source for irradiation. Annu Rep Res Reactor Inst Kyoto Univ 22:142
6. Yamamoto S et al (1996) Application of BGO scintillators to absolute measurement of neutron capture cross sections between 0.01eV to 10eV. J Nucl Sci Technol 33:815
7. Shibata K et al (2011) JENDL-4.0: a new library for nuclear science and engineering. J Nucl Sci Technol 48:1

Chapter 4
Development of Nondestructive Assay of Fuel Debris of Fukushima Daiichi NPP (2): Numerical Validation for the Application of a Self-Indication Method

Tadafumi Sano, Jun-ichi Hori, Yoshiyuki Takahashi, Hironobu Unesaki, and Ken Nakajima

Abstract To perform decommissioning of the Fukushima Daiichi NPP safely, it is very important to measure the components of the fuel debris. Therefore, a new nondestructive assay to identify and quantify the target nuclide in fuel debris using a pulsed-neutron source is under development in Kyoto University Research Reactor Institute.

We use the self-indication method for the nondestructive assay. This method is a neutron transmission method. The neutron transmission method is focused on resonance reactions (i.e., capture, fission) at the target nuclide. In the self-indication method, the transmitted neutrons from the sample are injected into an indicator. The indicator consists of a high-purity target nuclide. The transmitted neutrons are obtained by the time-of-flight (TOF) technique via resonance reactions in the indicator. The self-indication method has a high signal-to-noise (S/N) ratio compared to the conventional method.

In this study, numerical validation for the self-indication method to identify and quantify nuclides in a BWR-MOX pellet is described. The burn-up of the MOX pellet is 0 GWd/t, 10 GWd/t, 20 GWd/t, 30 GWd/t, 40 GWd/t, and 50 GWd/t. The 12-m measurement line in KUR-LINAC is simulated as a calculational geometry. Numerical calculations are carried out by continuous-energy Monte-Carlo code MVP2 with JENDL-4.0 as the nuclear data library. The burn-up calculations of the BWR-MOX pellet are performed by the deterministic neutronics code SARC 2006 with JENDL-4.0.

Numerical validation for application of the self-indication method is carried out. From the results, it is noted that the self-indication method has a good S/N ratio compared to the neutron transmission method for quantifying the amount of target nuclides in the fuel debris.

T. Sano (✉) • J. Hori • Y. Takahashi • H. Unesaki • K. Nakajima
Kyoto University Research Reactor Institute, 1010, Asashiro-nishi-2, Kumatori-cho, Sennan-gun, Osaka, Japan
e-mail: t-sano@rri.kyoto-u.ac.jp

Keywords Burn-up • KUR-LINAC • MOX pellet • Nondestructive assay • Numerical validation • Resonance • Self-indication method

4.1 Introduction

To perform decommissioning of the Fukushima Daiichi NPP safely, it is very important to measure the components of the fuel debris. Therefore, a new nondestructive assay to identify and quantify a target nuclide in the fuel debris using a pulsed-neutron source is under development in Kyoto University Research Reactor Institute.

We use the self-indication method for the nondestructive assay. This method is a neutron transmission method. The neutron transmission method is focused on resonance reactions (i.e., capture, fission) at the target nuclide. In the conventional neutron transmission method, a sample is irradiated by a pulsed-neutron beam and the energy distribution of transmitted neutrons from the sample is measured by the time-of-flight technique. Then, the target nuclide in the sample is identified and quantified by using the transmitted neutrons in the resonance energy region. This is a remarkably effective method to identify and quantify the target nuclide. However, if the energy spectrum of the transmitted neutron has many dips caused by resonance reactions of other nuclides, it is difficult to identify and quantify the target nuclide in the sample.

In the self-indication method, the transmitted neutrons from the sample are injected into an indicator, which consists of a high-purity target nuclide. The transmitted neutrons are obtained via resonance reactions in the indicator. The self-indication method has a high signal-to-noise (S/N) ratio compared to the conventional method.

In this chapter, numerical validation for application of the self-indication method is carried out. A calculational model and conditions are shown in Sect. 4.2 and the numerical results are shown in Sect. 4.3. From these results, some conclusions are drawn in Sect. 4.4.

4.2 Calculational Model and Condition

In this chapter, applicability of the self-indication method to identify and quantify nuclides in a BWR-MOX pellet is evaluated. The burnup of the MOX pellet is 0 GWd/t, 20 GWd/t, and 30 GWd/t. A plutonium vector in the fresh MOX pellet is employed as the OECD/NEA BWR MOX benchmark (Pu4) (^{235}U, 0.2 w/o; total Pu, 6.71 w/o; ^{238}Pu, 2.2 %; ^{239}Pu, 46.2 %; ^{240}Pu, 29.4 %; ^{241}Pu, 8.8 %) [1]. The burn-up calculations of the BWR-MOX pellet are carried out by using deterministic neutronics code SARC 2006 [2] with JENDL-4.0 [3]. The numerical validations are performed by using the MVP2.0 [4] with the JENDL-4.0. The MVP2.0 is a

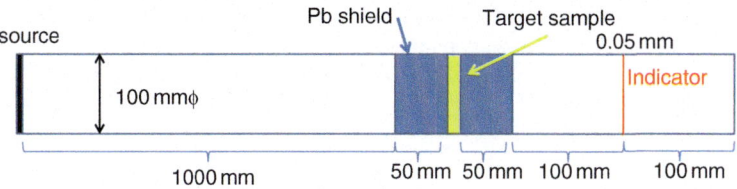

Fig. 4.1 Calculational geometry of 12-m measurement line in KUR-LINAC

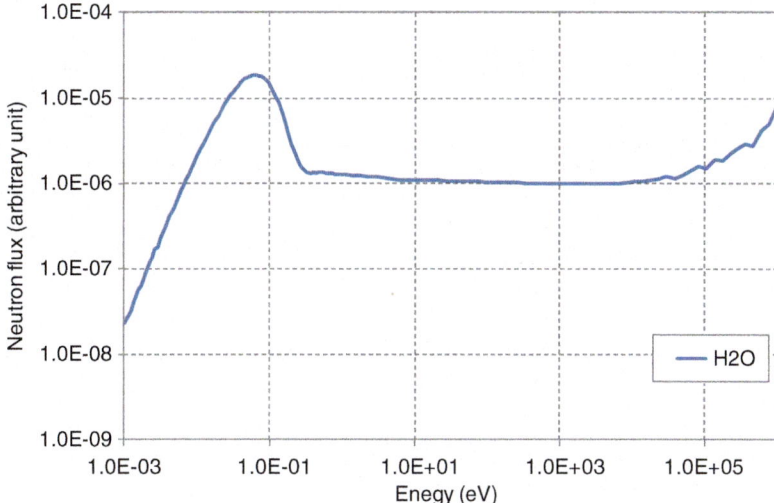

Fig. 4.2 Neutron spectrum in a Ta target of KUR-LINAC

continuous-energy Monte Carlo code developed by the Japan Atomic Energy Agency.

The 12-m measurement line in the KUR-LINAC is simulated as a calculational geometry shown in Fig. 4.1. Figure 4.2 shows a neutron spectrum in a tantalum target that is a neutron source of the KUR-LINAC. The spectrum is calculated by MVP2.0. Using the spectrum as the surface source, the validation is carried out.

4.3 Numerical Results and Discussion

The numerical validation for application of the self-indication method is discussed in this section. In the experiment, the transmitted neutron spectrum from the sample is measured via resonance reactions in the indicator whereas the reaction rates in the indicator are shown by numerical calculation. If the energy boundaries are made to have a finer division, the numerical result of the resonance reaction will have the same peak with dips as the measured data.

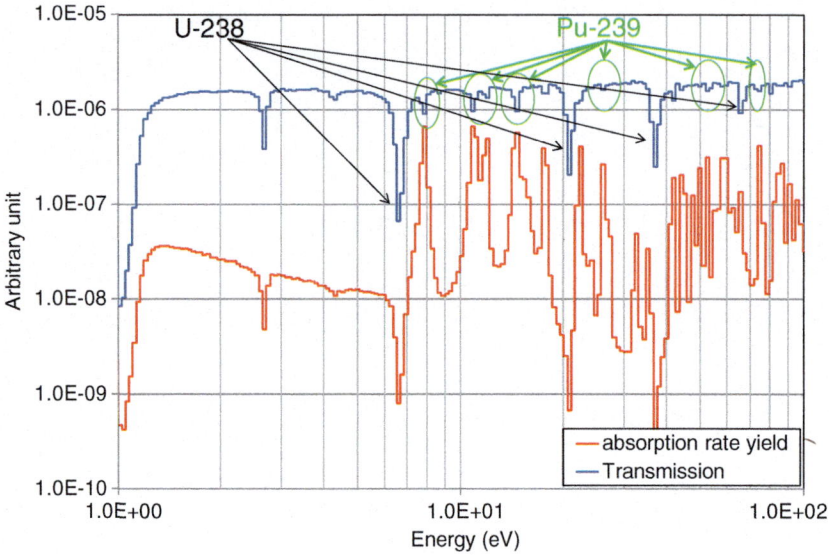

Fig. 4.3 Pu-239 absorption yield in an indicator (sample, 0 GWd/t)

Figure 4.3 shows the ^{239}Pu absorption rate yield by the present method (red) and the transmitted neutron spectrum by the conventional method (blue). The sample is the fresh (no burn-up) MOX pellet. Using the present method, one can easily obtain resonance absorption by ^{239}Pu. On the other hand, the transmitted neutron spectrum has many dips caused by resonance reaction of the other nuclides. Thus, if the sample is a burn-up pellet, it is difficult to quantify and identify by using the conventional method.

A numerical result to identify ^{129}I in the MOX pellet is described. The burn-up of the MOX pellet is 20 GWd/t. ^{129}I has only four resonances in the energy region of 0.1–100 eV: the resonance peaks are 41, 73, 75, and 97 eV. The transmitted neutrons are easily obtained via ^{129}I resonance absorption reactions in the indicator by the present method (Fig. 4.4).

Using the self-indication method, one cannot prepare a pure indicator to identify and quantify a target nuclide in a sample. Therefore, it is necessary to validate the application of the present method using an impure indicator. Figure 4.5 shows the numerical result of ^{239}Pu fission yield in the indicator, which has impure plutonium. The sample is a fresh MOX pellet, and the plutonium vector in the indicator is ^{239}Pu = 98.57 w/o, ^{239}Pu = 1.38 w/o, and ^{240}Pu = 0.05 w/o. In Fig. 4.5, the red line is a pure ^{239}Pu indicator, and the blue line shows that an indicator employed impure plutonium. Even in this case, as well as the result of using the pure ^{239}Pu as the resonance absorption in indicator is observed, it is shown to quantify and identify ^{239}Pu in the sample.

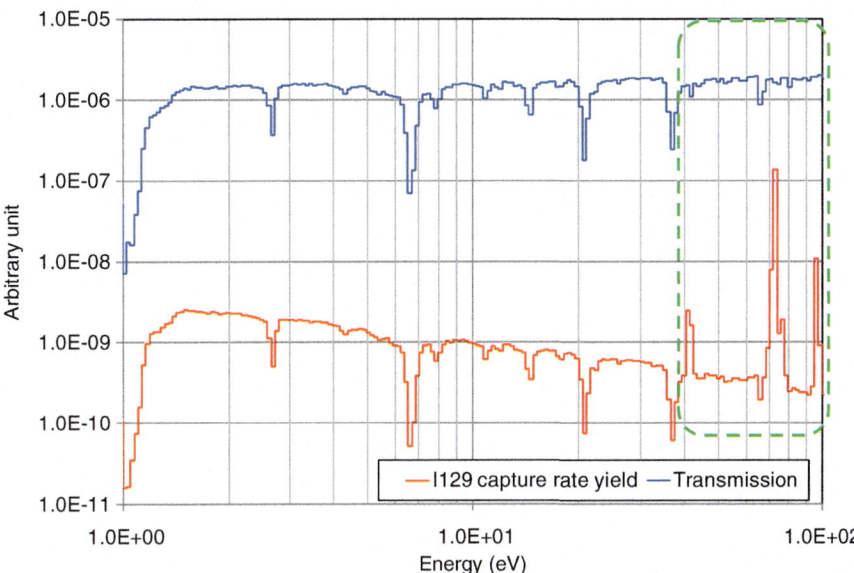

Fig. 4.4 I-129 absorption yield in an indicator (sample, 20 GWd/t)

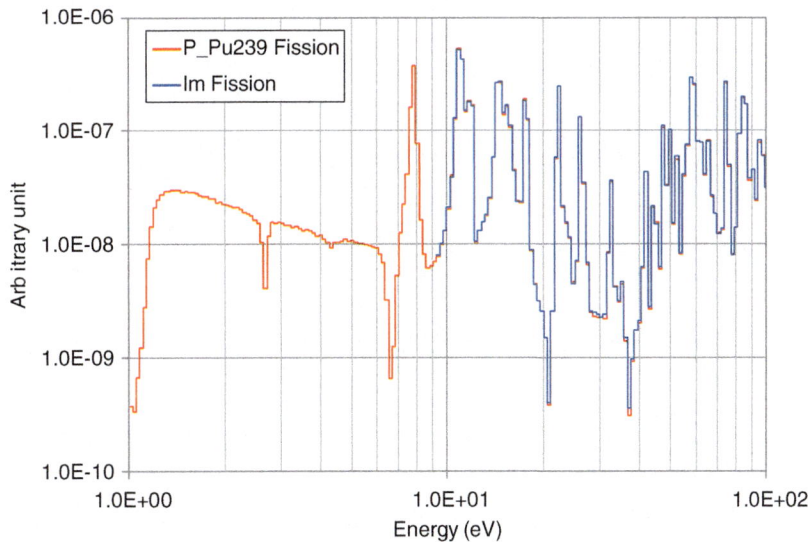

Fig. 4.5 Pu-239 absorption yield in an impure indicator (sample, 0 GWd/t)

Next, the applicability of the present method for fuel debris in Fukushima Daiichi NPP is examined. The fuel debris in Fukushima Daiichi NPP contains highly concentrated B-10, which has a large neutron absorption cross section. Thus, numerical validation of the present method and the conventional neutron transmission method for the sample with B-10 were carried out. The burn-up of the

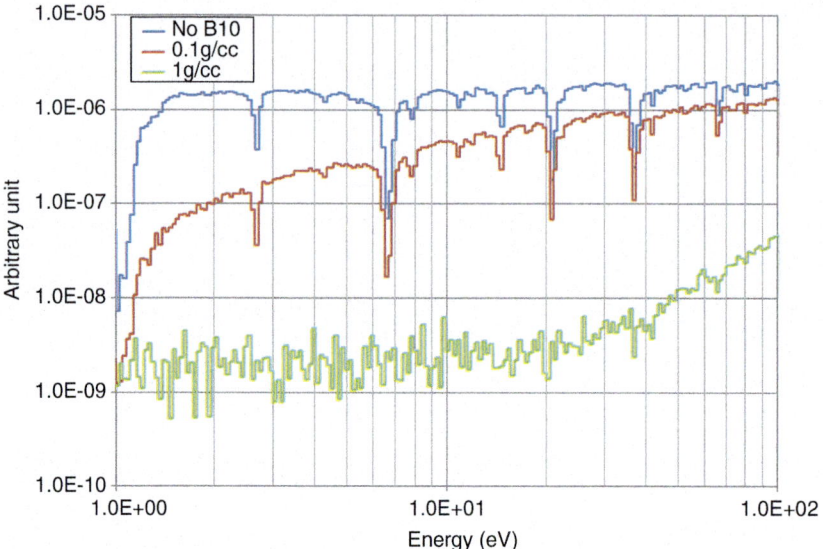

Fig. 4.6 Transmitted neutron spectrum from the sample (30 GWd/t) with B-10 (conventional method)

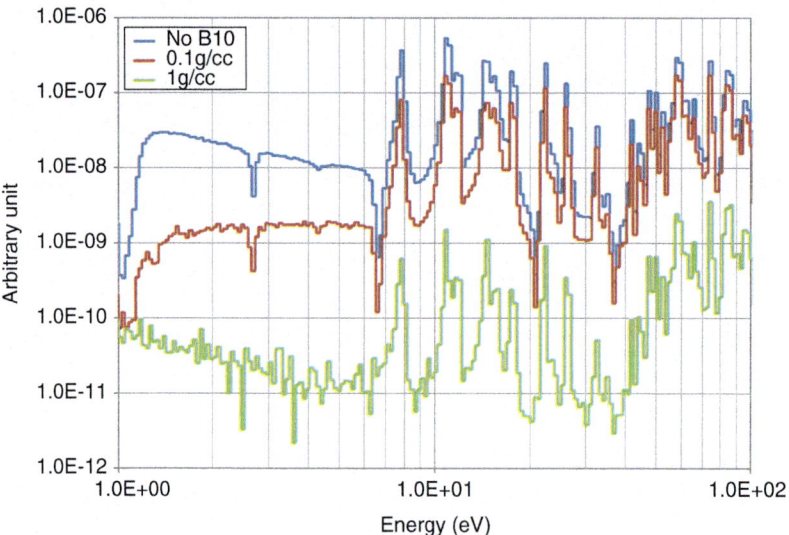

Fig. 4.7 Pu-239 absorption yield by self-indication method (sample, 30 GWd/t)

sample is 30 GWd/t. Using the transmission neutron method, it is difficult to obtain the dips caused by resonance reaction (Fig. 4.6) because neutron absorption by B-10 has a large contribution in the sample. On the other hand, one can obtain the neutron absorption rate yield in an indicator by the present method although the signal of the neutron is decreased (Fig. 4.7).

4.4 Conclusion

Numerical validation for application of the self-indication method has been carried out. As a result, the self-indication method is shown to have a better S/N than the neutron transmission method to quantify the amount of target nuclides.

The present method can be applied to identify and quantify a nuclide that has a small resonance, i.e., ^{129}I, and it is shown that one can measure an intended signal with good S/N by using an impure indicator. In addition, if the sample contains a highly concentrated neutron absorber, one can identify and quantify the target nuclide by using the self-indication method. Thus, the self-indication method can be applied to analyze the fuel debris in Fukushima Daiichi NPP.

Acknowledgments This work was supported by JSPS KAKENHI Grant Number 24760714.

Open Access This chapter is distributed under the terms of the Creative Commons Attribution Noncommercial License, which permits any noncommercial use, distribution, and reproduction in any medium, provided the original author(s) and source are credited.

References

1. Working Party on Physics of plutonium fuels and innovative fuel cycles (2003) Physics of Plutonium fuels BWR MOX benchmark specification and results. OECD/NEA, ISBN: 92-64-19905-5
2. Okumura K et al (2007) SRAC 2006: a comprehensive neutronics calculation code system. JAEA-Data/Code, 2007–004
3. Shibata K et al (2011) JENDL-4.0: a new library for nuclear science and engineering. J Nucl Sci Technol 48(1):1–30
4. Nagaya Y et al (2005) MVP/GMVP II: general purpose Monte Carlo codes for neutron and photon transport calculations based on continuous energy and multigroup methods. JAERI 1348, Japan Atomic Energy Research Institute

Chapter 5
Precise Measurements of Neutron Capture Cross Sections for LLFPs and MAs

S. Nakamura, A. Kimura, M. Ohta, T. Fujii, S. Fukutani, K. Furutaka,
S. Goko, H. Harada, K. Hirose, J. Hori, M. Igashira, T. Kamiyama,
T. Katabuchi, T. Kin, K. Kino, F. Kitatani, Y. Kiyanagi, M. Koizumi,
M. Mizumoto, M. Oshima, K. Takamiya, Y. Toh, and H. Yamana

Abstract To evaluate the feasibility of development of nuclear transmutation technology and an advanced nuclear system, precise nuclear data of neutron capture cross sections for long-lived fission products (LLFPs) and minor actinides (MAs) are indispensable. In this chapter, we present our research activities for the measurements of neutron capture cross sections for LLFPs and MAs.

Keywords Activation method • ANNRI • J-PARC • Long-lived fission products • Minor actinides • Neutron capture cross section • Time-of-flight method

---

S. Nakamura (✉) • A. Kimura • M. Ohta • K. Furutaka • S. Goko • H. Harada • K. Hirose
T. Kin • F. Kitatani • M. Koizumi • Y. Toh
Japan Atomic Energy Agency, 2-4 Shirane, Shirakata, Tokai-mura, Naka-gun,
Ibaraki 319-1195, Japan
e-mail: nakamura.shoji@jaea.go.jp

T. Fujii • S. Fukutani • J. Hori • K. Takamiya • H. Yamana
Research Reactor Institute, Kyoto University, 2-1010 Asashiro Nishi, Kumatori-cho,
Sennan-gun, Osaka 590-0494, Japan

M. Igashira • T. Katabuchi • M. Mizumoto
Research Laboratory for Nuclear Reactors, Tokyo Institute of Technology, O-okayama,
Meguro, Tokyo 152-8550, Japan

T. Kamiyama • K. Kino
Faculty of Engineering, Hokkaido University, Kita 13, Nishi 8, Kita-ku,
Sapporo 060-8628, Japan

Y. Kiyanagi
Faculty of Engineering, Hokkaido University, Kita 13, Nishi 8, Kita-ku,
Sapporo 060-8628, Japan

Nagoya University, Furo-cho, Chikusa-ku, Nagoya 464-8601, Japan

M. Oshima
Japan Atomic Energy Agency, 2-4 Shirane, Shirakata, Tokai-mura, Naka-gun,
Ibaraki 319-1195, Japan

Japan Chemical Analysis Center, 295-3 Sannou-cho, Inage-ku, Chiba-city,
Chiba 263-0002, Japan

© The Author(s) 2015
K. Nakajima (ed.), *Nuclear Back-end and Transmutation Technology for Waste Disposal*, DOI 10.1007/978-4-431-55111-9_5

5.1 Introduction

Associated with the social acceptability of nuclear power reactors, it is desirable to solve the problems of nuclear waste management of the long-lived fission products (LLFPs) and minor actinides (MAs) existing in spent nuclear fuels. A method of nuclear transmutation seems to be one of the solutions to reduce the radiotoxicity of nuclear wastes. The transmutation method makes it possible to reduce both the size of a repository for packages of nuclear wastes and the storage risks for the long term. To evaluate the feasibility of development of the nuclear transmutation method, precise nuclear data of neutron capture cross sections for LLFPs and MAs are indispensable.

This chapter presents joint research activities by JAEA and universities for measurements of the neutron capture cross sections for LLFPs and MAs by activation and neutron time-of-flight (TOF) methods.

5.2 Present Situation of Data for LLFPs and MAs

Although accurate data of neutron capture cross sections are necessary to evaluate reaction rates and burn-up times, there are discrepancies among the reported data for the thermal neutron capture cross sections for LLFPs and MAs. As an example of MA, Fig. 5.1 shows the trend of the thermal neutron capture cross section data for ^{237}Np: the discrepancies are about 10 %. Discrepancies between experimental and evaluated data still remain. As for LLFPs, e.g., ^{93}Zr, Fig. 5.2 shows that there

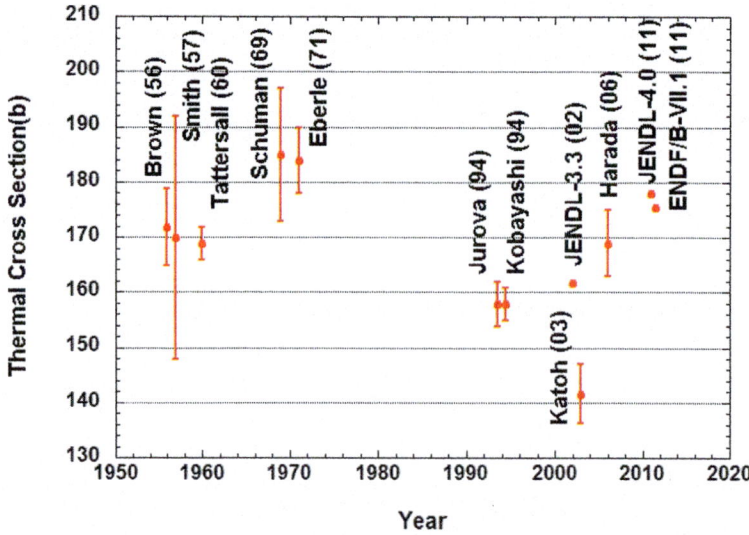

Fig. 5.1 Trend of thermal neutron capture cross section of ^{237}Np from the 1950s

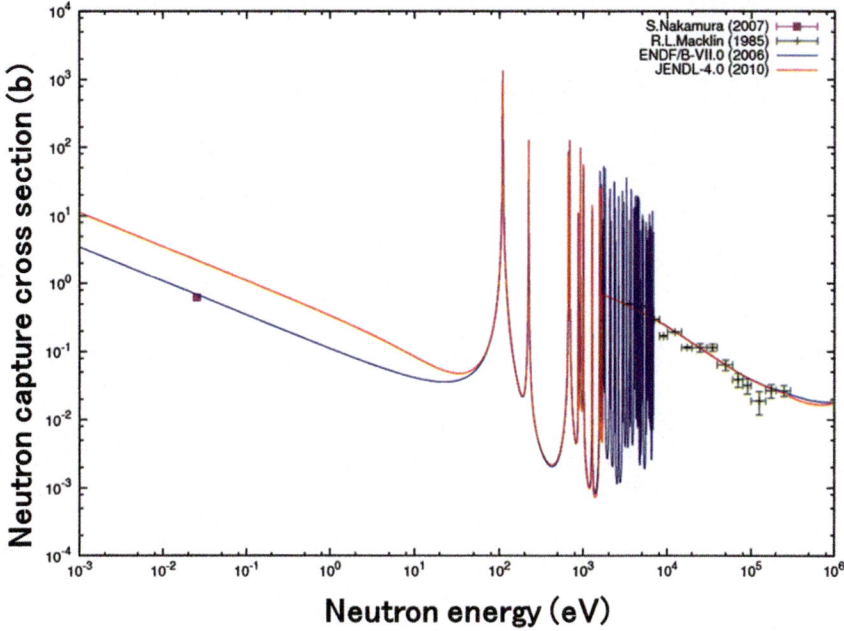

Fig. 5.2 Present situation of cross-section data for ^{93}Zr

are discrepancies between ENDF/B-VII.0 and JENDL-4.0 evaluations in the region of the thermal neutron energy. Thus, our concern was focused to remeasure the neutron capture cross sections of those LLFPs and MAs.

5.3 Measurement Activities by the Activation Method

Neutron capture cross sections were determined on the basis of Westcott's convention [1] by an activation method. The results for LLFPs [2–23] are listed in Table 5.1, and for MAs [24–31] in Table 5.2, together with previously reported data. Here, the symbols σ_{eff}, σ_0, and I_0 denote the effective cross section, the thermal neutron capture cross section, and the resonance integral, respectively; σ_0 is the cross section at the neutron energy of 25.3 meV.

Nuclear waste sometimes contains a large amount of stable nuclei having the same atomic number as that of long-lived fission products. These stable nuclei absorb thermal neutrons during the neutron irradiation of the nuclear waste and affect the neutron economics; the reaction rate of the target nuclei is reduced. Moreover, some of these stable nuclei breed more radioactive nuclei by the neutron capture process. It is also necessary for transmutation study to accurately estimate these influences caused by stable nuclei involved in the FP targets. The cross sections of the stable nuclei, such as ^{127}I [14] and ^{133}Cs [20], were also measured; the results are shown in Table 5.1.

Table 5.1 Results of thermal neutron capture cross sections and resonance integrals for long-lived fission products (LLFPs)

Nuclide (half-life)	Reported data (author, year)	JAEA data
^{137}Cs (30 years)	$\sigma_{\text{eff}} = 0.11 \pm 0.03$ b (Stupegia 1960 [2])	$\sigma_0 = 0.25 \pm 0.02$ b
		$I_0 = 0.36 \pm 0.07$ b (1990, 1993, 2000 [3–5])
^{90}Sr (29 years)	$\sigma_{\text{eff}} = 0.8 \pm 0.5$ b (Zeisel 1966 [6])	$\sigma_0 = 10.1 \pm 1.3/4.2$ mb
		$I_0 \leq 0.16$ b (1994 [7])
		$\sigma_0 = 10.1 \pm 1.3$ mb
		$I_0 = 104 \pm 16$ mb (2001 [8])
^{99}Tc (2.1×10^5 years)	$\sigma_0 = 20 \pm 2$ b	$\sigma_0 = 22.9 \pm 1.3$ b
	$I_0 = 186 \pm 16$ b (Lucas 1977 [9])	$I_0 = 398 \pm 38$ b (1995 [10])
^{129}I (1.6×10^7 years)	$\sigma_0 = 27 \pm 2$ b	$\sigma_0 = 30.3 \pm 1.2$ b
	$I_0 = 36 \pm 4$ b (Eastwood 1958 [11])	$I_0 = 33.8 \pm 1.4$ b (1996 [12])
^{127}I (Stable)	$\sigma_0 = 4.7 \pm 0.2$ b	$\sigma_0 = 6.40 \pm 0.29$ b
	$I_0 = 109 \pm 5$ b (Friedman 1983 [13])	$I_0 = 162 \pm 8$ b (1999 [14])
^{135}Cs (3×10^6 years)	$\sigma_0 = 8.7 \pm 0.5$ b	$\sigma_0 = 8.3 \pm 0.3$ b
	$I_0 = 61.7 \pm 2.3$ b (Baerg 1958 [15])	$I_0 = 38.1 \pm 2.6$ b (1997 [16])
^{134}Cs (2 years)	$\sigma_{\text{eff}} = 134 \pm 12$ b (Bayly 1958 [17])	$\sigma_{\text{eff}} = 141 \pm 9$ b (1999 [18])
^{133}Cs (Stable)	$\sigma_0 = 30.4 \pm 0.8$ b	$\sigma_0 = 29.0 \pm 1.0$ b
	$I_0 = 461 \pm 25$ b (Baerg 1960 [19])	$I_0 = 298 \pm 16$ b (1999 [20])
^{166m}Ho (1.2×10^3 years)	$\sigma_0 = 9{,}140 \pm 650$ b	$\sigma_{\text{eff}} = 3 \pm 1$ kb (2000 [22])
	$I_0 = 1{,}140 \pm 90$ b (Masyanov 1993 [21])	$\sigma_0 = 3.11 \pm 0.82$ kb
		$I_0 = 10.0 \pm 2.7$ kb (2002 [23])

Table 5.2 Results of thermal neutron capture cross sections and resonance integrals for minor actinides (MAs)

Nuclide (half-life)	Reported data (author, year)	JAEA data
^{237}Np (2.14×10^6 years)	$\sigma_0 = 158 \pm 3$ b	$\sigma_0 = 141.7 \pm 5.4$ b
	$I_0 = 652 \pm 24$ b (Kobayashi 1994 [24])	$I_0 = 862 \pm 51$ b (2003 [25])
		$\sigma_0 = 169 \pm 6$ b (2006 [26])
^{238}Np (2.1 days)	No data	$\sigma_{\text{eff}} = 479 \pm 24$ b (2004 [27])
^{241}Am (432 years)	$\sigma_{0,\,g} = 768 \pm 58$ b	$\sigma_{0,\,g} = 628 \pm 22$ b
	$I_{0,\,g} = 1{,}694 \pm 146$ b (Shinohara 1997 [28])	$I_{0,\,g} = 3.5 \pm 0.3$ kb (2007 [29])
^{243}Am (7,370 years)	$\sigma_{0,\,m} = 80$ b	$\sigma_{\text{eff}} = 174.0 \pm 5.3$ b (2006 [31])
	$\sigma_{0,\,g} = 84.3$ b (Ice 1966 [30])	

As seen in Table 5.1, the thermal cross section for ^{137}Cs is about twice as large as the previous data reported by Stupegia [2]. As for ^{90}Sr, its thermal cross section is found to be much smaller than the data reported by Zeisel [6]. As seen in Table 5.2, the cross section of ^{238}Np is obtained for the first time. Thus, the joint research activities of the Japan Atomic Energy Agency (JAEA) and universities have measured the cross sections for important LLFPs and MAs by the activation method.

5.4 Measurement Activities at J-PARC/MLF/ANNRI

A new experimental apparatus called the accurate neutron nucleus reaction measurement instrument (ANNRI) has been constructed on the beam line no. 4 (BL04) of the MLF in the J-PARC. The ANNRI has two detector systems. One of them is a large Ge detector array, which consists of two cluster-Ge detectors, eight coaxial-shaped Ge detectors, and BGO Compton suppression detectors; the other is a large NaI(Tl) spectrometer (Fig. 5.3). The ANNRI has an advantage for neutron cross-section measurements because the MLF facility can provide the strongest neutron intensity in the world.

The neutron capture cross sections of ^{237}Np [32, 33], ^{241}Am [34], ^{244}Cm [35], ^{93}Zr [36], ^{99}Tc [37], and ^{107}Pd [38] have been measured relative to the ^{10}B(n, $\alpha\gamma$) standard cross section by the TOF method. Some highlights of results obtained in our research activities are shown in Fig. 5.4 for ^{237}Np and in Fig. 5.5 for ^{93}Zr.

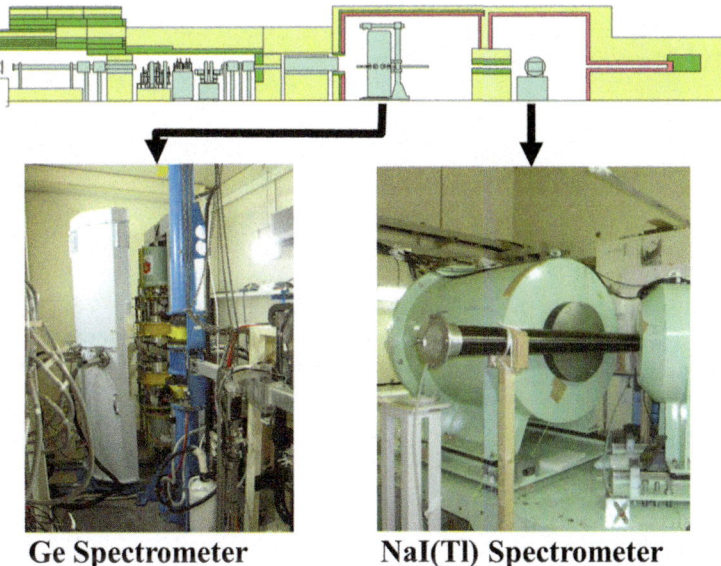

Ge Spectrometer   **NaI(Tl) Spectrometer**

Fig. 5.3 A new experimental apparatus called the accurate neutron nucleus reaction measurement instrument (ANNRI). The cross-sectional view of ANNRI is shown in the *upper panel*, the spectrometer is on the *left side*, and the NaI(Tl) spectrometer is on the *right side*

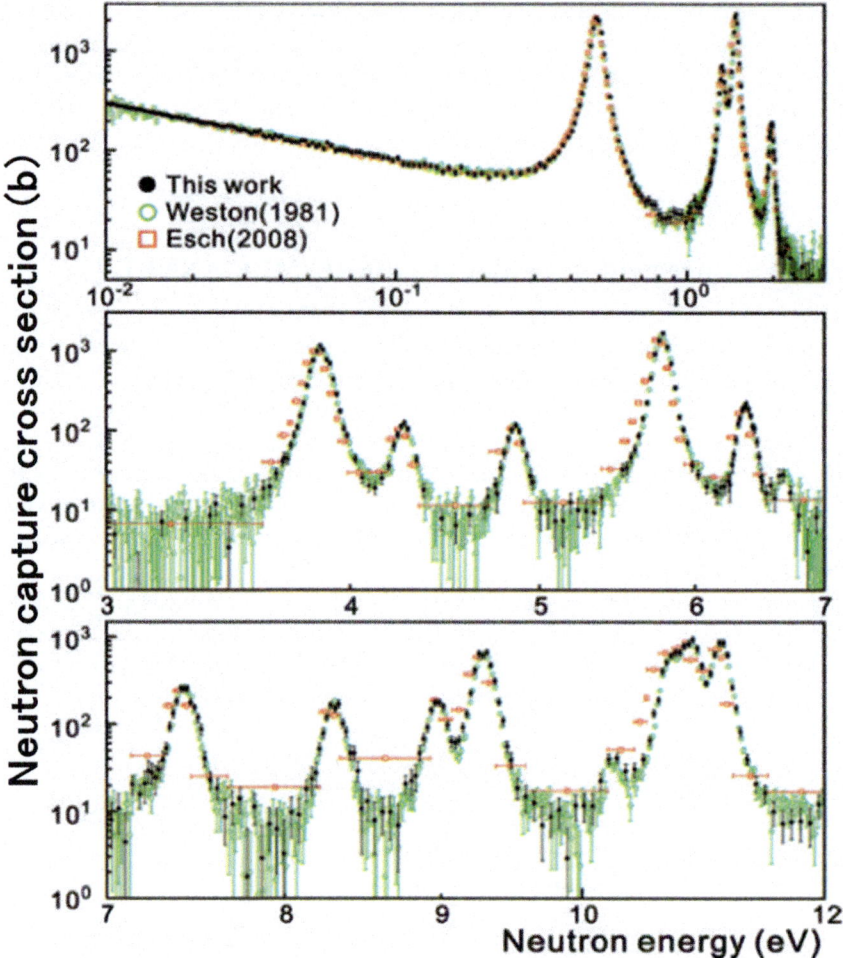

Fig. 5.4 ^{241}Am cross section in neutron energy from 0.01 to 10 eV

The results obtained at the ANNRI are good agreement with the data reported by Weston (Fig. 5.4). The ^{93}Zr cross sections in Fig. 5.5 present results greatly different from the evaluated data in the thermal neutron energy region. One finds that the present results support the value of the thermal cross section reported in 2007 [39].

5.5 Summary

This chapter described the JAEA research activities for the measurement of neutron capture cross sections for LLFPs and MAs by activation and neutron time-of-flight (TOF) methods. We summarized our results of the thermal neutron capture cross

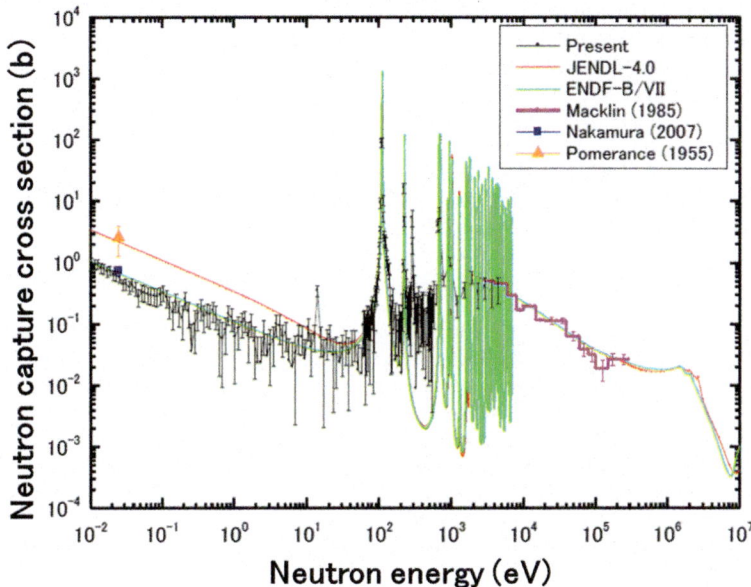

Fig. 5.5 ^{93}Zr cross section (tentative data) together with the evaluated data

section and the resonance integral for some of the important LLFPs and MAs by the activation method.

Operation of a new experimental apparatus called the accurate neutron nucleus reaction measurement instrument (ANNRI) in the MLF at J-PARC has been started for neutron capture cross-section measurements of MAs and LLFPs. Some of the highlights of our results have been shown here.

Acknowledgments The authors thank the staff at Kyoto University Reactor Institute, Rikkyo University Reactor and JRR-3M. A part of this work has been carried out under the Visiting Researcher's Program of the Research Reactor Institute, Kyoto University. Moreover, the authors appreciate the accelerator staff of J-PARC for their operation of the accelerator.

This work is supported by JSPS KAKENHI (22226016).

Open Access This chapter is distributed under the terms of the Creative Commons Attribution Noncommercial License, which permits any noncommercial use, distribution, and reproduction in any medium, provided the original author(s) and source are credited.

References

1. Westcott CH et al (1958) Proceedings of 2nd International Conference on Peaceful Uses of Atomic Energy, Geneva, vol 16, United Nations, New York, p. 70
2. Stupegia DC (1960) J Nucl Energ A12:16
3. Harada H et al (1990) J Nucl Sci Technol 27(6):577

4. Sekine T et al (1993) J Nucl Sci Technol 30(11):1099
5. Wada H et al (2000) J Nucl Sci Technol 37(10):827
6. Zeisel G (1966) Acta Phys Austr 23:5223
7. Harada H et al (1994) J Nucl Sci Technol 31(3):173
8. Nakamura S et al (2001) J Nucl Sci Technol 38(12):1029
9. Lucas M et al (1977) IAEA-TC-119/14, p 407–432
10. Harada H et al (1995) J Nucl Sci Technol 32(5):395
11. Eastwood TA et al (1958) Proceedings of 2nd International Conference on Peaceful Uses of Atomic Energy, Geneva, vol 16. United Nations, New York, p 54
12. Nakamura S et al (1996) J Nucl Sci Technol 33(4):283
13. Friedmann L et al (1983) Radiochim Acta 33:182
14. Nakamura S et al (1999) J Nucl Sci Technol 36(3):223
15. Baerg AP et al (1958) Can J Phys 36:863
16. Katoh T et al (1997) J Nucl Sci Technol 34(5):431
17. Bayly JG et al (1958) Inorg Nucl Chem 5:259
18. Katoh T et al (1999) J Nucl Sci Technol 36(8):635
19. Baerg AP et al (1960) Can J Chem 38:2528
20. Nakamura S et al (1999) J Nucl Sci Technol 36(10):847
21. Masyanov SM et al (1993) Atom Energ 73:673
22. Harada H et al (2000) J Nucl Sci Technol 37(9):821
23. Katoh T et al (2002) J Nucl Sci Technol 39(7):705
24. Kobayashi K et al (1994) J Nucl Sci Technol 31(12):1239
25. Katoh T et al (2003) J Nucl Sci Technol 40(8):559
26. Harada H et al (2006) J Nucl Sci Technol 43(11):1289
27. Harada H et al (2004) J Nucl Sci Technol 41(1):1
28. Shinohara N et al (1997) J Nucl Sci Technol 34(7):613
29. Nakamura S et al (2007) J Nucl Sci Technol 44(12):1500
30. Ice CH (1966) Du Pont: Savannah River reports, vol 66, p 69
31. Ohta M et al (2006) J Nucl Sci Technol 43(12):1441
32. Hirose K et al. (2014) Nuclear Data Sheets 119:48–51
33. Hirose K et al (2013) J Nucl Sci Technol 50:188–200
34. Harada H et al (2014) Nuclear Data Sheets 119:61–64
35. Kimura A et al (2012) J Nucl Sci Technol 49:708–724
36. Hori J et al (2011) J Korean Phys Soc 59(2):1777–1780
37. Kino K et al (2014) Nuclear Data Sheets 119:140–142
38. Nakamura S et al (2014) Nuclear Data Sheets 119:143–146
39. Nakamura S et al (2007) J Nucl Sci Technol 44(1):21

Chapter 6
Development of the Method to Assay Barely Measurable Elements in Spent Nuclear Fuel and Application to BWR 9 × 9 Fuel

Kenya Suyama, Gunzo Uchiyama, Hiroyuki Fukaya, Miki Umeda, Toru Yamamoto, and Motomu Suzuki

Abstract In fission products in used nuclear fuel, there are several stable isotopes that have a large neutron absorption effect. For evaluation of the neutronics characteristics of a nuclear reactor, the amount of such isotopes should be evaluated by using burn-up calculation codes. To confirm the correctness of such data obtained by calculation codes, it is important to assure the precision of the evaluation of the neutron multiplication factor of used nuclear fuel. However, it is known that there are several hardly measurable elements in such important fission products. Data for the amounts of the hardly measurable elements in used nuclear fuel are scarce worldwide.

The Japan Atomic Energy Agency (JAEA) had been developing a method to assess the amounts of these fission products that are hardly measurable and have a large neutron capture cross section, under the auspices of the Japan Nuclear Energy Safety Organization. In this work, a measurement method was developed combining a simple and effective chemical separation scheme of fission products from used nuclear fuel and an inductively coupled plasma mass spectrometry with high sensitivity and high precision. This method was applied to the measurement program for the used BWR 9 × 9 fuel assembly. This measurement method is applicable to the required measurements for countermeasures to the accident at the Fukushima Dai-ichi Nuclear Power Plant of Tokyo Electric Power Company (TEPCO). JAEA has a measurement plan for not only BWR but also PWR fuel.

K. Suyama (✉) • G. Uchiyama • H. Fukaya • M. Umeda
Japan Atomic Energy Agency, 2-4 Shirakata-Shirane, Tokai-mura, Ibaraki-ken 319-1195, Japan
e-mail: suyama.kenya@jaea.go.jp

T. Yamamoto
Secretariat of Nuclear Regulation Authority, Roppongi-First Bld. 14F, 1-9-9 Roppongi, Minato-ku, Tokyo 106-8450, Japan

M. Suzuki
Central Research Institute of Electric Power Industry, 2-11-1 Iwadokita, Komae-shi, Tokyo 201-8511, Japan

This presentation describes the measurement method developed in the study as well as the future measurement plan in JAEA.

Keywords Fission products • Isotopic composition • Post-irradiation examinations

6.1 Introduction

In fission products in used nuclear fuel, there are several stable isotopes that have a large neutron absorption effect. For evaluation of the neutronics characteristics of a nuclear reactor, the amounts of such isotopes should be evaluated by using burn-up calculation codes. For this purpose, a quantitative analytical method of uranium, plutonium, and fission products of spent fuels has been studied [1, 2]. However, it is known that there are several barely measurable elements in such important fission products.

To assay the amount of many fission products, radiation measurement is widely used. Cesium-134 and -137 are typical examples. However, this method is not applicable for isotopes that are important from the aspect of reactivity assessment because they are stable isotopes. For such isotopes, there is the possibility of adopting the isotopic dilution method (IDM), which has been used for measurement of actinides and a few fission products such as neodymium.

In the Japan Atomic Energy Agency (JAEA), thermal ionization mass spectrometry (TIMS) has been used for IDM to evaluate the burn-up value of the used fuel. TIMS is one of the most reliable instruments to determine the isotopic ratio and the obtained result is considered to be the reference. However, TIMS needs relatively large amounts of the fuel solution sample and a long time is required to obtain the final results after dissolution of the fuel and preparation of the measurement sample.

The most serious problem is that the important fission isotopes for the reactivity assessment belong to the rare earth elements. Because many of these have the same mass number, we need an efficient chemical separation method and high-performance instruments for measuring the isotopic composition that which should have high sensitivity and resolution. For this reason, the fission products important for reactivity assessment are barely measurable and available data for such fission products are scarce.

JAEA has been active in measuring the isotopic composition of the spent nuclear fuel from the 1980s and the obtained data have been archived in the SFCOMPO database [3], which has been supported by the OECD/NEA databank. Based on this past experience, JAEA launched a development program [4] of measurement of the fission products important for reactivity assessment under the auspices of the Japan Nuclear Energy Safety Organization (JNES) in 2008 and successfully finalized the program in 2012.

In this program, a combined method of chromatographic separation of uranium, plutonium, and fission products from irradiated nuclear fuels was developed. Furthermore, by the introduction of high-resolution inductively coupled plasma mass spectrometry (HR-ICP-MS), the IDM has been applied to lanthanide nuclides. The developed method was applied to the measurement of isotopic composition of used BWR 9×9 fuel and evaluation of the burn-up calculation code was carried out [5].

After the accidents at Fukushima Dai-ichi Nuclear Power Plants (hereafter referred to as 1F) of Tokyo Electric Power Company (TEPCO) in 2011, we need a confirmed method to assay the composition of the fuel irradiated in 1F to carry out decommissioning of the Fukushima site. For this purpose, JAEA has a further measurement plan of not only BWR but also PWR used fuel to obtain enough experience to measure the isotopic composition of the irradiated nuclear fuels.

This report summarizes the analytical procedure to measure the amount of fission products isotopes developed in JAEA and the future measurement program.

6.2 Analytical Procedure

The objective fission products required for reactivity assessment are samarium, europium and gadolinium. Cesium-133 is also required. Of the important fission products, several metallic isotopes exist in the barely dissolved residue: ^{97}Mo, ^{99}Tc, ^{101}Ru, ^{103}Ru, and ^{109}Ag.

We decided to adopt the isotopic dilution method (IDM) and the calibration curve method to measure the amounts of the stable fission products. For this purpose, we introduced the high-resolution inductively coupled plasma mass spectrometry (HR-ICP-MS), ELEMENT2 of Thermo Fisher Scientific (Photo 6.1). This instrument has very high sensitivity and enough precision and accuracy to measure the isotopic ratios of objective elements belonging to the rare earth elements.

In this technical development, five samples taken from ZN2 (average burn-up is 35.6 GWd/t) and ZN3 (average burn-up is 53.5GWd/t) fuel assemblies of used fuel of Fukushima Dai-ni Nuclear Power Plant Unit 1 (2F-1) were used for demonstrating the measurement method. Sample positions are shown in Figs. 6.1, 6.2, and 6.3. Five fuel samples taken from ZN2 and ZN3 fuel assemblies were dissolved initially in 3 M nitric acid solution at about 110 °C, then the dissolution residue was dissolved again in mixed solutions of nitric, hydrochloric, and sulfuric acid at 180 °C.

Before the measurement of isotopic ratio, the isobar should be separated to avoid contamination. Figure 6.4 shows a schematic of chemical separation. The dissolution solutions of spent fuels were filtrated and the filtrate solution was fed to an anion-exchange resin of UTEVA (Eichrom, USA) to separate U, Pu, and Nd individually. Figure 6.5 shows the yields of lanthanide in each fraction eluted from the Ln resin column in the separation experiment using a simulated dissolution solution of spent fuel. U and Pu in the solution were effectively separated from the solution with more than 95 % efficiency. The eluate solution from the UTEVA resin column was fed to the Ln resin column. Lanthanides elements were separated with hydrochloric acid solutions in the Ln resin column.

Photo 6.1 High-resolution inductively coupled plasma mass spectrometry (HR-ICP-MS) introduced in JAEA for the measurement of fission product nuclides [6]

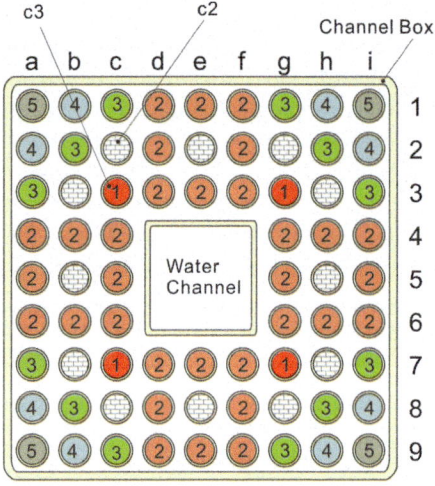

Fig. 6.1 Position of samples taken from ZN2 fuel assemblies (average burn-up, 35.6 GWd/t) [4]

6 Development of the Method to Assay Barely Measurable Elements in Spent...

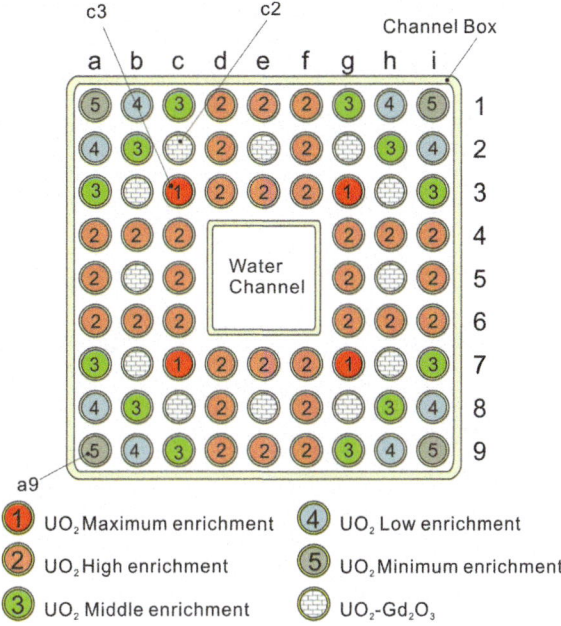

Fig. 6.2 Position of samples taken from ZN3 fuel assemblies (average burn-up, 53.5 GWd/t) [4]

By using HR-ICP-MS, the isotopic ratios of neodymium isotopes (^{142}Nd, ^{143}Nd, ^{144}Nd, ^{145}Nd, ^{146}Nd, ^{148}Nd, and ^{150}Nd), samarium isotopes (^{147}Sm, ^{148}Sm, ^{149}Sm, ^{150}Sm, ^{151}Sm, ^{152}Sm, and ^{154}Sm), europium isotopes (^{151}Eu and ^{153}Eu), and gadolinium isotopes (^{152}Gd, ^{154}Gd, ^{155}Gd, ^{156}Gd, ^{157}Gd, ^{158}Gd, and ^{160}Gd) in spent fuels were determined. The concentrations of ^{97}Mo, ^{99}Tc, ^{101}Ru, ^{103}Ru, and ^{109}Ag in the residue solution were analyzed by HR-ICP-MS by adopting the calibration curve method.

This new measurement method has the following merits.

1. It requires a sample of 10 μg, one-tenth of the amount required for use of TIMS.
2. We need 5 days to obtain results, which is one-sixth of the time required by the previous method using TIMS.
3. Obtained accuracy (1s) is approximately 1 %, which is sufficient considering the required measurement error.

Figures 6.6, 6.7, and 6.8 are examples of measurement results for ^{101}Ru, ^{103}Rh, and ^{150}Sm, respectively.

Fig. 6.3 Axial sampling position [4]

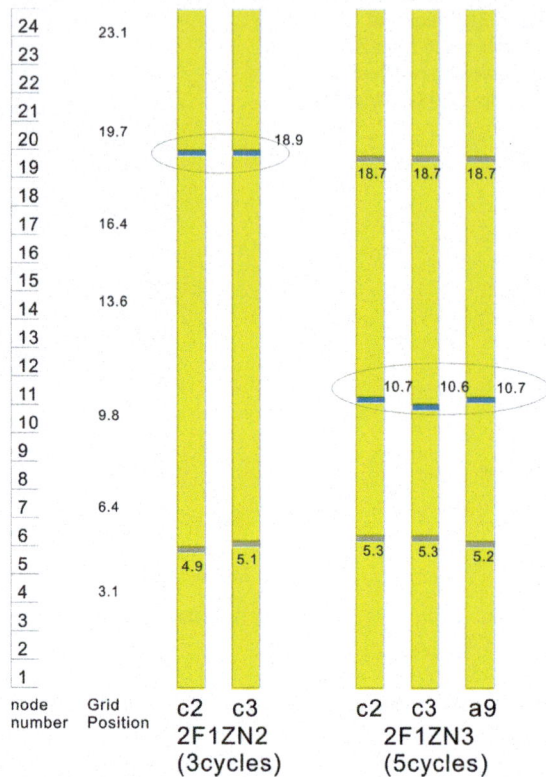

Fig. 6.4 Analytical procedure to separate fission products

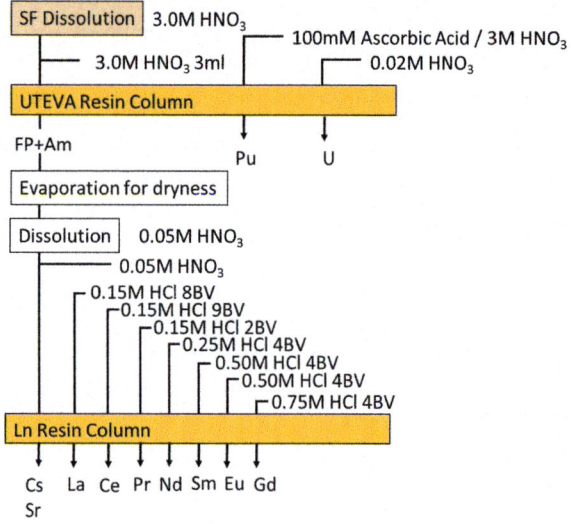

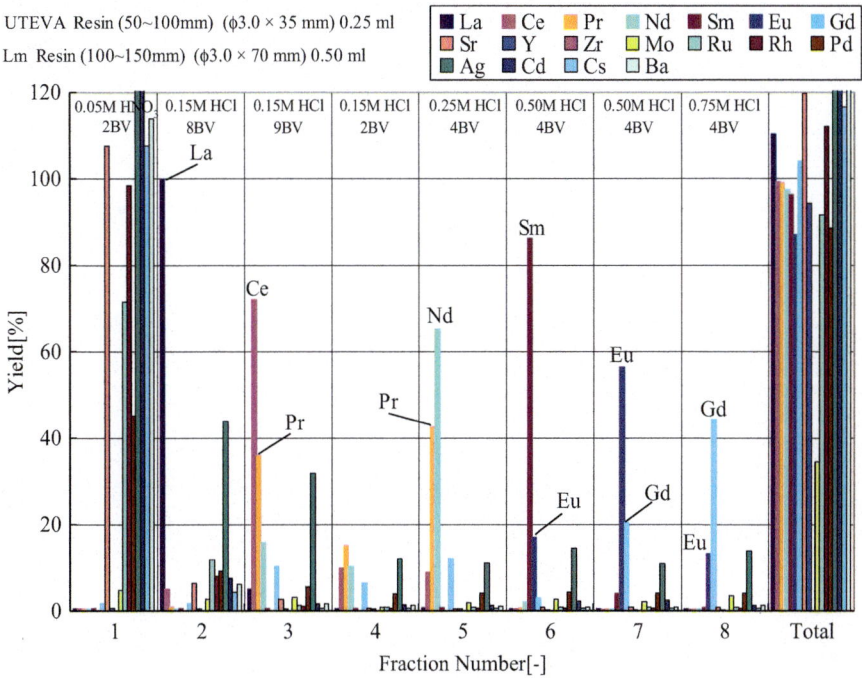

Fig. 6.5 Separation behavior of lanthanides elements with hydrochloric acid solutions as eluent in the Ln resin column [7]

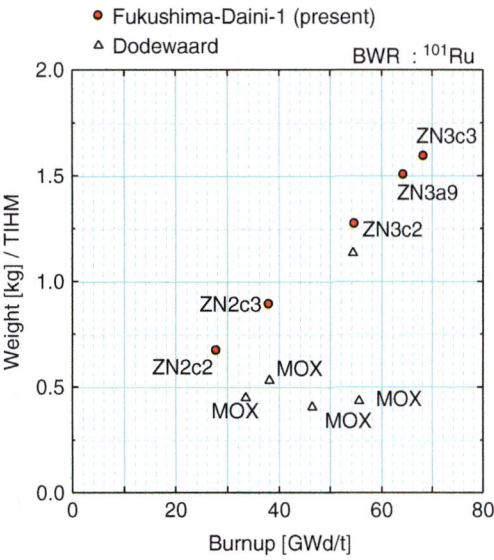

Fig. 6.6 Measurement results of ^{101}Ru [4]

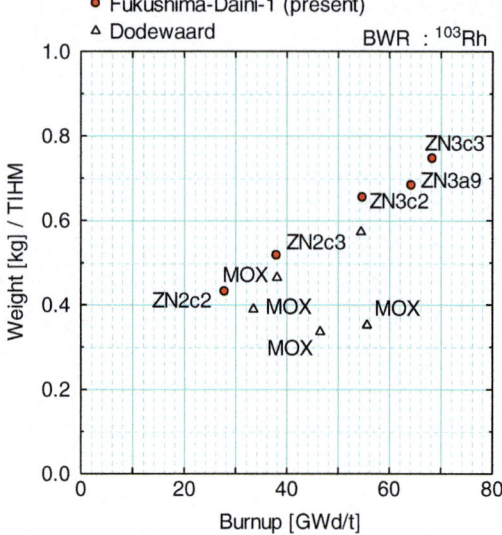

Fig. 6.7 Measurement results of ^{103}Rh [4]

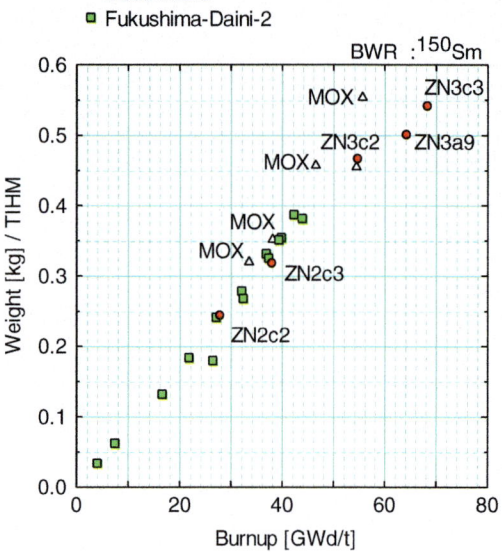

Fig. 6.8 Measurement results of ^{150}Sm [4]

6.3 Future Plans

Initially, this measurement technique and procedure have been developed to obtain experimental data for the demonstration of the neutronics calculation code. However, after the accident at the Fukushima Dai-ichi Nuclear Power Plant in 2011, it is recognized that the measurement of the isotopic composition of the used nuclear fuel is crucial to take countermeasures to the accident. We expect that measurement of the amounts of many varieties of isotopes is required for the decommissioning of the Fukushima Dai-ichi site.

To accumulate experience and recheck the measurement procedure, JAEA has already started the measurement campaign after the Fukushima accident. The first PIE sample was taken from the same fuel assembly used in the PIE campaign described in the earlier sections. In 2012, one fuel sample was taken from the ZN3 fuel assembly irradiated in Fukushima Dai-ni nuclear power plant unit 1 and the dissolution was conducted in 2013. It is expected that the measurement results will be obtained in F.Y. 2014.

JAEA will also assay the isotopic composition of spent nuclear fuel irradiated in PWR. It is planned to measure the isotopic compositions of nine samples taken from NO4F69 fuel assembly irradiated in Ohi Nuclear Power Station unit 4 of the Kansai Electric Power Co., Inc. (KEPCO). The nine PIE samples will be taken from three fuel rods including UO_2 and UO_2-Gd_2O_3 whose average burn-up values are estimated to be from 40 to 57 GWd/t approximately. This measurement campaign was started in 2013, and the first results are hoped to be seen in 2014.

6.4 Conclusion

The Japan Atomic Energy Agency had been active in the field of postirradiation examinations since the 1980s. Based on past experience and introducing the state-of-the-art technique, JAEA began a measurement program of fission products that are important for reactivity assessment. By this program, the quantitative analytical method based on isotopic dilution technology has been developed for fission products in spent fuels. JAEA will carry out the measurement program of isotopic composition of the used nuclear fuel for countermeasures to the accident at Fukushima Dai-ichi Nuclear Power Station. This program consists of the measurement of several samples not only from BWR but also PWR. The obtained results will be used for the evaluation of the burnup code system. Also, the experience of treating spent fuel and measuring its isotopic composition will strengthen the technical ability of JAEA for providing countermeasures for the Fukushima accident.

Acknowledgments The authors thank the following staff in charge of the measurement program in JAEA: K. Tonoike, M. Amaya, M. Umeda, T. Sonoda, K. Watanabe, N. Shinohara, M. Ito,

T. Ueno, M. Magara, J. Inagawa, S. Miyata, S. Sampei, K. Kamohara, M. Sato, H. Usami, K. Ohkubo, M. Totsuka, Y. Sakazume, and T. Kurosawa. The authors are deeply indebted to T. Ichihara and T. Nakai of the Kansai Electric Power Co., Inc., and M. Kawasaki and I. Hyodo of Tokyo Electric Power Co., Inc., who allowed us to use their spent fuel assemblies and to use owned technical data. The authors also like express their appreciation to Y. Taniguchi, H. Nagano, T. Ito, H. Kishita, Y. Kubo, and K. Kakiuchi of Nuclear Fuel Industries, Ltd., for their cooperation in using their spent fuel assemblies and technical data.

Open Access This chapter is distributed under the terms of the Creative Commons Attribution Noncommercial License, which permits any noncommercial use, distribution, and reproduction in any medium, provided the original author(s) and source are credited.

References

1. Ito M, Fukaya Y, Ueno T et al (2011) Isotopic composition measurement of FP nuclides in spent LWR UO_2 fuel and nucleonic analyses [2] measurement of FP nuclides. In: Proceedings of the 2011 annual meeting of the Atomic Energy Society of Japan, Fukui-shi, Japan
2. Suzuki M, Yamamoto T, Nakajima T et al (2012) Isotopic composition measurement of FP nuclides in spent LWR UO_2 fuel and nucleonic analyses [4]: Nucleonic analysis of metallic FP isotopes. In: Proceedings of the 2012 fall meeting of the Atomic Energy Society of Japan, Higashi-Hiroshima-shi, Japan
3. SFCOMPO - Spent Fuel Isotopic Composition Database, http://www.oecd-nea.org/sfcompo/, OECD/NEA
4. Fukaya H, Suyama K, Sonoda T, Okubo K, Umeda M, Uchiyama G (2013) Examination of measurement method of isotopic composition of fission products in spent fuel. JAEA-Research 2013-020, Japan Atomic Energy Agency [in Japanese]
5. Suzuki M, Yamamoto T, Fukaya H, Suyama K, Uchiyama G (2013) Lattice physics analysis of measured isotopic compositions of irradiated BWR 9×9 UO_2 fuel. J Nucl Sci Technol 50(12):1161–1176
6. JAEA R&D Review 2013, http://jolisfukyu.tokai-sc.jaea.go.jp/fukyu/mirai-en/2013/, Japan Atomic Energy Agency (2013)
7. Uchiyama G, Fukaya H, Suyama K, Ito M, Ueno T, Miyata S, Sonoda T, Usami H, Kawasaki Y, Sakadume Y, Kurosawa T, Tonoike K, Inagawa J, Umeda M, Magara M, Watanabe K, Shinohara N (2012) Development of quantitative analytical method of U, Pu and fission products in spent fuels by combined-chromatographic-mass-spectrometric-isotopic-dilution technology. In: Proceedings of ENC 2012, Manchester

Part II
Development of ADS Technologies: Current Status of Accelerator-Driven System Development

Chapter 7
Contribution of the European Commission to a European Strategy for HLW Management Through Partitioning & Transmutation

Presentation of MYRRHA and Its Role in the European P&T Strategy

Hamid Aït Abderrahim

Abstract MYRRHA (Multi-purpose hYbrid Research Reactor for High-tech Applications) is an experimental accelerator-driven system (ADS) currently being developed at SCK•CEN for replacement of material testing reactor BR2. The MYRRHA facility is conceived as a flexible fast-spectrum irradiation facility that is able to run in both subcritical and critical modes. The applications catalogue of MYRRHA includes fuel developments for innovative reactor systems, material developments for GEN IV systems and fusion reactors, doped silicon production, radioisotope production, and fundamental science applications, thanks to the high-power proton accelerator. Next to these applications, MYRRHA will demonstrate the ADS full concept by coupling a high-power proton accelerator, a multi-megawatts spallation target, and a subcritical reactor at reasonable power level to allow operational feedback, scalable to an industrial demonstrator, and to allow the study of efficient transmutation of high-level nuclear waste. Because MYRRHA is based on the heavy liquid metal technology, namely lead–bismuth eutectic (LBE), it will be able to significantly contribute to the development of Lead Fast Reactor (LFR) technology and will have the role of European Technology Pilot Plant in the roadmap for LFR. The current design of the MYRRHA ADS and its ability to contribute to the European Commission strategy for high-level waste management through Partitioning and Transmutation (P&T) are discussed in this chapter.

Keywords ADS • HLW Management • MYRRHA • P&T

H.A. Abderrahim (✉)
SCK•CEN, Boeretang 200, 2400 Mol, Belgium
e-mail: haitabde@sckcen.be; myrrha@sckcen.be

7.1 Introduction

When concerned with energy, one cannot avoid considering geostrategic questions and the international political situation. Indeed, major armed conflicts in the world in past decades are taking place in major fossil energy production countries or on the major roads connecting places of great production with those of large consumption. Therefore, Europe is very concerned about the security of its supply in terms of energy, especially when considering the limited energy fossil resources in the European Union (EU). As such, nuclear power remains a major energy source in the EU.

Presently, the EU relies, for 30 % of its electric power production, on generation II–III fission nuclear reactors, leading to the annual production of 2,500 t/year of used fuel, containing 25 t plutonium, and high-level wastes (HLW) such as 3.5 t of minor actinides (MA), namely, neptunium (Np), americium (Am), and curium (Cm), and 3 t of long-lived fission products (LLFPs). These MA and LLFP stocks need to be managed in an appropriate way. The reprocessing of used fuel (closed fuel cycle) followed by geological disposal, or direct geological disposal (open fuel cycle), are today the envisaged solutions in Europe, depending on national fuel cycle options and waste management policies. The required time scale for geological disposal exceeds our accumulated technological knowledge, and this remains the main concern of the public. Partitioning and Transmutation (P&T) has been pointed out in numerous studies as the strategy that can relax constraints on geological disposal and reduce the monitoring period to technological and manageable time scales. Therefore, a special effort is ongoing in Europe and beyond to integrate P&T in advanced fuel cycles and advanced options for HLW management. Transmutation based on critical or subcritical fast-spectrum transmuters should be evaluated to assess the technical and economic feasibility of this waste management option, which could ease the development of a deep geological storage.

Despite diverse strategies and policies pursued by European Member States concerning nuclear power and the envisaged fuel cycle policy ranging from the once-through without reprocessing to the double-strata fuel cycle ending with ADS as the ultimate burner or generation IV (Gen-IV) fast critical reactors multi-recycling all transuranic (TRUs), P&T requires an integrated effort at the European and even worldwide level. Even when considering the phase-out of nuclear energy, the combination of P&T and a dedicated burner such as ADS technologies, at a European scale, would allow meeting the objectives of both types of countries, those phasing out nuclear energy as well as countries favoring the continuation of nuclear energy development toward the deployment of new fast-spectrum systems.

The concept of partitioning and transmutation has three main goals: reduction of the radiological hazard associated with spent fuel by reducing the inventory of minor actinides, reduction of the time interval required to reach the radiotoxicity of

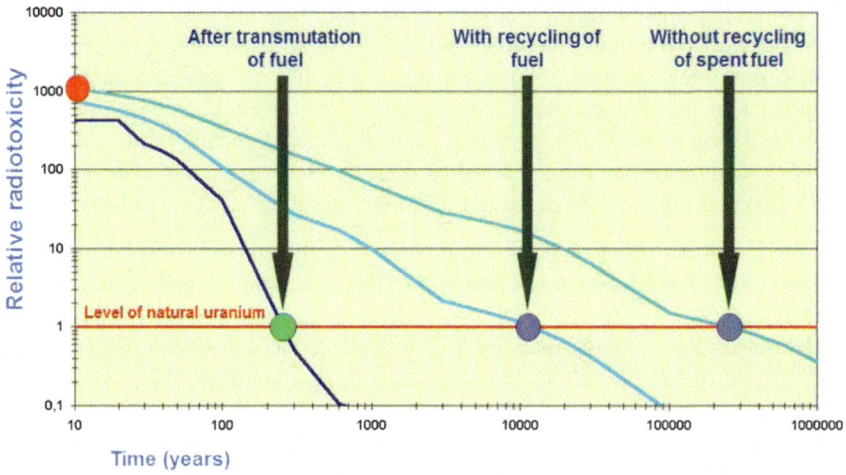

Fig. 7.1 Radiotoxicity of radioactive waste [4]

natural uranium, and reduction of the heat load of the HLW packages to be stored in geological disposal, leading to its efficient use.

Transmutation of high-level radioactive elements with a long half-life present in the nuclear waste reduces the radiological impact of the actinides (such as americium, curium, and neptunium) and fission products. The time scale (Fig. 7.1) needed for the radiotoxicity of the waste to drop to the level of natural uranium will be reduced from a 'geological' value (500,000 to 1 million years) to a value that is comparable to that of human activities (several hundreds of years) [1–3]. During transmutation, the nuclei of the actinides are fissioned into shorter-lived fission products.

To transmute the minor actinides in an efficient way, high intensity and high energy neutron fluences are necessary. Therefore, only nuclear fast fission reactors, being critical or subcritical, can be utilized.

If the aim is to transmute large amounts of minor actinides in the dedicated transmuter then it is necessary to use an accelerator-driven system. The subcriticality is mandatory because of the smaller delayed neutron fraction within the minor actinides (0.01–0.1 %) compared to uranium-235 (0.7 %) to allow the criticality variation control.

After nearly 20 years of basic research funded by national programs and EURATOM framework programs, the research community needs to be able to quantify indicators for decision makers, such as the proportion of waste to be channeled to this mode of management, but also issues related to safety, radiation protection, transport, secondary wastes, costs, and scheduling.

From 2005, the research community on P&T within the EU started structuring its research toward a more integrated approach. This effort resulted, during the FP6, into two large integrated projects, namely, EUROPART dealing with partitioning, and EUROTRANS dealing with accelerator driven system (ADS), design for transmutation, development of advanced fuel for transmutation, R&D activities related to the heavy liquid metal technology, innovative structural materials, and nuclear data measurement. This approach resulted in a European strategy, the so-called four building blocks at engineering level for P&T, as given next. The implementation of P&T of a large part of the high-level nuclear wastes in Europe needs the demonstration of its feasibility at an "engineering" level. The respective R&D activities could be arranged in these four "building blocks," as listed next:

1. Demonstration of the capability to process a sizable amount of spent fuel from commercial LWRs to separate plutonium (Pu), uranium (U), and minor actinides (MA),
2. Demonstration of the capability to fabricate, at a semi-industrial level, the dedicated fuel needed to load in a dedicated transmuter (JRC-ITU)
3. Design and construction of one or more dedicated transmuters
4. Provision of a specific installation for processing of the dedicated fuel unloaded from the transmuter, which can be of a different type than that used to process the original spent fuel unloaded from commercial power plants, together with the fabrication of new dedicated fuel

These "blocks will" result in identification of the costs and benefits of partitioning and transmutation for European society.

7.2 MYRRHA: A Flexible Fast-Spectrum Irradiation Facility

MYRRHA (Multi-purpose hYbrid Research Reactor for High-tech Applications) is the flexible experimental accelerator-driven system (ADS) in development at SCK•CEN. MYRRHA is able to work both in subcritical (ADS) and in critical mode. In this way, MYRRHA targets the following applications catalogue:

- To demonstrate the ADS full concept by coupling the three components (accelerator, spallation target, and subcritical reactor) at reasonable power level (50–100 MW$_{th}$) to allow operation feedback, scalable to an industrial demonstrator;
- To allow the study of the efficient technological transmutation of high-level nuclear waste, in particular, minor actinides that would require high fast flux intensity ($\Phi_{>0.75 \text{MeV}} = 10^{15}$ n/cm^2 s);
- To be operated as a flexible fast-spectrum irradiation facility allowing for
  - Fuel developments for innovative reactor systems, which need irradiation rigs with a representative flux spectrum, a representative irradiation temperature, and high total flux levels ($\Phi_{tot} = 5 \cdot 10^{14}$ to 10^{15} n/cm^2 s); the main target will be fast-spectrum GEN IV systems, which require fast-spectrum conditions;
  - Material developments for GEN IV systems, which need large irradiation volumes with high uniform fast flux level ($\Phi_{>1 \text{ MeV}} = 1 \sim 5 \cdot 10^{14}$ n/cm^2 s) in

various irradiation positions, representative irradiation temperature, and representative neutron spectrum conditions; the main target will be fast-spectrum GEN IV systems;
- Material developments for fusion reactors, which need also large irradiation volumes with high constant fast flux level ($\Phi_{>1\,MeV} = 1 \sim 5 \cdot 10^{14}$ n/cm^2 s), a representative irradiation temperature, and a representative ratio appm He/dpa(Fe) = 10;
- Radioisotope production for medical and industrial applications by
  - Holding a backup role for classical medical radioisotopes;
  - Focusing on R&D and production of radioisotopes requiring very high thermal flux levels ($\Phi_{thermal} = 2$ to $3 \cdot 10^{15}$ n/cm^2 s) because of double-capture reactions;
- Industrial applications, such as Si-doping, need a thermal flux level depending on the desired irradiation time: for a flux level $\Phi_{thermal} = 10^{13}$ n/cm^2 s, an irradiation time in the order of days is needed, and for a flux level of $\Phi_{thermal} = 10^{14}$ n/cm^2 s, an irradiation time in the order of hours is needed to obtain the required specifications.

Further in this section, we discuss some basic characteristics of the accelerator and of the core and primary system design.

7.3 The MYRRHA Accelerator

The accelerator is the driver of MYRRHA because it provides the high-energy protons that are used in the spallation target to create neutrons, which in turn feed the core. In the current design of MYRRHA, the machine must be able to provide a proton beam with energy of 600 MeV and an average beam current of 3.2 mA. The beam is delivered to the core in continuous wave (CW) mode. Once per second, the beam is shut off for 200 μs so that accurate on-line measurements and monitoring of the subcriticality of the reactor can take place. The beam is delivered to the core from above through a beam window.

Accelerator availability is a crucial issue for the operation of the ADS. A high availability is expressed by a long mean time between failure (MTBF), which is commonly obtained by a combination of overdesign and redundancy. In addition to these two strategies, fault tolerance must be implemented to obtain the required MTBF. Fault tolerance will allow the accelerator to recover the beam within a beam trip duration tolerance after failure of a single component. In the MYRRHA case, the beam trip duration tolerance is 3 s. Within an operational period of MYRRHA, the number of allowed beam trips exceeding 3 s must remain under 10. Shorter beam trips are allowed without limitations. The combination of redundancy and fault tolerance should allow obtaining a MTBF value in excess of 250 h.

At present, proton accelerators with megawatt-level beam power in CW mode only exist in two basic concepts: sector-focused cyclotrons and linear accelerators (linacs). Cyclotrons are an attractive option with respect to construction costs, but they do not have any modularity, which means that a fault tolerance scheme cannot be implemented. Also, an upgrade of its beam energy and intensity for industrial application presently is not a realistic option. A linear accelerator, especially if made superconducting, has the potential for implementing a fault tolerance scheme and offers a high modularity, resulting in the possibility to recover the beam within a short time and increasing the beam energy and intensity toward industrial application of ADS technology.

7.4 Design of the Core and Primary System

Because MYRRHA is a pool-type ADS, the reactor vessel houses all the primary systems. In previous designs of MYRRHA, an outer vessel served as secondary containment in case the reactor vessel leaks or breaks. In the current design, the reactor pit fulfills this function, improving the capabilities of the reactor vault air cooling system. The vessel is closed by the reactor cover, which supports all the in-vessel components. A diaphragm, inside the vessel, acts to separate the hot and cold LBE plenums; it supports the in-vessel fuel storage (IVFS) and provides a pressure separation. The core is held in place by the core support structure consisting of a core barrel and a core support plate. Figure 7.2 shows vertical cut sections of the MYRRHA reactor showing its main internal components.

At the present state of the design, the reactor core (Fig. 7.3) consists of mixed oxide (MOX) fuel pins, typical for fast reactors. In subcritical mode, the central hexagon houses a window beam tube-type spallation target. Thirty-seven positions can be occupied by in-pile test sections (IPS) or by the spallation target (the central one of the core in subcritical configuration) or by control and shutdown rods (in the core critical configuration). This design gives a large flexibility in the choice of the more suitable position (neutron flux) for each experiment.

The requested high fast flux intensity has been obtained by optimizing the core configuration geometry (fuel rod diameter and pitch) and maximizing the power density. We will be using, for the first core loadings, 15-15Ti stabilized stainless steel as cladding material instead of T91 ferritic-martensitic steel that will be qualified progressively further on during MYRRHA operation for a later use. The use of lead–bismuth eutectic (LBE) as coolant permits lowering the core inlet operating temperature (down to 270 °C), decreasing the risk of corrosion and allowing increasing the core ΔT. This design, together with the adoption of reliable and passive shutdown systems, will allow meeting the high fast flux intensity target.

In subcritical mode, the accelerator (as described in the previous section) is the driver of the system. It provides the high-energy protons that are used in the spallation target to create neutrons which in their turn feed the subcritical core. In subcritical mode the spallation target assembly, located in the central position of the

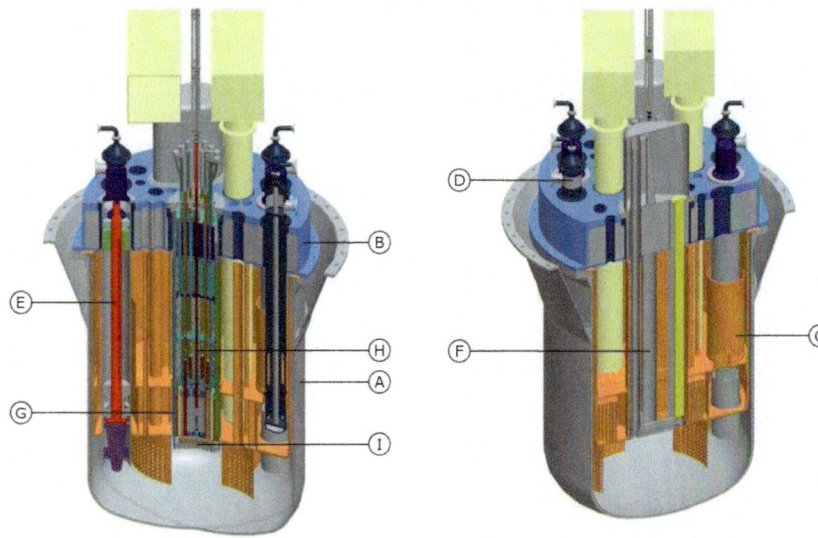

A. Reactor Vessel
B. Reactor Cover
C. Diaphragm
D. Primary Heat Exchanger
E. Primary Pump
F. In-vessel Fuel Handling Machine
G. Core
H. Above Core Structure
I. Core Restraint System

Fig. 7.2 Section of the MYRRHA-FASTEF reactor

core, brings the proton beam via the beam tube into the central core region. The spallation heat deposit is dissipated to the reactor primary circuit. The spallation module guarantees the barrier between the reactor LBE and the reactor hall and ensures optimal conditions for the spallation reaction. The spallation module assembly is conceived as an IPS and is easily removable or replaceable.

The primary, secondary, and tertiary cooling systems have been designed to evacuate a maximum thermal core power of 110 MW. The 10 MW more than the nominal core power account for the power deposited by the protons, for the power of in-vessel fuel, and for the power deposited in the structures by γ-heating. The average coolant temperature increase in the core in nominal conditions is 140 °C with a coolant velocity of 2 m/s. The primary cooling system consists of two pumps and four primary heat exchangers (PHX).

The interference of the core with the proton beam, the fact that the room located directly above the core will be occupied by much instrumentation and IPS penetrations, and core compactness result in insufficient space for fuel handling to (un)load the core from above. Since the very first design of MYRRHA, fuel handling has been performed from underneath the core. Fuel assemblies are kept by buoyancy under the core support plate.

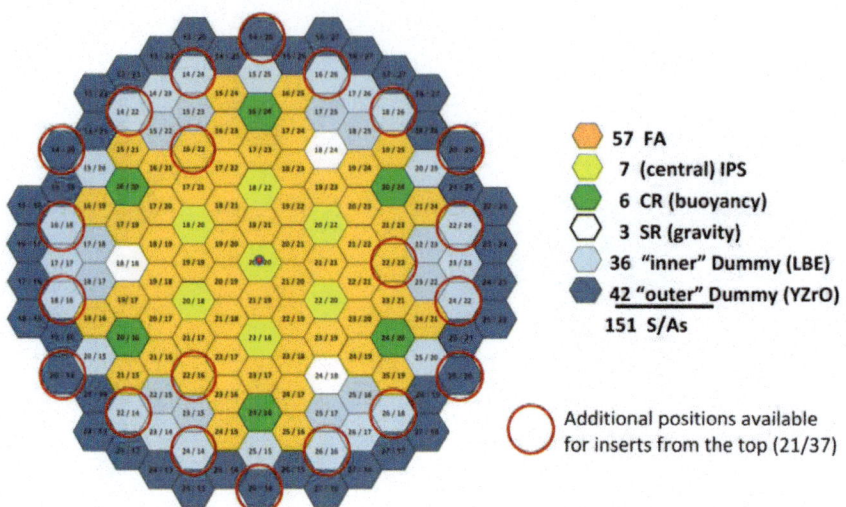

Fig. 7.3 Cut of the MYRRHA-FASTEF core, showing the central target, the different types of fuel assemblies, and dummy components

Two fuel-handling machines are used, located at opposite sides of the core (Fig. 7.4). Each machine covers one side of the core. The use of two machines provides sufficient range to cover the necessary fuel storage positions without the need of an increase for the reactor vessel when only one fuel-handling machine is used. Each machine is based on the well-known fast reactor technology of the 'rotating plug' concept using SCARA (Selective Compliant Assembly Robot Arm) robots. To extract or insert the fuel assemblies, the robot arm can move up or down for about 2 m. A gripper and guide arm is used to handle the FAs: the gripper locks the FA, and the guide has two functions, namely to hold the FA in the vertical orientation and to ensure neighboring FAs are not disturbed when a FA is extracted from the core. An ultrasonic (US) sensor is used to uniquely identify the FAs.

The in-vessel fuel-handling machine will also perform in-vessel inspection and recovery of an unconstrained FA. Incremental single-point scanning of the diaphragm can be performed by an US sensor mounted at the gripper of the IVFHM. The baffle under the diaphragm is crucial for the strategy as it limits the work area where inspection and recovery are needed. It eliminates also the need of additional recovery and inspection manipulators, prevents items from migrating into the space between the diaphragm and the reactor cover, and permits side scanning.

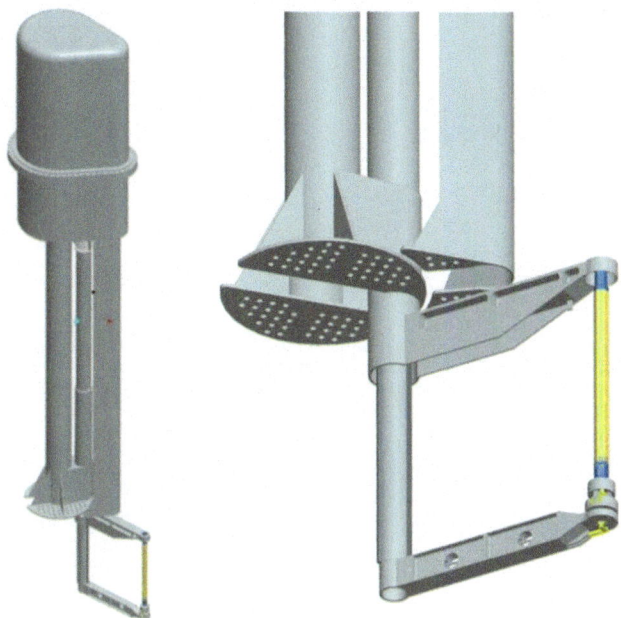

Fig. 7.4 The in-vessel fuel-handling machine

7.5 MYRRHA, A Research Tool in Support of the European Roadmap for P&T

Spent nuclear fuel from light water reactors (LWR) contains a mixture of uranium and plutonium (up to 95 % of the initial uranium mass), fission products, and minor actinides such as neptunium, americium, and curium. In the shorter term, the highly active but short-lived fission products will dominate the activity of this spent fuel. However, the transuranics including plutonium and the minor actinides (together with a few long-lived fission products) are largely responsible for the long-term radiotoxicity and heat production of LWR spent fuel.

The principle behind Partitioning and Transmutation (P&T) is to isolate the minor actinides from this LWR spent fuel and transmute them. As for these isotopes the fission to capture ratio increases with increasing neutron energy, a fast neutron spectrum facility is required. By burning the minor actinides, the long-lived, heat-producing component of spent fuel can be strongly reduced, which decreases the radiotoxicity of the spent fuel and its heat load. Both conditions will ease the design and construction of a long-term storage solution (geological disposal) from the engineering point of view.

Partitioning & Transmutation requires the development of an advanced fuel cycle. Currently, two major options for P&T are being studied worldwide: the single-stratum approach wherein the minor actinides are burned in fast reactors that are deployed for electricity production and the double-strata approach where the Pu

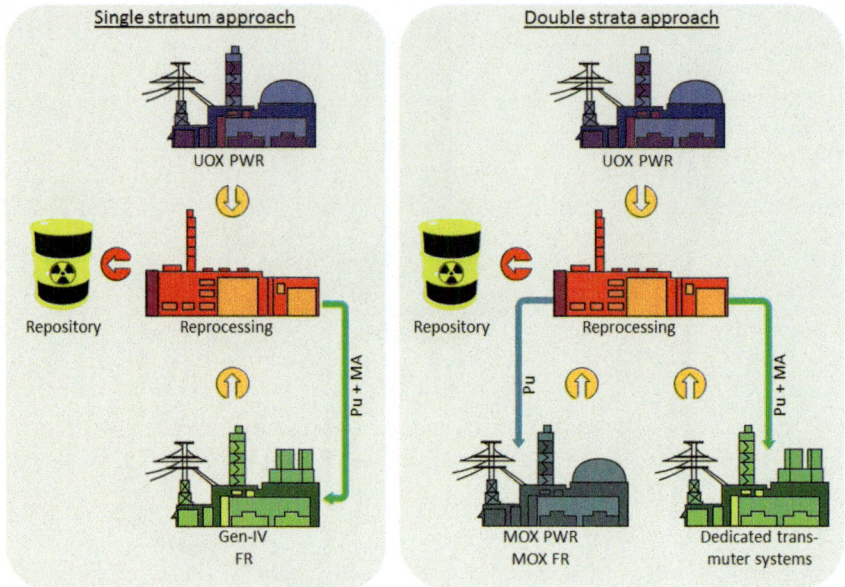

Fig. 7.5 Single-stratum vs. double-strata approach

is burned for electricity production in LWRs and FRs whereas the minor actinides are burned in a dedicated facility (Fig. 7.5).

In the single-stratum approach, the minor actinides can be mixed homogeneously in the fast reactor fuel or can be loaded in dedicated targets. In the homogeneous option, care must be taken in the analysis of the change in the core safety parameters such as delayed neutron fraction, Doppler constant, and void coefficient. By increasing the concentration of minor actinides in the fuel mixture, these safety parameters typically go in the wrong direction and hence pose a threat to the reactor safety. Because of this, one expects a maximum of 4–5 % minor actinide loading in the fuel.

Also, the fabrication and reprocessing of this "spiked" fast reactor fuel or the dedicated minor actinide target requires extra care because the presence of the minor actinides increases heat production during these fabrication processes. The presence of Cm-244 will pose a shielding problem because of its spontaneous fission and hence neutron emission.

Given the fact that only small amounts of minor actinides can be loaded per reactor, limited by a maximum concentration in case of the homogeneous option or limited by the number of target positions in the heterogeneous option, a large number of reactors will be required to use this minor actinide-spiked fuel or house these dedicated targets; this will certainly be the case when nations decide to also treat their legacy LWR waste and not only the minor actinides produced in this future advanced fuel cycle. Implied are a large number of transports of these fuels and targets from reprocessing site to fuel fabrication site and to transmutation sites and back.

In the double-strata approach, a dedicated transmutation facility is foreseen in the form of an accelerator-driven system. Because of the reactor physics properties of such an ADS (one does not rely on a subtle equilibrium such as the chain reaction, but the ADS subcritical core acts merely as a multiplier of a primary neutron source), one can devise fuels that have a very high minor actinide content. The EC-FP6 program IP-EUROTRANS delivered the conceptual design of such an industrial transmuter (EFIT). In EFIT, 400 MW^{th} core designs were made with uranium-free inert matrix fuels having a mixture of plutonium and minor actinides. In EFIT, the so-called 42–0 approach core was developed, meaning a core design that would be as plutonium neutral as possible (no burning nor breeding of plutonium) and which could in optimal conditions burn 42 kg minor actinides per TWh power produced. This system was used in the EC-FP6 program PATEROS, which produced a roadmap for the development of Partitioning and Transmutation at the European level. The deployment of such an industrial transmuter as EFIT would be very difficult for small nuclear countries and hence this scheme is optimal in a regional approach.

Because the burning of the minor actinides is done in a very concentrated manner, these industrial transmuters can be located near a fuel reprocessing and transmuter fuel fabrication facility, limiting the transportation of hazardous materials. Calculations have indicated that the support ratio, that is, the ratio of the total power of industrial transmuters to the total power of electricity-generating systems, is about 6 %. Also with this "concentrated" approach, one can much easier envisage the burning of the LWR legacy waste in a reasonable amount of time without impacting the regular electricity production installations.

Within the PATEROS project, a number of nuclear fuel cycle scenarios have been studied. Different regions have been identified: a group of countries that are stagnant with respect to nuclear energy production or in phase-out ("Group A," typically Belgium, Czech Republic, Germany, Spain, Sweden, Switzerland) and a group of countries which are developing an advanced fuel cycling with the deployment of fast reactors ("Group B," typically France). Different objectives were set concerning the burning of the minor actinides. Within the EC-F7 ARCAS project, which continues on the work done in PATEROS, it was estimated that to burn the minor actinides present in Group A in a reasonable time frame (less than 100 years), the group would need to deploy 7 EFIT-like facilities. If also Group B wants to stabilize their minor actinide inventory, 15 EFIT-like installations would be needed, and if total minor actinide elimination is required in Groups A and B, 20 EFIT-like installations are to be built.

At the European level, four building block strategies for partitioning and transmutation have been identified. Each block poses a serious challenge in research and development to reach an industrial-scale deployment. These blocks are as follows.

- Demonstration of advanced reprocessing of spent nuclear fuel from LWRs, separating uranium, plutonium, and minor actinides;

- Demonstrate the capability to fabricate at semi-industrial level dedicated transmuter fuel heavily loaded in minor actinides;
- Design and construct one or more dedicated transmuters;
- Demonstration of advanced reprocessing of transmuter fuel together with the fabrication of new transmuter fuel.

MYRRHA will support this roadmap by playing the role of an accelerator-driven system prototype (at reasonable power level) and as a flexible irradiation facility providing fast neutrons for the qualification of materials and fuel for an industrial transmuter. MYRRHA will be capable of irradiating samples of this inert matrix fuels, but it is also foreseen to house fuel pins or even a limited number of fuel assemblies heavily loaded with MAs for irradiation and qualification purposes.

7.6 Conclusions

SCK•CEN is proposing to replace its aging flagship facility, the Material Testing Reactor BR2, by a new flexible irradiation facility, MYRRHA. Considering international and European needs, MYRRHA is conceived as a flexible fast spectrum irradiation facility able to work in both subcritical and critical mode. Despite several nonobvious design challenges, such as the use of LBE, the increased level of seismic loading (consequence of Fukushima), or the choice of passive mode for decay heat removal in emergency conditions, we found no significant showstopper in the design. The R&D program that is running in parallel has taken into account international recommendations from experts concerning the remaining technological challenges as mentioned in Section VI (above).

MYRRHA is now foreseen to be in full operation by 2025, and it will be able to be operated in both operation modes, subcritical and critical. In subcritical mode, it will demonstrate the ADS technology and the efficient demonstration of MA in subcritical mode. As a fast spectrum irradiation facility, it will address fuel research for innovative reactor systems, material research for GEN IV systems and for fusion reactors, radioisotope production for medical and industrial applications, and industrial applications, such as Si-doping.

The MYRRHA design has now entered into the Front End Engineering Phase, covering the period 2012–2015. The engineering company that handles this phase has currently started the work. At the end of this phase, the purpose is to have

- Progressed in such a way in the design of the facility that the specifications for the different procurement packages of the facility can be written,
- Adequately addressed the remaining outstanding R&D issues,
- obtained the construction and exploitation permits, and
- Formed the international members' consortium for MYRRHA.

Belgium and SCK•CEN have opened participation in the MYRRHA to EU member states and to the European Commission but also to worldwide participation, as the issue of safe and efficient management of high-level nuclear waste is a worldwide issue, whatever the policy adopted or to be adopted by the countries that have industrialized nuclear power generation and want to phase it out, those willing to continue its use, and those willing to start nuclear power generation.

Open Access This chapter is distributed under the terms of the Creative Commons Attribution Noncommercial License, which permits any noncommercial use, distribution, and reproduction in any medium, provided the original author(s) and source are credited.

References

1. OECD (2012) OECD Factbook 2011–2012: economic, environmental and social statistics. OECD. OECD Publications, France. doi:10.1787/factbook-2011-en
2. OECD/NEA (2006) Physics and safety of transmutation systems, a status report. OECD Publications, France.
3. Martinez-Val J (2009) PATEROS P&T Roadmap proposal for advanced fuel cycles leading to a sustainable nuclear energy: syntheses report.
4. Lecomte M (2008) Le traitement-recyclage du combustible nucléaire usé. CEA, Editions Le Moniteur. CEA Saclay et Groupe Moniteur, Paris, France.

Chapter 8
Design of J-PARC Transmutation Experimental Facility

Toshinobu Sasa

Abstract After the Fukushima accident caused by the Great East Japan Earthquake, nuclear transmutation acquired much interest as an effective option of nuclear waste management. The Japan Atomic Energy Agency (JAEA) proposes the transmutation of minor actinides by an accelerator-driven system (ADS) using lead–bismuth eutectic alloy (Pb-Bi) as a spallation target and a coolant of the subcritical core. The current ADS design has 800 MWth of rated power, which is driven by a 20 MW proton LINAC, to transmute minor actinides generated from 10 units of standard light water reactors.

To obtain the data required for ADS design, including the European MYRRHA project, JAEA plans to build a Transmutation Experimental Facility (TEF) within the framework of the J-PARC project. TEF consists of two buildings: one is an ADS target test facility (TEF-T), in which will be installed a high-power Pb-Bi spallation target, and the other is the Transmutation Physics Experimental Facility (TEF-P), which will set up a fast critical/subcritical assembly driven by a low-power proton beam. TEF will be located at the end of the 400 MeV LINAC of J-PARC and accept a 250-kW proton beam with repetition rate of 25 Hz. As major research and development items of TEF-T, irradiation tests for structural materials and engineering tests for Pb-Bi applications to determine the effective lifetime of the proton beam window will be performed. The reference design parameter, that considers operating conditions of the ADS transmutor, was determined by thermal-hydraulic analyses and structural analyses. When the target operates with full-power beam, a fast neutron spectrum field is formed around the target, and it is possible to apply multipurpose usage. Various research plans have been proposed, and layout of the experimental hall surrounding the target is under way. Basic physics application such as measurements of nuclear reaction data is considered as one of the major purposes.

Keywords Accelerator-driven system • J-PARC • Transmutation • Transmutation Experimental Facility

---

T. Sasa (✉)
Transmutation Section, J-PARC Center, Japan Atomic Energy Agency, 2-4, Shirakata-Shirane, Tokai-mura, Ibaraki 319-1195, Japan
e-mail: sasa.toshinobu@jaea.go.jp

8.1 Introduction

After the Fukushima accident caused by the Great East Japan Earthquake, public interest in the management of radioactive wastes and spent nuclear fuels has increased. The Science Council of Japan recommends prioritizing research and developments to reduce the radiological burden of high-level wastes by transmutation technology.

The Japan Atomic Energy Agency (JAEA) proceeded with R&D to reduce the radiological hazard of high-level wastes by partitioning and transmutation (P-T) technology [1]. In the framework of the J-PARC project, JAEA also promoted constructing the Transmutation Experimental Facility (TEF) to study minor actinide (MA) transmutation by both fast reactors and accelerator-driven systems [2]. TEF is located at the end of the LINAC, which is also an important component to be developed for future ADS, and shares the proton beam with other experimental facilities used for material sciences, life sciences, and high-energy nuclear physics.

The TEF (Fig. 8.1) consists of two buildings, the Transmutation Physics Experimental Facility (TEF-P) [3] and the ADS Target Test Facility (TEF-T) [4]. Two facilities are connected by the proton beam line with a low-power beam extraction mechanism using a laser beam [5]. TEF-P is a facility with zero-power critical assembly wherein a low-power proton beam is available to study the reactor physics and the controllability of accelerator-driven systems (ADS). It also has availability for measuring the reaction cross sections of MA and structural materials, for example. TEF-T is planned as an irradiation test facility that can accept a maximum 400 MeV–250 kW proton beam to the lead-bismuth (Pb-Bi) spallation target. Using these two facilities, the basic physical properties of a subcritical system and engineering tests of a spallation target are to be studied.

R&Ds for important technologies required to build the facilities are also performed, such as laser charge exchange technique to extract a very low power proton beam for reactor physics experiments, a remote handling method to load MA-bearing fuel into the critical assembly, and a spallation product removal method especially for the polonium. The objectives and construction schedule of the facilities, the latest design concept, and key technologies to construct TEF are under way.

8.2 Outline of the Transmutation Experimental Facility

8.2.1 Outline of TEF-T

For the JAEA-proposed ADS, Pb-Bi is a primary candidate of coolant and spallation target. To solve technical difficulties for Pb-Bi utilization, construction of TEF-T is planned to complete the data sets that are required for the design of ADS.

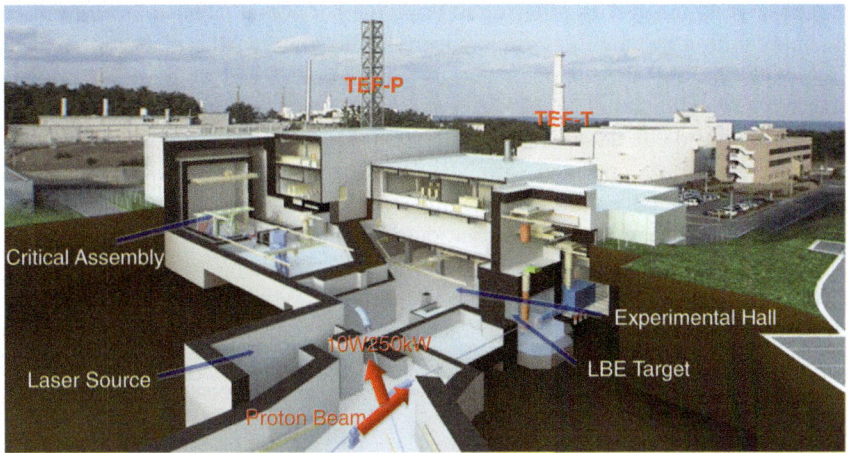

Fig. 8.1 Transmutation Experimental Facility

The experiments to obtain the material irradiation data for the beam window are the most important mission of TEF-T.

A high-power spallation target, which will be mainly used for material irradiation of candidate materials for a beam window of full-scale ADS, is an essential issue to realize a TEF-T. To set up the beam parameters, future ADS concepts are taken into account. In the reference case of the target, proton beam current density of 20 µA/cm^2, which equals the maximum beam current density of the JAEA-proposed 800 MWth ADS, was assumed.

8.2.2 Outline of TEF-P

Several neutronic experiments for ADS have been performed in both Europe [6, 7] and Japan. In Japan, subcritical experiments with fast neutron spectrum core were performed at the Fast Critical Assembly (FCA) in JAEA/Tokai, and subcritical experiments with thermal subcritical core driven by 100 MeV protons are performed at Kyoto University Research Reactor Institute. Many experimental studies also have been performed on the neutronics of the spallation neutron source with various target materials such as lead, tungsten, mercury, and uranium. These experiments for spallation targets are also useful to validate the neutronic characteristics of ADS. However, there are no experiments combined with a spallation source installed inside the subcritical fast-neutron core. The purpose of the TEF-P is divided roughly into three subjects: (1) reactor physics aspects of the subcritical core driven by a spallation source, (2) demonstration of the controllability of the subcritical core including power control by the proton beam power adjustment, and (3) investigation of the transmutation performance of the subcritical core using a certain amount of MA and LLFP.

TEF-P is designed with referring to FCA, the horizontal table–split type critical assembly with a rectangular lattice matrix. In this concept, the plate-type fuel for FCA with various simulation materials such as lead and sodium for coolant, tungsten for solid target, ZrH for moderator, B_4C for absorber, and AlN for simulating nitride fuel can be commonly used at TEF-P. Therefore, previous experiments can be correlated with TEF-P experiments. The proton beam will be introduced horizontally at the center of the fixed half assembly, and various kinds of spallation targets can be installed at various axial position of the radial center of the subcritical core. Application of MA fuel is one of the promising characteristics of TEF-P. Installation of a partial mock-up region of MA fuel with air cooling is considered to measure the physics parameters of the transmutation system. R&D to utilize MA fuel by remote handling systems is under way.

8.3 Design of Spallation Target for TEF-T

To evaluate the feasibility of a designed beam window of TEF target, numerical analysis with a three-dimensional (3D) model was performed. The analysis was done by considering the current density and shape of the incident proton beam to the target and the thermal fluid behavior of Pb-Bi around the beam window as a function of flow rate and inlet temperature. The thickness of the beam window is also considered from 2 to 3 mm. After the temperature distribution analysis, structural strength of the beam window is determined to evaluate soundness of the target. A concave shape beam window was used for this analysis. The prototype design of the beam window for TEF target system is shown in Fig. 8.2.

The material of the beam window would be a type 316 stainless steel. The concave section in the center part of the target was connected to the convex section in the terminal part, and then it was connected to the straight tube. A straight tube part has coaxially arranged annular and tube-type channels. The inner diameters of the outside tube and inside tube were set to 150 and 105 mm, respectively. The total length of the analysis region was 600 mm, which corresponds to an effective target depth for the 400 MeV proton. An irradiation sample holder, which was installed in the inner tube, holds eight irradiation specimens in the horizontal direction. The size of each specimen was $40 \times 145 \times 2$ mm. The rectification lattice having the aperture of the plural squares type was installed at the front end of the sample holder. A slit 2 mm in width was arranged along the side of the rectification lattice to cool the sample holder by flowing Pb-Bi.

The thermal-fluid behavior of the target was analyzed by the STAR-CD. The quarter-part model was set to tetra metric type and the divided face was set to a reflected image condition. At first, Pb-Bi flowed through the annular region and joined in the center of the beam window, and then, turned over and flowed in the inner tube after having passed a rectification lattice and an irradiation sample. In a default condition, flow rate at the inlet of annulus region was set to 1 l/s, and this was equivalent to the flow velocity of 0.125 m/s. Because the Pb-Bi flow forms a

Fig. 8.2 Prototype LBE spallation target for TEF-T

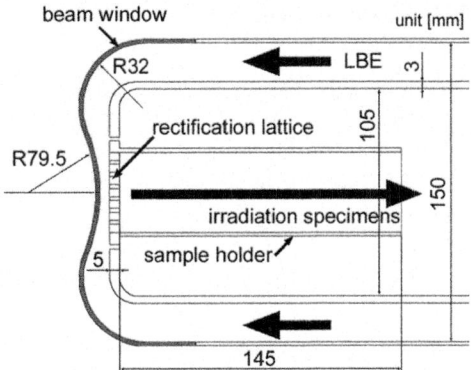

complicated turbulent flow, the standard k-ε model for high Re number type was used for a turbulence model. Heat deposition distribution by the primary proton beam, which was calculated by a hadronic cascade code PHITS [8], was used for the analysis. The internal pressure to the inside of the beam window was set to 0.3 MPa in consideration of the flowing Pb-Bi and the cover gas. On the outer side of the beam window and the border of the atmosphere, release of the radiant heat was considered. Embrittlement of the structural materials by irradiation was not considered.

The analyses were performed by changing flow rates from 1 to 4 l/s. In each case, a dead region was commonly formed in the center of the inside of the beam window. The maximum velocity of Pb-Bi was confirmed at the rectification lattice part and was approximately 1.2 m/s in the case of the inlet flow rate of 1 l/s. When the inlet flow rate increased to 4 l/s, the maximum velocity in the target reached 4.8 m/s, which is too high to apply to the Pb-Bi target. The maximum temperature is 544 °C in the case of a 3-mm-thick window. The peak temperature can be decreased to 477 °C in the case of 2-mm-thick window. The temperature differences between outside and inside at the center of the window were 65 and 37 °C in the case of the 3-mm-thick window and the 2-mm-thick window, respectively. From these results, it was determined that a condition of 2 mm was desirable.

Based on the results provided by STAR-CD, analysis to verify the feasibility of the beam window was performed by ABAQUS code. The operating conditions for the first stage of material irradiation in TEF were decided by a result of the analysis on each condition. The temperature and thermal stress for the steady state were estimated using ABAQUS code, the computational code for the finite-element method. In the ABAQUS code, only a beam window was modeled as the cylinder-slab geometry. From the analysis result for the 2-mm-thick window, the stress strength reached the maximum value of 190 MPa on the outer surface of the beam window. When the maximum temperature of the beam window is adopted to 470 °C from the result of STAR-CD, maximum stress is lower than the tolerance level of the materials for fast reactor, and hence, the feasibility of a designed beam window was confirmed.

8.4 Conclusion

To perform the design study for the transmutation system of long-lived nuclides, the construction of TEF, which consists of two buildings, TEF-T and TEF-P, is proposed under the J-PARC Project. According to the current construction schedule, TEF-T will be built at the first phase and TEF-P will be constructed at the latter phase. Licensing procedures for TEF-P construction will be processed simultaneously with TEF-T construction.

TEF-T is a facility to prepare the database for engineering design of an ADS using a 400 MeV–250 kW proton beam and the Pb-Bi spallation target. The purposes of TEF-T are R&D for the structural strength of the beam window, which is irradiated by both high-energy protons and neutrons, compatibility of the structural material with flowing liquid Pb-Bi, and operation of the high-power spallation target. Several kinds of target head can be installed according to the experimental requirement. It was shown that the reference case of injected proton beam condition (400 MeV–250 kW and 20 µA/cm^2 of beam current density) was applicable to the TEF-T target. Further studies to improve irradiation performance are under way.

TEF-P is a critical assembly, which can accept the 400 MeV–10 W proton beam for the spallation neutron source. The purposes of TEF-P are the experimental validation of the data and method to predict neutronics of the fast subcritical system with spallation neutron source, demonstration of the controllability of a subcritical system driven by an accelerator, and basic research of reactor physics for transmutation of MA and LLFP. The distinguishing points of the TEF-P in comparison with existing experimental facilities can be summarized as follows: (1) both the high-energy proton beam and the nuclear fuel are available, (2) the maximum neutron source intensity of about 10^{12} n/s is strong enough to perform precise measurements even in the deep subcritical state (e.g., $k_{eff} = 0.90$) and is low enough to easily access the assembly after the irradiation, (3) a wide range of pulse width (1 ns–0.5 ms) is available by the laser charge exchange technique, (4) MA and LLFP can be used as a shape of foil, sample, and fuel by installing an appropriate shielding and remote handling devices.

Along with the design study of the TEF, R&D for the components required for TEF, such as the laser charge exchange technique to extract a very low power proton beam, test manufacturing of MA fuel-handling devices, and operation of lead-bismuth test loops are under way. From the experimental results of the laser charge exchange technique, beam extraction in the magnetic field is successfully demonstrated. Mockup of the coolant simulator block and remote handling mechanism for pin-type fuel loading has been done. An effective method to remove polonium with a standard stainless mesh filter was established through the hot experiments. Significant improvement of analysis accuracy of actual ADS was expected by critical experiment with MA fuel at TEF-P.

When the target of TEF-T operates with a full power beam, a fast neutron spectrum field is formed around the target and it is possible to apply multipurpose usage. Various research plans have been proposed, and layout of the experimental hall surrounding the target is under way. Basic physics application such as measurements of nuclear reaction data is considered as one of the major purposes. We called for a preliminary letter of intent to encourage the project. Requests for multipurpose usage will be taken into account in the facility design of TEF.

Open Access This chapter is distributed under the terms of the Creative Commons Attribution Noncommercial License, which permits any noncommercial use, distribution, and reproduction in any medium, provided the original author(s) and source are credited.

References

1. Sasa T et al (2004) Research and development on accelerator-driven transmutation system at JAERI. Nucl Eng Des 230:209–222
2. The Joint Project Team of JAERI and KEK (2000) The joint project for high-intensity proton accelerators (in Japanese). JAERI-Tech 2000-003
3. Oigawa H et al (2001) Conceptual design of transmutation experimental facility. In: Proceedings of the Global 2001, Paris (CD-ROM)
4. Sasa T et al (2005) Conceptual study of transmutation experimental facility. (2) Study on ADS target test facility (in Japanese). JAERI-Tech 2005–021
5. Tomisawa T et al (2005) Investigation of photo neutralization efficiency of high intensity H-beam with Nd:YAG laser in J-PARC. In: Proceedings of the 7th European workshop on beam diagnostics and instrumentation for particle accelerators (DIPAC 2005), Lyon, p 275–277
6. Soule R et al (2004) Neutronic studies in support to ADS: the muse experiments in the MASURCA facility. Nucl Sci Eng 148:124–152
7. Uyttenhove W et al (2011) The neutronic design of a critical lead reflected zero-power reference core for on-line subcriticality measurements in accelerator driven systems. Ann Nucl Energy 38(7):1519–1526
8. Niita K et al (2010) PHITS: Particle and Heavy Ion Transport code System, version 2.23. JAEA-Data/Code 2010–022

Chapter 9
Accelerator-Driven System (ADS) Study in Kyoto University Research Reactor Institute (KURRI)

Cheol Ho Pyeon

Abstract Experimental studies on the uranium- and thorium-loaded accelerator-driven system (ADS) are being conducted for basic research of nuclear transmutation analyses with the combined use of the core at the Kyoto University Critical Assembly (KUCA) and the fixed-field alternating gradient (FFAG; 100 MeV protons) accelerator in the Kyoto University Research Reactor Institute. The ADS experiments with 100 MeV protons were carried out to investigate the neutronic characteristics of ADS, and the static and kinetic parameters were accurately analyzed through both the measurements and the Monte Carlo simulations of reactor physics parameters. An upcoming ADS at KUCA could be composed of highly enriched uranium fuel and Pb-Bi material, and the reaction rate ratio analyses (^{237}Np and ^{241}Am) of nuclear transmutation could be conducted in the ADS (hard spectrum core) at KUCA. The neutronic characteristics of Pb-Bi are expected to be examined through reactor physics experiments at KUCA with the use of solid Pb-Bi materials at the target and in the core.

Keywords 100 MeV protons • ADS • FFAG accelerator • KUCA • Spallation neutrons • Tungsten target

9.1 Introduction

The accelerator-driven system (ADS) has been considered as an innovative system for the nuclear transmutation of minor actinides and long-lived fission products with the use of spallation neutrons obtained from the injection of high-energy protons into a heavy metal target. At the Kyoto University Critical Assembly (KUCA), a series of ADS experiments [1–5] was carried out by coupling with the fixed-field alternating gradient (FFAG) accelerator [6–8], and the spallation

---

C.H. Pyeon (✉)
Nuclear Engineering Science Division, Research Reactor Institute, Kyoto University, Asashiro-nishi, Kumatori-cho, Sennan-gun, Osaka 590-0494, Japan
e-mail: pyeon@rri.kyoto-u.ac.jp

neutrons generated by 100 MeV protons from the FFAG accelerator were successfully injected into uranium- [1, 2, 4] and thorium-loaded [5, 7] cores.

In the ADS facility at KUCA, reactor physics experiments are being carried out to study the neutronic characteristics through the measurements of reactor physics parameters, including reaction rates, neutron spectrum, neutron multiplication, subcriticality, and neutron decay constant. Among these, neutron multiplication was considered as an important index to recognize the number of fission neutrons in the core induced by the external neutron source.

The mockup experiments [5] of thorium-loaded ADS carried out by varying the neutron spectrum and the external neutron source were aimed at investigating the influence of different neutron profiles on thorium capture reactions and the prompt and delayed neutron behaviors in the subcritical system. The results provided important effects of the neutron spectrum and the external neutron source on both static and kinetic parameters: the effect of the neutron spectrum was investigated by varying the moderator material in the fuel region, and that of external neutron source by injecting separately 14 MeV neutrons and 100 MeV protons into the thorium-loaded core varying the moderator. Before the subcritical experiments, a thorium plate irradiation experiment was carried out in the KUCA core to analyze the thorium capture and fission reactions in the critical system as a reference of the subcritical system, although the feasibility of ^{232}Th capture and ^{233}U fission reactions could be examined in the subcritical state.

In this chapter, experimental results of the uranium- and thorium-loaded ADS are shown. Accuracy was evaluated through the comparison between the experiments and the calculations of the Monte Carlo analyses through the MCNPX [9] code with ENDF/B-VII.0 [10], JENDL/HE-2007 [11], and JENDL/D-99 [12] libraries. The ADS static and kinetic experiments at KUCA are presented in Sect. 9.2, the results and discussion of the experiments and calculations in Sect. 9.3, and the conclusion of the study in Sect. 9.4.

9.2 Experimental Settings

9.2.1 Uranium-Loaded ADS Experiments

KUCA comprises two solid polyethylene-moderated thermal cores designated A and B and one water-moderated thermal core designated C. The A-core is mainly used for experiments of ADS basic research. The three cores are operated at a low mW power in the normal operating state; the maximum power is 100 W. The constitution and the configuration of the cores can be altered easily, and the coupling with the conventional Cockcroft-Walton type accelerator and with the FFAG accelerator has allowed conducting experiments separately with the use of 14 MeV neutrons from deuteron–tritium fusion reactions and 100 MeV protons with the heavy metal target, respectively.

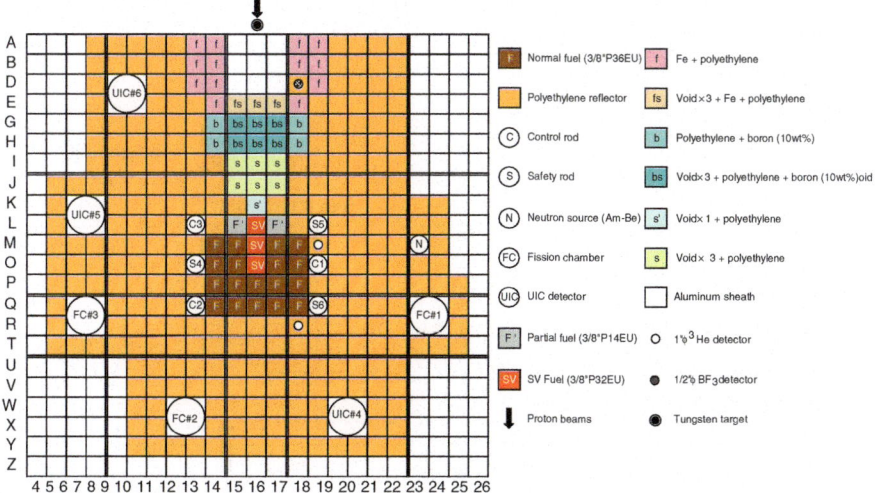

Fig. 9.1 *Top view* of the configuration of the A-core in the accelerator-driven system (ADS) experiments with 100 MeV protons

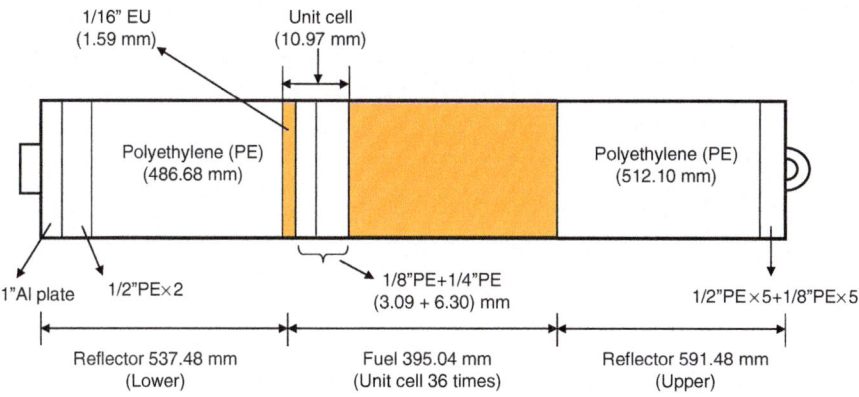

Fig. 9.2 *Side view* of of 3/8P″36EU fuel assembly (*F*, Fig. 9.1) in the A-core

The A-core (Fig. 9.1) employed in the ADS experiments was essentially a thermal neutron system composed of a highly enriched uranium fuel and the polyethylene moderator/reflector. In the fuel region, a unit cell is composed of the highly enriched uranium fuel plate 1/16 in. thick and polyethylene plates 1/4 in. and 1/8 in. thick (Fig. 9.2). The SV assembly is composed of a 5.08 × 5.08 × 5.08 cm center void region, 32 fuel unit cells, and the polyethylene blocks. In these ADS experiments, three types of fuel rods designated as the normal, partial, and special fuel SV were employed. For reasons of the safety regulations for KUCA, the heavy metal target was located not at the center of the core but outside the critical assembly. As in the previous ADS experiments with 14 MeV neutrons,

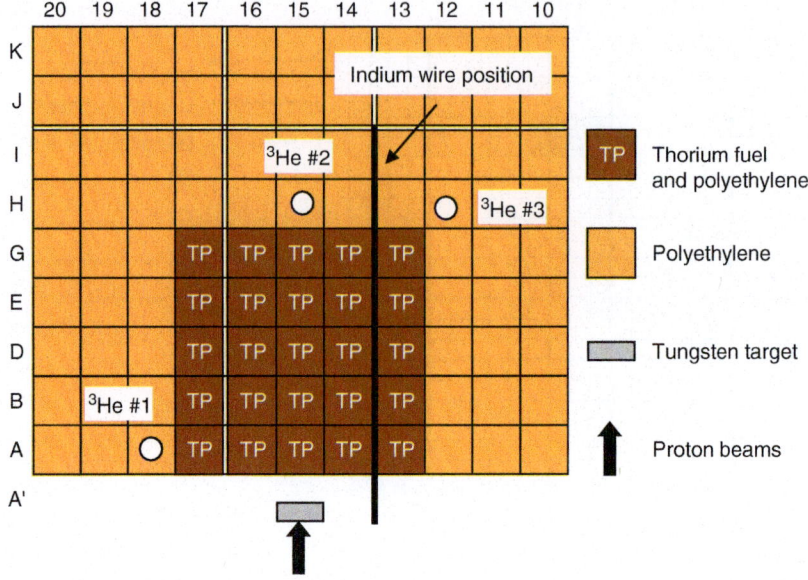

Fig. 9.3 *Top view* of thorium-loaded ADS core with 100 MeV protons

the introduction of a neutron guide and a beam duct is requisite to lead the high-energy neutrons generated from the heavy metal target to the center of the core as much as possible. In the uranium-loaded ADS experiments, the proton beam parameters were 100 MeV energy, 0.01 nA intensity, 30 Hz pulsed frequency, 100 ns pulsed width, and 80-mm diameter spot size at the tungsten target (100 mm diameter and 9 mm thick). The level of the neutron yield generated at the target was around 1.0×10^6 1/s by the injection of 100 MeV protons onto the tungsten target.

9.2.2 Thorium-Loaded ADS Benchmarks

In the ADS with 100 MeV protons (Fig. 9.3), the fuel rod was composed of a thorium metal plate and a polyethylene (PE), graphite (Gr), or beryllium (Be) moderator arranged in the A-core. Other components were selected from HEU and natural uranium (NU; $2 \times 2 \times 1/8$ in.) plates. The cores were composed of Th-PE (Fig. 9.4), Th-Gr, Th-Be, Th-HEU-PE, and NU-PE, according to a selection of moderator materials: PE, Gr, Be, HEU-PE, and NU-PE, respectively, and spallation neutrons were generated outside the core after injection onto the tungsten target. The thorium-loaded ADS experiments were conducted especially to investigate the relative influence of different neutron profiles on capture reactions of ^{232}Th and ^{238}U: the reaction of ^{238}U was taken as reference data for evaluating the validity of ^{232}Th capture cross sections.

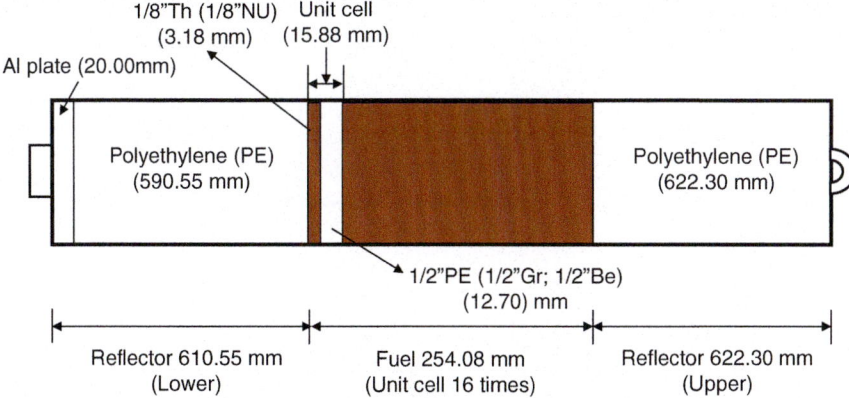

Fig. 9.4 *Side view* of Th-PE fuel assembly (TP) in thorium-loaded ADS core in Fig. 9.3

The main parameters of the proton beams were 100 MeV energy, 0.3 nA intensity, 20 Hz pulsed frequency, 100 ns pulsed width, and 40-mm-diameter spot size at the tungsten target (50 mm diameter and 9 mm thick). The level of the neutron yield generated at the target was around 1.0×10^7 1/s by the injection of 100 MeV protons onto the tungsten target.

Prompt and delayed neutron behavior was monitored by placing three ^{3}He detectors (20 mm diameter and 300 mm long) at three locations. Throughout the time evolution of the prompt and delayed neutrons, the prompt neutron decay constant was deduced by least-squares fitting to an exponential function over the optimal duration. Subcriticality was deduced by the extrapolated area ratio method [13] on the basis of prompt and delayed neutron behaviors. For 100 MeV protons, neutron detectors (^{3}He detectors: #1, #2, and #3) were set at three locations.

9.3 Results and Discussion

9.3.1 Uranium-Loaded ADS Experiments

9.3.1.1 Static Experiments

Thermal neutron flux distribution was estimated through the horizontal measurement of the ^{115}In$(n, \gamma)^{116m}$In reaction rate distribution by the foil activation method using an indium (In) wire 1.0 mm in diameter. The wire was set in an aluminum guide tube, from the tungsten target to the center of the fuel region [from the position of (13, 14 – A) to that of (13, 14 – P) (Fig. 9.1)], at the middle height of the fuel assembly. The experimental and numerical results of the reaction rates were normalized using an In foil ($20 \times 20 \times 2$ mm) emitted by ^{115}In$(n, n')^{115m}$In at the target. In this static experiment, the subcritical state (0.77 % $\Delta k/k$) was also attained

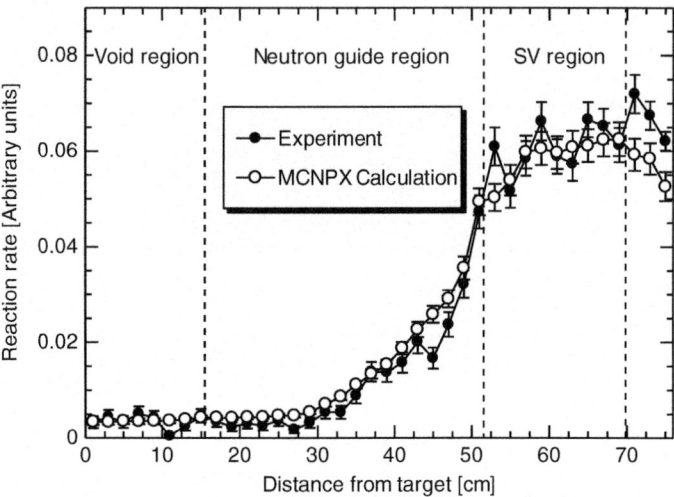

Fig. 9.5 Comparison of measured and calculated reaction rate distributions along the horizontal from (13, 14 – A) to (13, 14 – P) in Fig. 9.1 [2]

by the full insertion of C1, C2, and C3 rods. The numerical calculation was performed by MCNPX based on ENDF/B-VII.0. The generation of the spallation neutrons was included in the MCNPX calculation bombarding the tungsten target with 100 MeV proton beams. Because the reactivity effect of the In wire is considered to be not negligible, the In wire was taken into account in the simulated calculation: the reaction rates were deduced from tallies taken in the In wire setting region. The result of the fixed source calculation for the reaction rates was obtained after 2,000 active cycles of 100,000 histories, which led to a statistical error less than 10 % in the reaction rates. As shown in Fig. 9.5, the measured and the calculated reaction rate distributions were compared to validate the calculation method. The calculated reaction rate distribution agreed approximately with the experimental results within the statistical errors in the experiments, although these experimental errors were rather larger than those of the calculations. These larger errors in the experiments were attributed to the status of the proton beams described in Sect. 9.2.1, including the weak beam intensity and the poor beam shaping at the target.

9.3.1.2 Kinetic Experiments

To obtain information on the detector position dependence of the prompt neutron decay measurement, the neutron detectors were set at three positions as shown in Fig. 9.1: near the tungsten target [position of (17, D): 1/2-in.-diameter BF_3 detector]; and around the core [positions of (18, M) and (17, R): 1-in.-diameter ^{3}He detectors]. The prompt and delayed neutron behaviors (Fig. 9.6) were

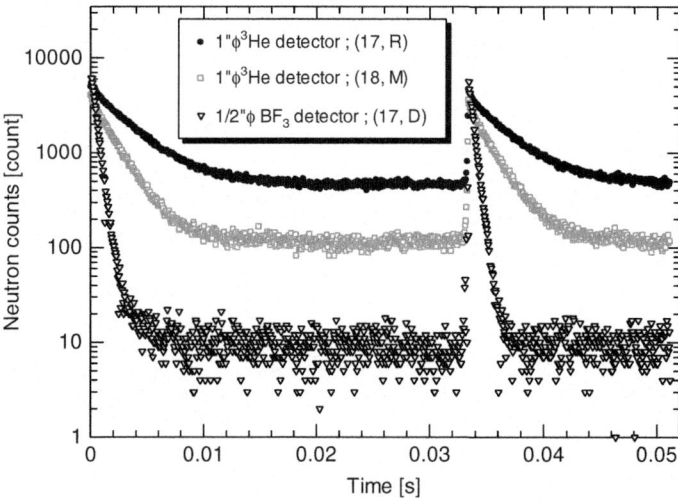

Fig. 9.6 Measured prompt and delayed neutron behaviors obtained from BF_3 and ^3He detectors in the A-core in Fig. 9.1 [2]

experimentally confirmed by observing the time evolution of neutron density in ADS, an exponential decay behavior and a slowly decreasing one, respectively. These behaviors clearly indicated that the neutron multiplication was caused by an external source: the sustainable nuclear chain reactions were induced in the subcritical core by the spallation neutrons through the interaction of the tungsten target and the proton beams from the FFAG accelerator. In these kinetic experiments, the subcriticality was deduced from the prompt neutron decay constant by the extrapolated area ratio method. The difference of measured results of 0.74 %$\Delta k/k$ and 0.61 %$\Delta k/k$ at the positions of (17, R) and (18, M) in Fig. 9.1, respectively, from the experimental evaluation of 0.77 %$\Delta k/k$, which was deduced from the combination of both the control rod worth by the rod drop method and its calibration curve by the positive period method, was within about 20 %. Note that the subcritical state was attained by a full insertion of C1, C2, and C3 control rods into the core.

9.3.2 Thorium-Loaded ADS Experiments

9.3.2.1 Static Experiments

the profile of neutron flux for the ^{232}Th capture reactions was estimated through the horizontal measurement of $^{115}In(n, \gamma)^{116m}In$ reaction rate distribution, as well as described in Sect. 9.3.1.1. The wire was set in an aluminum guide tube, from the tungsten target to the center of the fuel region [from the position of (13, 14 – A′) to that of (13, 14 – I); Fig. 9.3], at the middle height of the fuel assembly. The absolute values of the measured reaction rates (Fig. 9.7) revealed differently the variation of

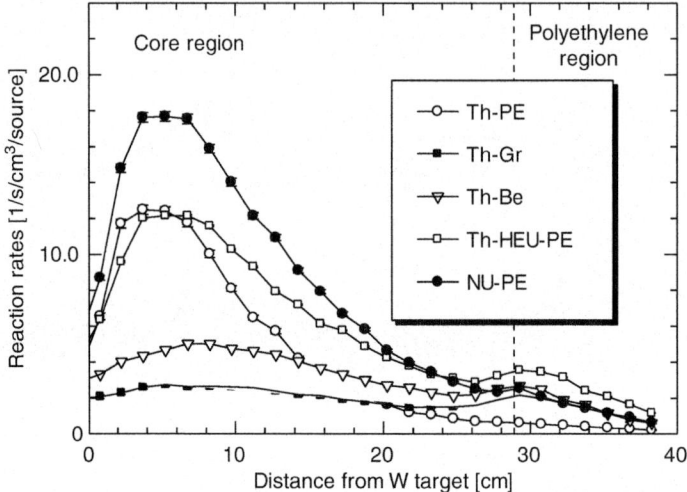

Fig. 9.7 Measured ^{115}In$(n, \gamma)^{116m}$In reaction rates obtained from the thorium-loaded ADS experiments with 100 MeV protons [5]

reaction rates attributed to varying the neutron spectrum in the core, when the spallation neutrons generated by 100 MeV protons were injected into the core. The moderating effect of the high-energy neutrons in some cores (Th-PE, Th-HEU-PE, and NU-PE: $k_{eff} = 0.00613$, 0.58754, and 0.50867, respectively) was observed around the boundary between the core and polyethylene regions. The ^{115}In$(n, \gamma)^{116m}$In reaction rates in the NU-PE core were higher than in other cores, demonstrating, that the reaction rates of ^{238}U in the NU-PE core were larger than those of ^{232}Th in the thorium cores with the use of 100 MeV protons. Additionally, the effect of the neutron spectrum on the reaction rates was observed with 100 MeV protons by comparing the measured results of reaction rates shown in Fig. 9.5. Thus, an expected physical effect was indeed observed as a result of the neutron spectrum change obtained by varying the moderator materials in the fuel assembly. Additionally, the accuracy [5] of experimental and numerical analyses was compared successfully with the ratio (C/E) of calculations and experiments around the relative difference of 10 %, through the subcritical parameter of neutron multiplication M.

9.3.2.2 Kinetic Experiments

The time evolution of prompt and delayed neutron behaviors was examined through the injection of an external neutron source (Fig. 9.8). In the Th-HEU-PE core, the prompt neutron decay constant (Table 9.1) at ^{3}He detector #1 was different from those at the others by the least-squares fitting, regardless of the kind of external neutron source and the position of neutron detection. It was considered overestimated, especially at detector #1, which was located near the external

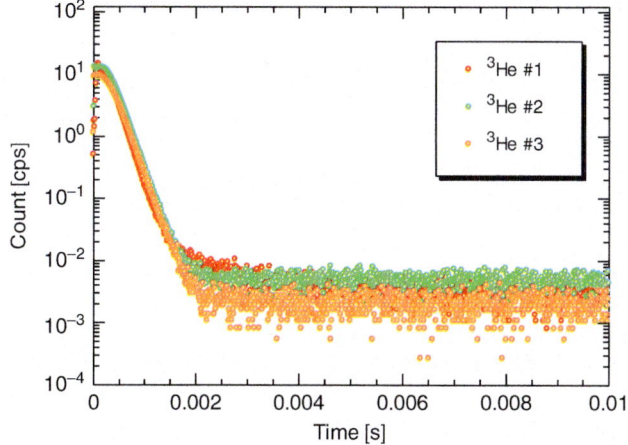

Fig. 9.8 Measured prompt and delayed neutron behaviors obtained from the thorium-loaded ADS experiments with 100 MeV protons (Th-HEU-PE)

Table 9.1 Measured results of prompt neutron decay constant (α [1/s]) in Th-HEU-PE core

Source	^{3}He #1	^{3}He #2	^{3}He #3
14 MeV neutrons	5,735 ± 5	5,155 ± 4	5,161 ± 4
100 MeV protons	5,788 ± 5	5,338 ± 5	5,229 ± 5

Table 9.2 Measured results of subcriticality (dollar units) in Th-HEU-PE core by the extrapolated area ratio method

Source	^{3}He #1	^{3}He #2	^{3}He #3
14 MeV neutrons	12.36 ± 0.51	29.70 ± 0.03	61.28 ± 0.09
100 MeV protons	31.19 ± 0.15	26.00 ± 0.10	43.14 ± 0.24

neutron source. The two different neutron sources provided different delayed neutron backgrounds. Subcriticality in dollar units was deduced by the extrapolated area ratio method with the use of prompt and delayed neutron components, and experimentally evaluated according to the kind of external neutron source and the location of neutron detection. As is well known, these results revealed subcriticality dependence on both the kind of external neutron source and the location of neutron detection, although the value of subcriticality was theoretically unchanged, regardless of the external neutron source and the location of the detector. Consequently, the experimental results (Table 9.2) showed that the subcriticality in pcm units for 14 MeV neutrons ($k_{\mathrm{eff}} = 0.6577$) was different from that for 100 MeV protons ($k_{\mathrm{eff}} = 0.7319$); remarkably, the discrepancy was also observed between the experiments and calculations ($k_{\mathrm{eff}} = 0.5876$), although the calculated value was evaluated with the use of MCNPX eigenvalue calculations (total number of histories, 1×10^8).

9.4 Conclusions

At KUCA, the ADS experiments with 100 MeV protons were carried out with the combined use of the KUCA A-core and the FFAG accelerator. The neutronic characteristics of ADS were investigated through experimental and numerical analyses of reaction rate distribution and subcriticality.

The thorium-loaded ADS study was conducted as observed by prompt neutron behavior and reaction rates through kinetic and static experiments, respectively. Further, experiments of thorium-loaded ADS were successfully carried out in the subcritical states with the use of an external neutron source (14 MeV neutrons and 100 MeV protons, respectively).

In the future, upcoming ADS experiments with 100 MeV protons could be carried out at the highly enriched uranium-fueled and Pb-Bi-zoned core of KUCA to investigate the neutronic characteristics of Pb-Bi solid materials used in the core and at the target. Also, irradiation experiments of ^{237}Np and ^{241}Am could be conducted in the hard-spectrum core at KUCA to examine the feasibility of reaction rate ratio (capture/fission) conversion analyses of nuclear transmutation.

Acknowledgments This work was supported by the KUR Research Program for Scientific Basis of Nuclear Safety from the Ministry of Education, Culture, Sports, Science and Technology (MEXT) of Japan. The authors are grateful to all the technical staff of KUCA for their assistance during the experiments.

Open Access This chapter is distributed under the terms of the Creative Commons Attribution Noncommercial License, which permits any noncommercial use, distribution, and reproduction in any medium, provided the original author(s) and source are credited.

References

1. Lim JY, Pyeon CH, Yagi T et al (2012) Subcritical multiplication parameters of the accelerator-driven system with 100 MeV protons at the Kyoto University Critical Assembly. Sci Technol Nucl Install 395878:9
2. Pyeon CH, Misawa T, Lim JY et al (2009) First injection of spallation neutrons generated by high-energy protons into the Kyoto University Critical Assembly. J Nucl Sci Technol 46:1091–1093
3. Pyeon CH, Lim JY, Takemoto Y et al (2011) Preliminary study on the thorium-loaded accelerator-driven system with 100 MeV protons at the Kyoto University Critical Assembly. Ann Nucl Energy 38:2298–2302
4. Pyeon CH, Azuma T, Takemoto Y et al (2013) Experimental analyses of spallation neutrons generated by 100 MeV protons at the Kyoto University Critical Assembly. Nucl Eng Technol 45:81–88
5. Pyeon CH, Yagi T, Sukawa K et al (2014) Mockup experiments on the thorium-loaded accelerator-driven system at the Kyoto University Critical Assembly. Nucl Sci Eng 177:156 (in press)
6. Lagrange JB, Planche T, Yamakawa E et al (2013) Straight scaling FFAG beam line. Nucl Instrum Methods A 691:55–63

7. Planche T, Lagrange JB, Yamakawa E et al (2011) Harmonic number jump acceleration of muon beams in zero-chromatic FFAG rings. Nucl Instrum Methods A 632:7–17
8. Yamakawa E, Uesugi T, Lagrange JB et al (2013) Serpentine acceleration in zero-chromatic FFAG accelerators. Nucl Instrum Methods A 716:46–53
9. Pelowitz DB (2005) MCNPX User's manual, version 2.5.0. LA-CP-05-0369. Los Alamos National Laboratory
10. Chadwick MB, Obložinský P, Herman M et al (2006) ENDF/B-VII.0: next generation evaluated nuclear data library for nuclear science and technology. Nucl Data Sheets 107:2931–3060
11. Takada H, Kosako K, Fukahori T (2009) Validation of JENDL high-energy file through analyses of spallation experiments at incident proton energies from 0.5 to 2.83 GeV. J Nucl Sci Technol 46:589–598
12. Kobayashi K, Iguchi T, Iwasaki S et al (2002) JENDL dosimetry file 99 (JENDL/D-99). JAERI Report 1344
13. Gozani T (1962) A modified procedure for the evaluation of pulsed source experiments in subcritical reactors. Nukleonik 4:348

Part III
Mechanical and Material Technologies for ADS: Development of Mechanical Engineering or Material Engineering-Related Technologies for ADS and Other Advanced Reactor Systems

Chapter 10
Heat Transfer Study for ADS Solid Target: Surface Wettability and Its Effect on a Boiling Heat Transfer

Daisuke Ito, Kazuki Hase, and Yasushi Saito

Abstract In relationship to a solid target cooling system of an accelerator-driven system (ADS), wettability effect on boiling heat transfer has been experimentally investigated by irradiation with ultraviolet and gamma rays (γ-rays). The experimental apparatus consists of a copper heater block, a rectangular container, and a thermostat bath. Two copper heater blocks were fabricated: one is for radiation-induced surface activation (RISA) and the other is for photoelectric reaction by ultraviolet whose heat transfer surface is coated by a TiO_2 film. These copper heater blocks were irradiated by ultraviolet or by γ-rays to change the surface wettability. Boiling heat transfer under subcooling conditions was measured before and after the irradiations to study the wettability effect. Experimental results show that nucleate boiling curves are shifted to the higher wall superheated side with the irradiated surface because of the decrease of the active nucleation sites. Heat transfer enhancement was found in both the critical heat flux and microbubble emission boiling (MEB) regions under these experimental conditions.

Keywords Microbubble emission boiling • Photocatalysis • Proton beam • Radiation-induced surface activation • Surface wettability

10.1 Introduction

An accelerator-driven system (ADS) is a hybrid-type nuclear system consisting of a proton accelerator, a spallation target, and a subcritical assembly in which high-energy particles and high heat density are generated in the target and subcritical assembly by the spallation and fission reactions. Lead-bismuth is considered the leading candidate for the liquid-metal spallation target for nuclear transmutation

---

D. Ito (✉) • Y. Saito
Research Reactor Institute, Kyoto University, 2-1010 Asashiro-nishi,
Kumatori-cho, Sennan-gun, Osaka 590-0494, Japan
e-mail: itod@rri.kyoto-u.ac.jp

K. Hase
Power Systems Company, Toshiba Corporation, Kawasaki, Japan

[1–3], whereas a solid target such as tungsten or tantalum should be also developed for a water-cooled ADS neutron source [4, 5].

High-energy radiation affects the surface wettability and boiling heat transfer of the solid target. Wettability on a solid surface can be changed by using ultraviolet radiation or γ-rays, and recently the authors have found that the surface wettability can be also changed by proton-beam irradiation [6]. Applying the wettability change resulting from ultraviolet irradiation to titanium dioxide (TiO_2), heat transfer research has been carried out to evaluate the wettability effect [7]. In addition, radiation-induced surface activation (RISA) enhances the surface wettability by irradiating a metal oxide layer with γ-rays. Takamasa et al. [8] have applied RISA to heat-transfer experiments and reported that boiling heat transfer could be enhanced by changing the wettability of the heating surface. However, there has been no research to investigate surface wettability effect on boiling heat transfer at a solid target cooling system, where microbubble emission boiling (MEB) [9] might occur. MEB can take place when the heat transfer area is small (about 1 cm^2) with subcooling conditions. In the target cooling system, the target should be cooled by subcooled water, and the heat-transfer area can be small when the proton beam is focused to a small area. Thus, MEB should be investigated for thermal hydraulic design and safety analysis of the solid target system, and also the effect of wettability on boiling heat transfer should be studied.

The purpose of this study is to investigate wettability change by ultraviolet, γ-ray, and proton beam and to study the wettability effect on subcooled boiling heat transfer with a small heat-transfer area, and finally to obtain knowledge on the heat-transfer mechanism of the MEB phenomena.

10.2 Surface Wettability Change by Irradiation

10.2.1 Sample and Irradiation Facility

To investigate surface wettability change by irradiation, samples are irradiated by using an ultraviolet lamp, a ^{60}Co γ-ray source, and a proton accelerator. In this study, a TiO_2 sample, which is a typical photocatalyst [10], is used to compare the irradiation effects of ultraviolet, γ-ray, and proton beam. TiO_2 is prepared through anodizing a 0.1-mm-thick titanium plate [11]. Details of the experimental procedure with TiO_2 samples and irradiation facilities are described as follows.

10.2.1.1 Ultraviolet

Ultraviolet irradiates TiO_2 by using a commercial UV lamp. Irradiation intensity is measured by an ultraviolet meter and is controlled by changing the distance between the lamp and the sample. The intensity is varied at a range from 0.01

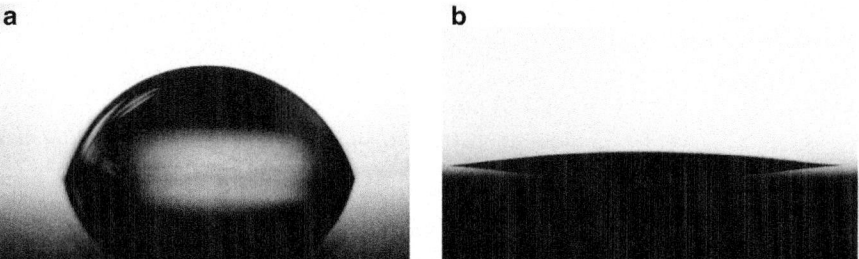

Fig. 10.1 Water droplets on the TiO$_2$ surface before and after ultraviolet irradiation. (**a**) Before ultraviolet irradiation. (**b**) After ultraviolet irradiation with 1 mW/cm^2 for 1 h

to 5 mW/cm^2. The center wavelength of the ultraviolet from this lamp is 365 nm. Figure 10.1 shows a typical irradiation effect on surface wettability change before and after ultraviolet irradiation.

10.2.1.2 Gamma Rays (γ-Rays)

The ^{60}Co γ-ray irradiation facility in the Research Reactor Institute, Kyoto University, is utilized for γ-ray irradiation. The integrated irradiation dose is estimated by an irradiation time and a distance from the γ-ray source. The γ-ray energy of this facility is about 1 MeV (1.17 and 1.33 MeV) and the maximum dose rate is about 15 kGy/h.

10.2.1.3 Proton Beam

The FFAG (fixed-field alternating gradient) accelerator in the Research Reactor Institute, Kyoto University, is utilized for proton-beam irradiation. The energy of the proton beam is set at about 100 or 150 MeV. The maximum beam current of this facility is about 10 nA.

10.2.2 Contact Angle Measurement

The wettability of the sample before and after irradiation is evaluated by measuring the contact angle of a water droplet on a sample surface. The measurement system (Fig. 10.2) consists of a digital video camera, a stage (with a biaxial stage and a goniometer), a backlight, and a PC. Pure water of 2 μl is dropped onto the horizontal surface of the sample using a micropipette. The water droplet is imaged by the camera and the images are processed to obtain the contact angle. In the image processing, it is assumed that the droplet is a part of a sphere (Fig. 10.3), and the contact angle is estimated by the following equation:

Fig. 10.2 Contact angle measurement system

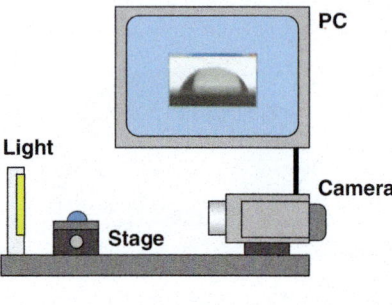

Fig. 10.3 Estimation of contact angle

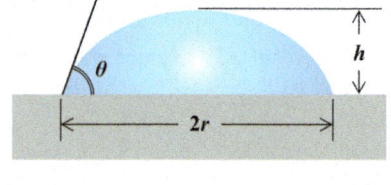

$$\theta = \sin^{-1}\frac{2rh}{r^2 + h^2} \qquad (10.1)$$

where r and h are obtained by using an image processing software (ImageJ).

10.2.3 Effect of Irradiations on Surface Wettability

Figure 10.4a–c shows the wettability change from ultraviolet, γ-ray, and proton-beam irradiation, respectively. The horizontal axis denotes an integrated irradiation dose or irradiation time and the vertical axis denotes the measured contact angle. As shown in these figures, the contact angle decreases with the irradiation dose. In these experiments, the ambient effect is also studied during the irradiations, which are performed in air or water. As shown in Fig. 10.4a, the ambient effect on the contact angle change is not obvious in the ultraviolet irradiation. However, the ambient effect is very distinct both in the γ-ray and the proton-beam irradiations. It is suggested that the wettability enhancement by the radiations may be attributed to the radiolysis of water.

10.3 Effect of Boiling Heat Transfer on Surface Wettability

10.3.1 Experimental Setup and Procedure

A schematic view of the experimental apparatus for pool boiling experiments with small heat transfer area is shown in Fig. 10.5. The apparatus consists of a copper block, a furnace, a rectangular container, and a heat exchanger. The copper block has a cylindrical part (10 mm in diameter, 15 mm in height). Several cartridge

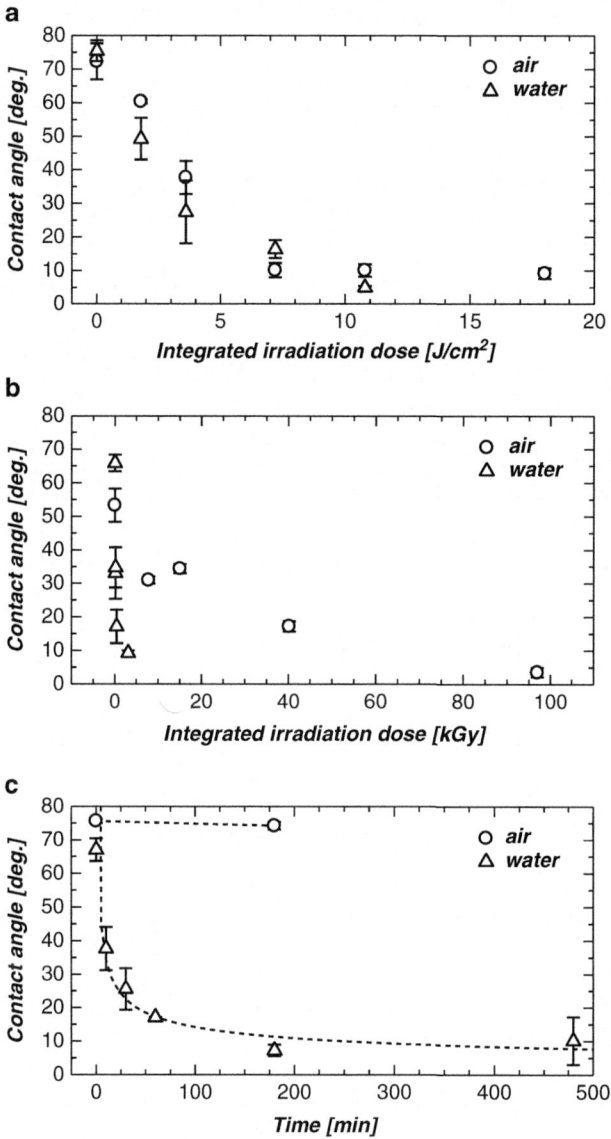

Fig. 10.4 Change of contact angle from ultraviolet, γ-ray, and proton beam irradiation to TiO_2 sample. (**a**) Ultraviolet irradiation. (**b**) γ-Ray irradiation. (**c**) Proton beam irradiation

heaters are installed in the copper block (3 kW in total). Two different copper blocks were fabricated to investigate the wettability effect on the heat transfer: one is a test section for ultraviolet irradiation and the other is for γ-ray irradiation. Both test sections are illustrated in Fig. 10.5: three thermocouples are inserted into the cylindrical part of each copper block to estimate heat flux and surface temperature.

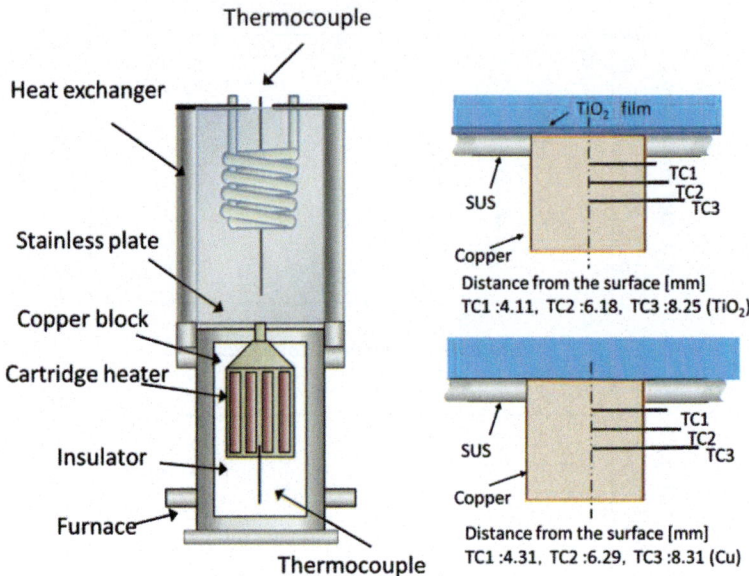

Fig. 10.5 Schematic of experimental apparatus

Table 10.1 Irradiation conditions

Irradiation source	Ultraviolet	γ-ray
Surface material	TiO_2	Copper oxide
Irradiation condition	3 mW/cm², 30 min	220 kGy

The heat transfer surface for ultraviolet experiments is coated by a TiO_2 thin film by a sputtering process in Kyushu University. The surface for γ-ray experiments is polished with #400 emery paper and then heated in air at 200 °C for 1 h. After the thermal oxidation of the copper surface, it is irradiated in the ^{60}Co facility where the integrated dose is 220 kGy. The irradiation conditions for heat-transfer experiments are summarized in Table 10.1.

10.3.2 Results and Discussion

Measured boiling curves with and without ultraviolet irradiation are shown in Fig. 10.6. Calculated lines denote existing correlations for nucleate boiling [12–14] and critical heat flux [15, 16]. As shown in this figure, the boiling curve after irradiation moves to the higher wall superheated side in the nucleate boiling region, which may be caused by inactivation of nucleation sites on the heat transfer surface resulting from hydrophilicity. In this experiment the maximum heat flux after the

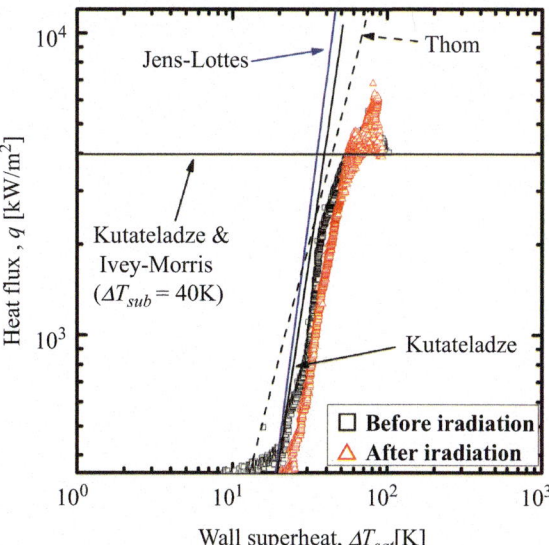

Fig. 10.6 Boiling curve before and after ultraviolet irradiation

nucleate boiling region is defined as a critical heat flux (CHF). In both boiling curves, the wall temperature rises rapidly just after the CHF region and the heat flux increases with decreasing wall temperature, resulting in heat flux being larger than CHF. These phenomena can be considered as a microbubble emission boiling (MEB) because many small bubbles are observed in this region. In the MEB region the boiling curve after irradiation moves to the lower wall superheated side, in contrast to the nucleate boiling region.

Measured boiling curves with and without γ-ray irradiation are shown in Figs. 10.7, 10.8, and 10.9. Table 10.2 shows the contact angles before and after irradiation to the oxidized copper surface. As shown in these figures, MEB phenomena are observed with or without irradiation for $\Delta T_{sub} = 20, 40$, and 60 K. The boiling curves in the nucleate boiling region (low superheat region) are shifted to the higher superheat side similar to the ultraviolet irradiation experiments. However, in contrast, the boiling curves in the MEB region (higher superheat region) are shifted to the lower superheat side after irradiation, which may be caused by enhancement of thin liquid film at the re-wetting phenomena.

CHF and maximum heat fluxes in the MEB region are plotted against liquid subcooling in Fig. 10.10. Solid and dashed lines denote the predicted values of existing CHF correlations by Kutatekadze [15] and Haramura and Katto [17], respectively. The measured CHF agree well with the predicted values by the aforementioned existing correlations. Measured CHF after the irradiations are slightly larger than those before the irradiations; however, the effect of the irradiation on the CHF is not obvious in the present experimental conditions. Maximum heat flux in the MEB region is almost 50 % larger than the CHF value, and the effect of the irradiation on the maximum heat flux is also not distinct, similar to the CHF within present experimental conditions.

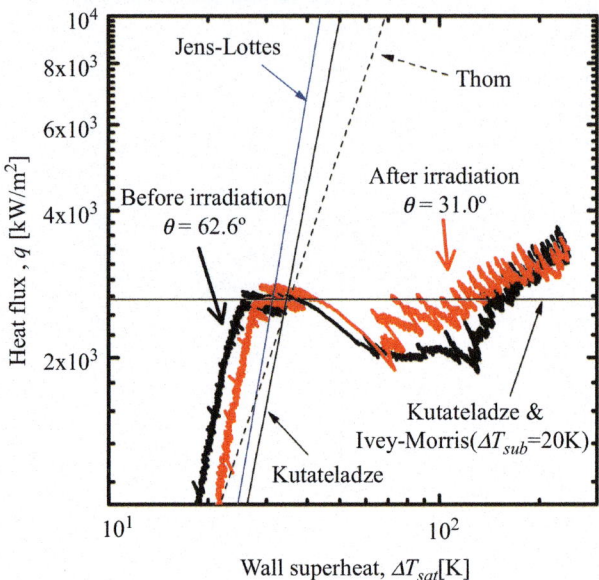

Fig. 10.7 Boiling curve before and after γ-ray for $\Delta T_{sub} = 20$ K

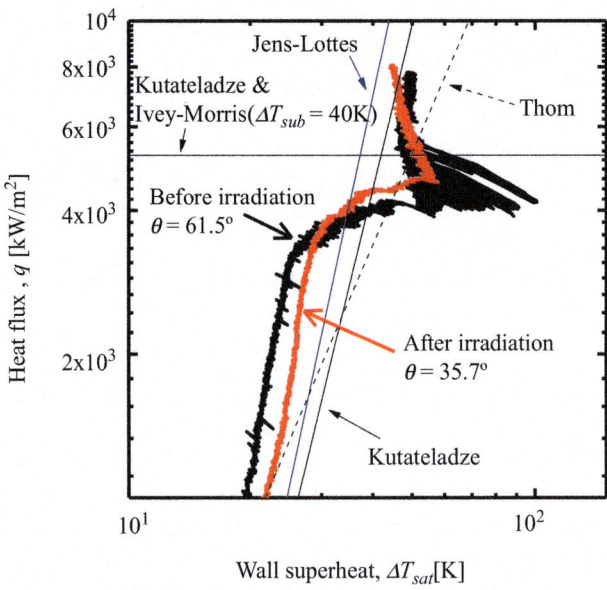

Fig. 10.8 Boiling curve before and after γ-ray for $\Delta T_{sub} = 40$ K

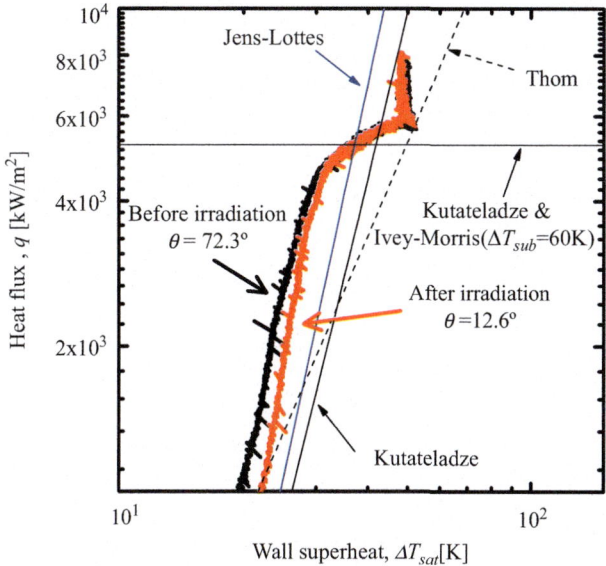

Fig. 10.9 Boiling curve before and after γ-ray for $\Delta T_{sub} = 60$ K

Table 10.2 Contact angles before and after γ-ray irradiations

Subcooling (K)	Before irradiation (°)	After irradiation (°)
20	62.6	31.0
40	61.5	35.7
60	72.3	12.6

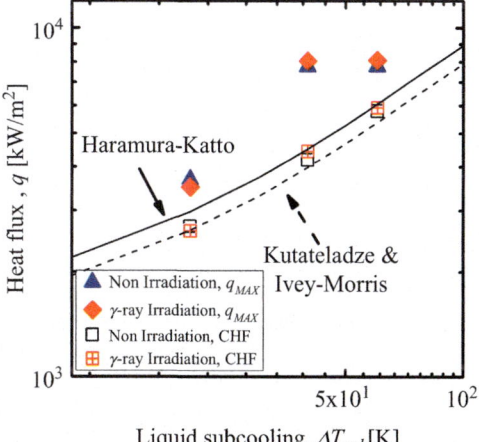

Fig. 10.10 Critical heat flux (CHF) and maximum heat flux before and after γ-ray irradiation

10.4 Conclusions

Effect of surface wettability on boiling heat transfer was studied by applying photocatalysis effect and radiation-induced surface activation. The main conclusions are as follows.

- Wettability enhancement was observed by proton-beam irradiation as well as the ultraviolet and γ-ray irradiation.
- In comparing the irradiation between that in air and and that in water, no influence of the radiation environment is observed for ultraviolet irradiation. However, the wettability was well enhanced by γ-ray and proton-beam irradiation in water rather than in air.
- The boiling curve with ultraviolet irradiation moves to the higher wall superheated side in the nucleate boiling region. In contrast, the boiling curve with irradiation moves to the lower wall superheated side in the MEB region.
- A similar tendency of the boiling curve was observed with γ-ray irradiation in comparison to the ultraviolet irradiation.
- The effect of irradiation on the CHF was not obvious at present experimental conditions.

Acknowledgments The authors express their sincere gratitude to Prof. Y. Takata and Dr. S. Hidaka of Kyushu University for their valuable comments and technical assistance. A part of this study is the result of "Research and Development for an Accelerator-Driven Sub-Critical System using a FFAG Accelerator" carried out under the Strategic Promotion Program for Basic Nuclear Research by the Ministry of Education, Culture, Sports, Science and Technology of Japan.

Open Access This chapter is distributed under the terms of the Creative Commons Attribution Noncommercial License, which permits any noncommercial use, distribution, and reproduction in any medium, provided the original author(s) and source are credited.

References

1. Cinotti L, Giraud B, Aït Abderrahim H (2004) The experimental accelerator driven system (XADS) designs in the EURATOM 5th framework programme. J Nucl Mater 335:148–155
2. Maes D (2006) Mechanical design of the small-scale experimental ADS: MYRRHA. Energ Convers Manag 47:2710–2723
3. Oigawa H et al (2011) Role of ADS in the back-end of the fuel cycle strategies and associated design activities: the case of Japan. J Nucl Mater 415:229–236
4. Cheng X et al (2006) Thermal-hydraulic analysis of the TRADE spallation target. Nucl Instrum Methods Phys Res A 562:855–858
5. Hao J-H et al (2013) Target thickness optimization design of a spallation neutron source target cooling system. Appl Therm Eng 61:641–648
6. Ito D, Nishi D, Saito Y (2014) Effect of ultraviolet and radiation on surface wettability. Multiphase Flow 30:209–220 (in Japanese)

7. Takata Y, Hidaka S, Cao JM, Nakamura T, Yamamoto H, Masuda M, Ito T (2005) Effect of surface wettability on boiling and evaporation. Energy 30:209–220
8. Takamasa T, Hazuku T, Okamoto K, Mishima K, Furuya M (2005) Radiation-induced surface activation on Leidenfrost and quenching phenomena. Exp Therm Fluid Sci 29:267–274
9. Suzuki K et al (2005) Enhancement of heat transfer in subcooled flow boiling with microbubble emission. Exp Therm Fluid Sci 29:827–832
10. Fujishima A, Honda K (1972) Electrochemical photolysis of water at a semiconductor electrode. Nature (Lond) 238:37–38
11. Masahashi N, Semboshi S, Ohtsu N, Oku M (2008) Microstructure and superhydrophilicity of anodic TiO_2 films on pure titanium. Thin Solid Films 516:7488–7496
12. Kutateladze SS (1952) Heat transfer in condensation and boiling, 2nd edn. AEC-trans-3770. U.S. Atomic Energy Commission, Technical Information Service, Washington
13. Jens WH, Lottes PA (1962) Analysis of heat transfer, burnout, pressure drop and density data for high pressure water. USAEC Report ANL-4627
14. Thom JRS, Walker WM, Fallon TA, Reising GFS (1966) Boiling in subcooled water during flow up heated tubes or annuli. Proc Inst Mech Eng 180:226–246, Part 3C
15. Kutateladze SS (1953) Heat transfer in condensation and boiling, 2nd edn. AEC-trans-3405, U. S. AEC Technical Information Service,
16. Ivey HJ, Morris DJ (1962) On the relevance of the vapor-liquid exchange mechanism for subcooled boiling heat transfer at high pressure. AEEW Report 137, UKAEA
17. Haramura Y, Katto Y (1983) A new hydrodynamic model of critical heat flux, applicable widely to both pool and forced convection boiling on submerged bodies in saturated liquids. Int J Heat Mass Transfer 26-3:389–399

Chapter 11
Experimental Study of Flow Structure and Turbulent Characteristics in Lead–Bismuth Two-Phase Flow

Gen Ariyoshi, Daisuke Ito, and Yasushi Saito

Abstract In a severe accident of a lead–bismuth-cooled accelerator-driven system, a gas–liquid two-phase flow with a large liquid-to-gas density ratio might appear, such as a steam leakage into hot lead–bismuth flow. It is still difficult to predict such phenomena because there are no available flow models for two-phase flow with a large density ratio compared to ordinary two-phase flows such as an air–water two-phase flow. Therefore, a two-phase flow model should be developed based on experimental data of two-phase flows with a large density ratio. In this study, a liquid–metal two-phase flow was measured by using a four-sensor electrical conductivity probe and a miniature electromagnetic probe to establish an experimental database for lead–bismuth flow structure. In measurements with the four-sensor probe, the radial profiles of void fraction and interfacial area concentration were measured at different axial positions. Experiments were also performed to understand the turbulent structure in a liquid–metal two-phase flow by using the electromagnetic probe. From the data measured by both four-sensor and electromagnetic probes, it is shown that the turbulence intensity at the pipe center was proportional to the void fraction to the power of 0.8 for higher void fraction. These results represented a similar tendency as previous data in air–water two-phase flows.

Keywords Accelerator-driven system • Electromagnetic probe • Four-sensor probe • Lead–bismuth • Turbulence characteristics • Two-phase flow • Void fraction

---

G. Ariyoshi (✉)
Graduate School of Energy Science, Kyoto University, Kyoto, Japan
e-mail: ariyoshi.gen.46n@st.kyoto-u.ac.jp

D. Ito • Y. Saito
Kyoto University Research Reactor Institute, Osaka, Japan

11.1 Introduction

The accelerator-driven system (ADS) has been developed as the next-generation nuclear energy system and is expected to be used as a nuclear transmutation process [1]. ADS is a hybrid system that consists of a high-intensity proton accelerator, a nuclear spallation target, and a subcritical core. Lead–bismuth eutectic (LBE) is considered as an option of the spallation target and can also be used as the coolant of the reactor. Neutrons are produced by a nuclear spallation reaction between the protons supplied from the accelerator and the LBE target, and a chain reaction of nuclear fission can then be maintained by the contribution of spallation neutrons. The chain reaction in the core will stop when the supply of protons stops. Therefore, the ADS has a higher safety margin, in principle, than other nuclear energy systems.

As research toward the development of the ADS, a subcritical reactor physics study, a reactor thermal-hydraulics study, and studies on the material, accelerator, and fuel for the ADS have been carried out. However, safety assessment is very important in preparation for a possible severe accident. A pipe rupture in a steam generator is one of the severe accidents of a LBE-cooled ADS. In this case, the direct contact between the LBE and the water ejected from the ruptured pipe of the steam generator might lead to LBE–steam two-phase flow in the reactor pool. If the gas bubble comes into the fuel region, the core reactivity might be affected. Thus, the gas–liquid two-phase flow appearing in the ADS core should be understood in taking measures for such an accident. The gas–liquid two-phase flow in an ADS has density ratio that is an order larger than that of air–water two-phase flow. Although flow models of gas–liquid two-phase flow with a large liquid-to-gas density ratio are required for severe accident analysis, there are fewer studies on two-phase flow in the physical property range of large density ratio mixture. Thus, an experimental database on two-phase flow properties in two-phase flow with a large density ratio should be built and the two-phase flow model should be developed based on the database. In this study, an LBE two-phase flow was measured by using a four-sensor electrical conductivity probe and a miniature electromagnetic probe, and knowledge of the flow structure and the turbulent characteristics in two-phase flow with a large density ratio was obtained.

11.2 Measurement Techniques

11.2.1 Four-Sensor Probe

The four-sensor probe [2] used in this study consists of a central front sensor and three peripheral rear sensors (Fig. 11.1a). Tungsten acupuncture needles with a maximum diameter of 0.1 mm were coated with epoxy resin varnish except the tip; the diameter of the tip is less than 1 μm. The insulated needles were inserted into a

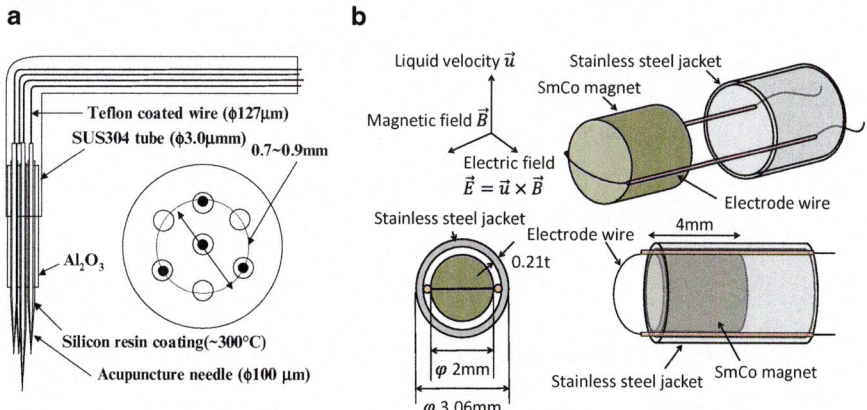

Fig. 11.1 Schematics of four-sensor probe (**a**) and electromagnetic probe (**b**)

seven-bore insulating tube made of Al_2O_3. The output signals were acquired at a sampling frequency of 10 kHz and then processed on a PC.

11.2.2 Electromagnetic Probe

The schematic of the electromagnetic probe [3, 4] used in this study is shown in Fig. 11.1b. The probe consists of a SmCo magnet, electrode wires, and a stainless steel jacket. Because a small cylindrical magnet with a diameter of 2 mm was used to miniaturize probe size, the induced potential between the electrodes at the tip of the probe was rather small. Therefore, the detected signal was amplified by a low-noise pre-amplifier and a DC amplifier. The signals digitized by an A/D converter were processed on a PC. The sampling frequency was 10 kHz.

The principle of the electromagnetic probe is based on Faraday's law. When the conducting fluid passes across a magnetic field, potential is induced in a direction normal to the magnetic field and the fluid velocity. Here, the induced potential is proportional to the velocity. In this study, the calibration of the electromagnetic probe was carried out using a rigid rotating setup (Fig. 11.2a) that consists of a cylindrical tank, a rotating system, and a heater. The tank was filled with LBE and rotated at a constant speed. The probe was inserted into the molten LBE rotating rigidly in the tank. The voltage corresponding to the tangential velocity component was measured, and this calibration was performed for all probes used in this study. Typical calibration results are shown in Fig. 11.2b.

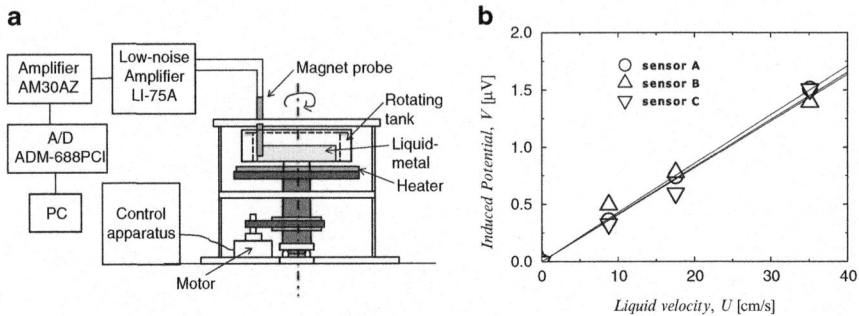

Fig. 11.2 Schematic of calibration system (**a**) and typical calibration results (**b**)

11.3 Experimental Setup

The schematic diagram of the LBE test loop is illustrated in Fig. 11.3. The test loop consists of a test section, a gas injector, an electromagnetic pump, a flow meter, and a drain tank. The test section is a stainless steel pipe with an inner diameter of 50 mm and a length of 2,000 mm. The working fluids are molten LBE and nitrogen gas. The flow rate of LBE was measured by the magnetic flow meter. Nitrogen gas was injected into LBE flow by the gas injector, which consists of 101 stainless steel needle tubes 0.58 mm in inner diameter. The gas flow rate was controlled by a mass flow controller. The operating temperature of this loop was maintained at 200 °C and the heating power was controlled by a temperature controller unit. The flow rate, differential pressure, temperature, and liquid level were monitored by a data acquisition unit connected to a PC. In the experiments, the superficial gas and liquid velocities were varied. Three four-sensor probes or electromagnetic probes were installed at three different axial positions ($z/D = 3.2$, 17.6, and 32.4) of the test section to investigate the axial development of two-phase flow structure. In addition, these probes were traversed at 12 radial points to obtain the radial profiles.

11.4 Results and Discussion

11.4.1 Radial Profiles of Two-Phase Flow Properties

The radial profiles of void fraction, interfacial area concentration, liquid velocity, and turbulence intensity are shown in Fig. 11.4. The horizontal axis is the radial position from the pipe center to the pipe wall, and the vertical axis is the measured data. The void fraction increases along the flow direction, mainly as a result of the static pressure change. The void fraction profile at $z/D = 3.2$ seems to be uniform and shows a flat shape. As increasing z/D, the void fraction profile changes to core peak because the large bubble moves to the core region. The interfacial area

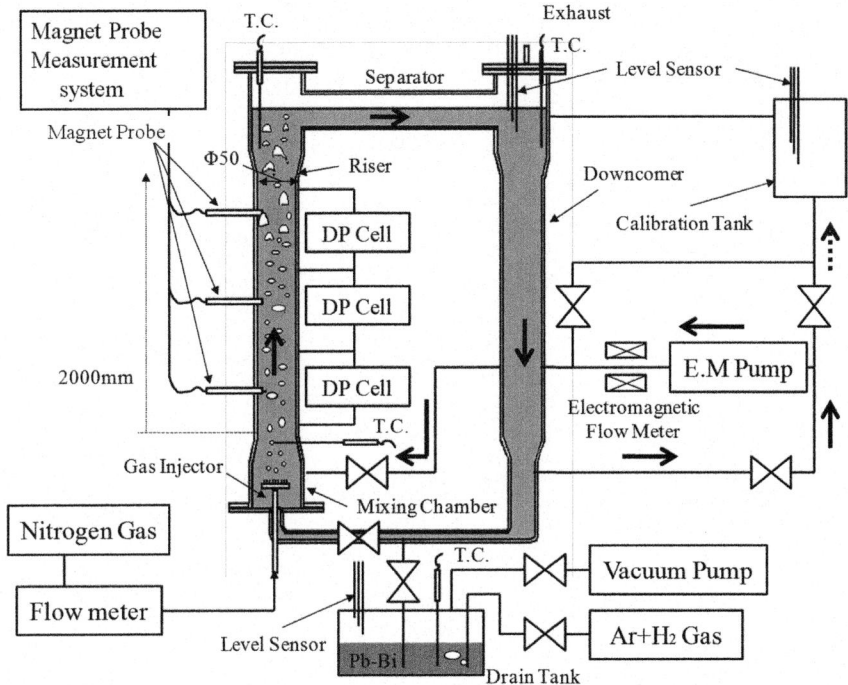

Fig. 11.3 Experimental setup for lead–bismuth-eutectic (LBE) two-phase flow measurement

concentration also increases as increasing z/D, as shown in Fig. 11.4b. The axial liquid velocity profiles are shown in Fig. 11.4c. The dashed lines are the calculated value of seventh power law, as follows:

$$U(r) = U_{\max}\left(1 - \frac{r}{R}\right)^{1/7} \qquad (11.1)$$

where $U_{\max}$ is the measured velocity at the pipe center. The measured liquid velocity profile at $z/D = 32.4$ has a core peak, although the profiles at $z/D = 3.2$ and 17.6 have a wall peak. Thus, the liquid velocity profiles were also developed axially and the gas–liquid interfacial drag force caused by the rising bubbles might act on the liquid phase. The turbulence intensity profiles in the liquid metal two-phase flow have a wall peak and they increase with the increase of z/D. These profiles have a shape similar to that in an air–water two-phase flow. However, the nondimensional turbulence intensity was much larger than the measured result in air–water flow systems; this might be attributed to the smaller liquid velocity in the present experimental condition, where the bubble-induced turbulence may be dominant.

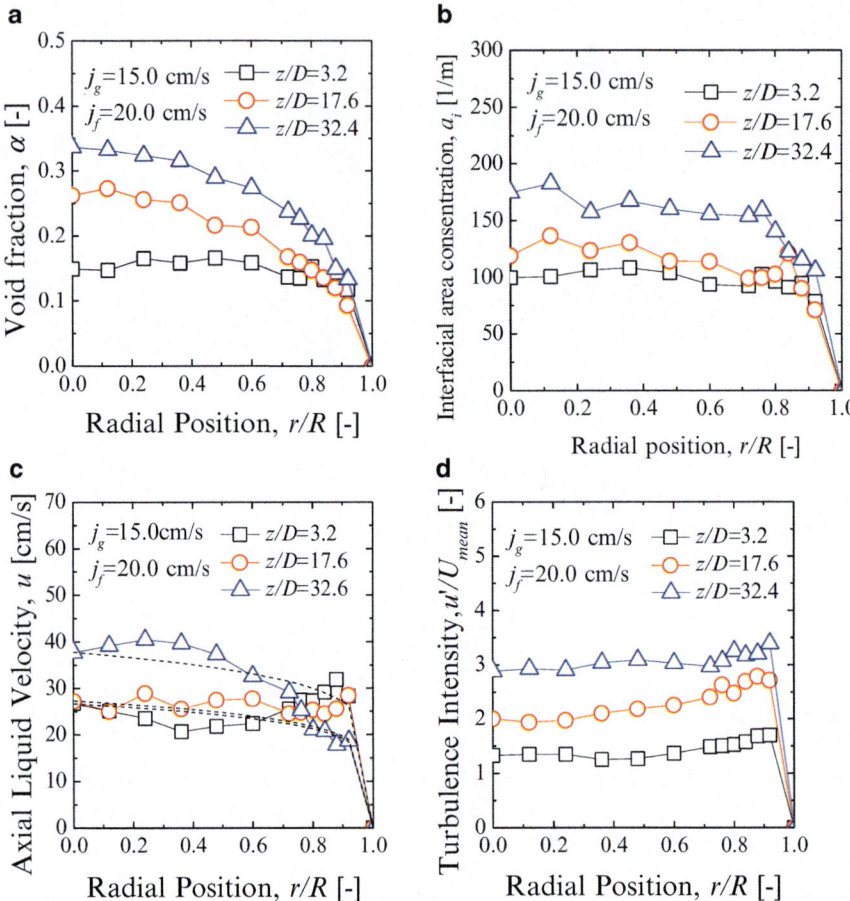

Fig. 11.4 Typical measurement results: void fraction (**a**), interfacial area concentration (**b**), axial liquid velocity (**c**), and turbulence intensity (**d**)

11.4.2 Comparison of Interfacial Area Concentration

Interfacial area concentration measured by the four sensor probes was compared with existing correlations (Fig. 11.5). The vertical axis shows the estimation error between the measured and calculated interfacial area concentration. All the correlation overestimates the interfacial area concentration by 50–90 %, which might be caused by the differences in bubble size and shape. Most of the correlations were formulated with air–water two-phase flow data for a bubbly flow regime. However, the bubble shape in a liquid metal two-phase flow might be strongly distorted by the momentum exchange at the gas–liquid interface. Thus, a more appropriate expression of the interfacial area concentration for liquid metal two-phase flow should be developed based on the experimental database.

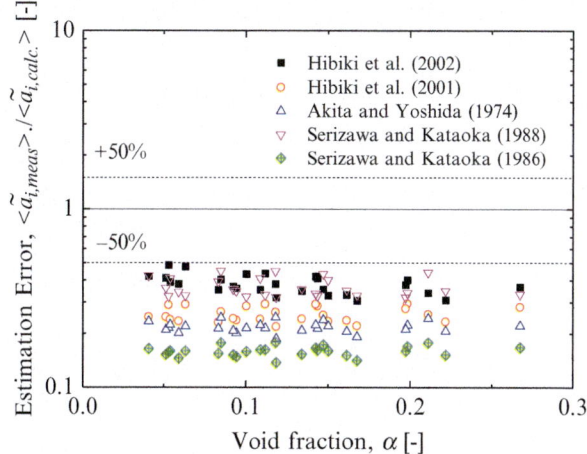

Fig. 11.5 Comparison of interfacial area concentration with existing correlations

11.4.3 Bubble-Induced Turbulence

The turbulence intensity measured in this study could be divided into wall turbulence and bubble-induced turbulence. However, turbulent production from bubbles is dominant at the pipe center. Thus, the turbulence intensity at $r/R = 0$ was plotted against the void fraction measured by the four-sensor probe (Fig. 11.6). In addition, the present results were compared with the previous experimental data of the bubble-induced turbulence in an air–water two-phase flow system. The solid line in this figure denotes the calculated value by the following semi-theoretical equation [5].

$$u' = u_r \alpha^{0.5}. \tag{11.2}$$

In this equation, the velocity field around the bubble is assumed as potential flow and the rotational component of the wake is ignored. In addition, the value calculated by the empirical equation for air–water two-phase flow [6] is also drawn as the dashed line in Fig. 11.6; the equation is represented as follows:

$$u' = 0.85 \alpha^{0.8}. \tag{11.3}$$

Although the fluid properties are different with the air–water two-phase flow, the measured turbulence intensity agrees with Eq. (11.3) and the previous data [7–9], except the result at $z/D = 3.2$. However, Eq. (11.3) was derived for an air–water flow system and its applicability to liquid metal flow was not clear. Therefore, the mechanism of turbulence production in liquid metal two-phase flow should be investigated in more detail. On the other hand, the turbulence intensity at $z/D = 3.2$ was slightly larger than other plots and Eq. (11.3). The measurement

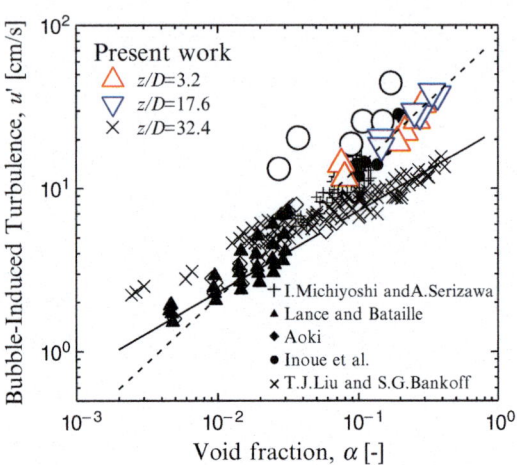

Fig. 11.6 Bubble-induced turbulence

position at $z/D = 3.2$ was relatively close to the gas injector, so it is expected that the flow was not fully developed.

11.5 Conclusions

A liquid metal two-phase flow was investigated by using a four-sensor probe and an electromagnetic probe. From the measurement results of two-phase flow structure and turbulence characteristics, the following knowledge was obtained.

- Radial profile of void fraction changes from wall peak to core peak along the flow direction.
- Axial development of the liquid velocity field shows different tendency for the void fraction profiles.
- Existing correlations for interfacial area concentration overestimate interfacial area concentrations at present experimental conditions, which might be attributed to the difference in bubble size. A new correlation should be modeled with further consideration of bubble size and the wall conditions.
- Bubble-induced turbulence at the pipe center in lead–bismuth two-phase flow agrees well with the previous experimental data for air–water flows. However, the mechanism should be clarified by measuring the liquid–metal two-phase flow in a wide range of flow conditions.

Open Access This chapter is distributed under the terms of the Creative Commons Attribution Noncommercial License, which permits any noncommercial use, distribution, and reproduction in any medium, provided the original author(s) and source are credited.

References

1. Tsujimoto K, Sasa T, Nishihara K, Oigawa H, Takano H (2004) Neutronics design for lead–bismuth cooled accelerator-driven system for transmutation of minor actinide. J Nucl Sci Technol 41(1):21–36
2. Saito Y, Mishima K (2012) Bubble measurements in liquid-metal two-phase flow by using a four-sensor probe. Multiphase Sci Technol 24:279–297
3. Ricou R, Vives C (1982) Local velocity and mass transfer measurements in molten metals using an incorporated magnet probe. Int J Heat Mass Transfer 25(10):1579–1588
4. Iguchi M, Tokunaga H, Tatemichi H (1997) Bubble and liquid flow characteristics in a Wood's metal bath stirred by bottom helium gas injection. Metall Mater Trans B 28B:1053–1061
5. Larce M, Bataille J (1983) Turbulence in the liquid phase of a bubbly air-water flow. In: Advances in two-phase flow and heat transfer, fundamentals and applications, vol 1. Martinus Nijhoff Publishers, Boston, pp 403–427
6. Michiyoshi I, Serizawa A (1986) Turbulence in two-phase bubbly flow. Nucl Eng Des 95:253–267
7. Aoki S (1982) Eddy diffusivity of momentum in bubbly flow. M.S. thesis, Department of Nuclear Engineering, Kyoto University
8. Inoue A, Aoki S, Koga T, Yaegashi H (1976) Void-fraction, bubble and liquid velocity profiles of two-phase bubble flow in a vertical pipe. Trans JSME 42:2521–2531
9. Liu JT, Bankoff GS (1993) Structure of air-water bubbly flow in a vertical pipe. I. Liquid mean velocity and turbulence measurements. Int J Heat Mass Transfer 36(4):1049–1060

Part IV
Basic Research on Reactor Physics of ADS: Basic Theoretical Studies for Reactor Physics in ADS

Chapter 12
Theory of Power Spectral Density and Feynman-Alpha Method in Accelerator-Driven System and Their Higher-Order Mode Effects

Toshihiro Yamamoto

Abstract This chapter discusses the theory of higher-order modes in the Feynman Y function and cross-power spectral density (CPSD) in an accelerator-driven system (ADS) where pulsed spallation neutrons are injected at a constant time interval. Theoretical formulae that consider the higher-order modes of the correlated and uncorrelated components in the Feynman Y function and CPSD for an ADS were recently derived in a paper published by the author. These formulae for the Feynman Y function and CPSD are applied to a subcritical multiplying system with a one-dimensional infinite slab geometry in this chapter. The Feynman Y functions and CPSD calculated with the theoretical formulae are compared with the Monte Carlo simulations of these noise techniques. The theoretical formulae reproduce the Monte Carlo simulations very well, thereby substantiating the theoretical formulae derived in this chapter. The correlated and uncorrelated components of the Feynman Y functions and CPSD are decomposed into the sum of the fundamental mode and higher-order modes. This chapter discusses the effect of subcriticality on the higher-order mode effects.

Keywords ADS • Feynman-α method • Higher-order mode • Monte Carlo • Neutron noise • Power spectral density

12.1 Introduction

In accelerator-driven systems (ADS), fission chain reactions are driven by spallation neutrons emitted from a proton beam target. An ADS is quite different from an ordinary nuclear reactor in that it is always operated at a subcritical state. Thus, the safety requirements for reactivity control can be eased in ADSs. The

---

T. Yamamoto (✉)
Kyoto University, Research Reactor Institute, 2-1010 Asashiro Nishi,
Kumatori-cho, Sennan-gun, Osaka 590-0494, Japan
e-mail: tyama@rri.kyoto-u.ac.jp

subcriticality of an ADS, however, needs to be continuously monitored to maintain its criticality safety. A reactor noise technique such as the Feynman-α method and the power spectral density method can be a potential candidate for monitoring the subcriticality of ADSs. The noise theory in ADSs is different from the classical reactor noise theory in that multiple neutrons are injected from the proton beam target at a single spallation event and pulsed neutrons are emitted deterministically at a constant period. Many theoretical and experimental studies on the noise theory in ADSs have been performed thus far. The theoretical formula for the Feynman-α method or Rossi-α method in ADSs was studied by, for example, Pázsit et al. [1], Pázsit et al. [2], Kitamura et al. [3], and Muñoz-Cobo et al. [4]. Another technique that uses the auto-power spectral density (APSD) or cross-power spectral density (CPSD) was studied by, for example, Muñoz-Cobo et al. [5], Rugama et al. [6], Ballester and Muñoz-Cobo [7], and Degweker and Rana [8]. Sakon et al. recently carried out a series of power spectral analyses in a thermal subcritical reactor system driven by a periodically pulsed 14 MeV neutron source at the Kyoto University Critical Assembly (KUCA) [9].

Both the Feynman-α method and the power spectral density method are intended to measure a prompt neutron time-decay constant α of the fundamental mode because the subcriticality is directly related to the fundamental mode α. The measured results, however, are inevitably contaminated by the higher-order mode components. To obtain an accurate knowledge of the subcriticality, the effect of the higher-order modes needs to be quantified in detail.

Endo et al. [10] derived a theoretical formula of the Feynman Y function that considers the higher order modes. Muñoz-Cobo et al. [11] also derived a similar theoretical formula from a different approach. Using these formulae, Yamamoto [12, 13] demonstrated quantitative analyses of the spatial- and energy-higher order modes in Feynman Y functions, respectively. In these two works, the Feynman Y functions were successfully resolved into spatial- or energy-higher order modes. These discussions, however, involved subcritical multiplying systems driven by a neutron source with Poisson character. They did not account for either a periodically pulsed neutron source or its non-Poisson character. Some previous work that considered the higher-order modes in the noise techniques for ADSs has been published (e.g., [6], [7]). In these previous publications, however, the effects of the higher-order modes have not been quantitatively investigated. Yamamoto [14, 15] presented the formulae of the Feynman Y function and CPSD for ADSs that consider the higher-order mode effects. Yamamoto [15] resolved the Feynman Y functions and power spectral densities into the mode components. Verification of the formulae was demonstrated by comparing the theoretical predictions with the Monte Carlo simulations of the subcriticality measurement in an ADS.

The purpose of the present chapter is to investigate how the subcriticality would affect Feynman Y function and power spectral density. The subcriticality of an ADS differs from design to design. The smaller the subcriticality, the larger the neutron multiplication that can be gained, which, on the other hand, decreases the margin of criticality safety. The subcriticality undergoes a gradual change as the fuel burn-up proceeds. The Feynman Y function and power spectral density emerge differently as the subcriticality changes. This chapter shows the dependence of subcriticality measurement on its subcriticality, which will contribute to the design of ADSs and planning of subcriticality measurements in the future.

12.2 Theory of Feynman-α Method in ADS

This section reviews the theory on the higher-order modes in the Feynman-α method in an ADS based on the work of Yamamoto [15]. Neglecting the energy- and spatial dependence of neutrons in a subcritical system driven by a neutron source with Poisson character, we obtain the Feynman Y function (the variance-to-mean ratio of neutron counts minus unity) as

$$Y(\Delta) = \frac{\langle C_1(\Delta)C_1(\Delta)\rangle - \langle C_1(\Delta)\rangle^2}{\langle C_1(\Delta)\rangle} - 1 \propto \left(1 - \frac{1 - e^{-\alpha_0\Delta}}{\alpha_0\Delta}\right) \quad (12.1)$$

where Δ = counting gate width, $C_1(\Delta)$ = neutron counts in Δ, α_0 = fundamental mode prompt neutron time-decay constant. When considering the energy and spatial dependence in an ADS, however, the Feynman Y function is more involved, as shown next.

The formula for the Feynman Y function in an ADS where q spallation neutrons are emitted from the beam target at a constant period T is given by this expression [15]:

$$Y(\Delta) = \frac{\langle C_1(\Delta)C_1(\Delta)\rangle - \langle C_1(\Delta)\rangle^2}{\langle C_1(\Delta)\rangle} - 1 = Y_C(\Delta) + Y_{CS}(\Delta) + Y_{UN}(\Delta), \quad (12.2)$$

where

$$Y_C(\Delta) = \frac{2}{C_R}\sum_{\ell=0}^{\infty}\sum_{m=0}^{\infty}\sum_{n=0}^{\infty} \frac{S_\ell F_{\ell \to mn} D_{1,m} D_{1,n}}{\alpha_\ell(\alpha_m + \alpha_n)\alpha_n}\left(1 - \frac{1 - e^{-\alpha_n\Delta}}{\alpha_n\Delta}\right), \quad (12.3)$$

$$Y_{CS}(\Delta) = -\frac{2q}{C_R T}\sum_{m=0}^{\infty}\sum_{n=0}^{\infty} \frac{D_{1,m} D_{1,n} \Psi_m^* \Psi_\ell^*}{(\alpha_m + \alpha_n)\alpha_n}\left(1 - \frac{1 - e^{-\alpha_n\Delta}}{\alpha_n\Delta}\right), \quad (12.4)$$

$$Y_{UN}(\Delta) = \frac{4q^2}{C_R \Delta T^2}\sum_{n=0}^{\infty}\sum_{\ell=0}^{\infty}\sum_{m=1}^{\infty} D_{1,n} D_{1,\ell} \Psi_n^* \Psi_\ell^* \frac{A_{\ell m n}(1 - \cos(\omega_m\Delta))}{\omega_m^2(A_{\ell m n}^2 + B_{\ell m n}^2)}, \quad (12.5)$$

the angle brackets denote the ensemble-averaging operator, C_R = count rate, and α_m = time-decay constant of the mth-order mode. (Refer to Yamamoto [14, 15] for other nomenclature.) Equation (12.3) represents the correlated component of the Y function, which also appears in a subcritical system with Poisson source. The correlation in Eq. (12.3) results from the multiple neutron emissions per fission

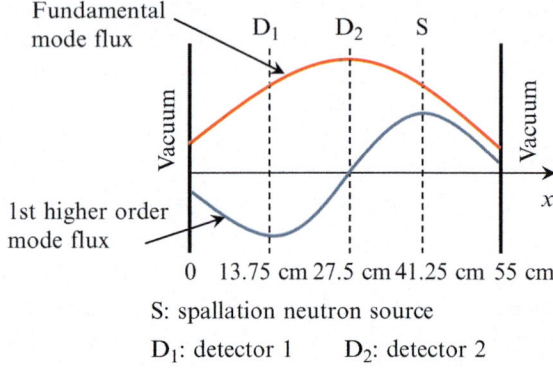

Fig. 12.1 Configuration of detector and neutron source in the one-dimensional infinite slab for test calculations

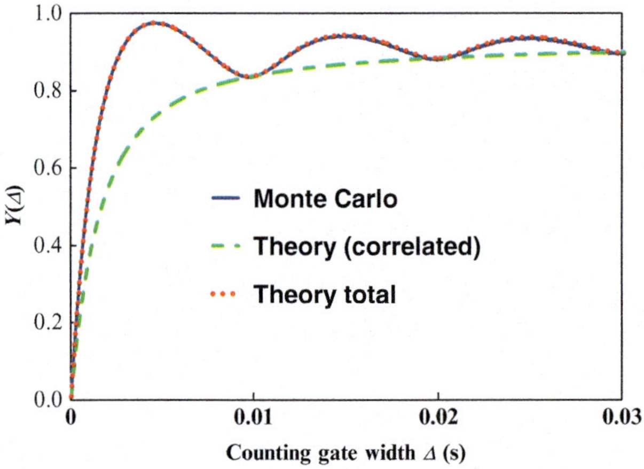

Fig. 12.2 Feynman Y function versus counting gate width by Monte Carlo simulation and theoretical value at the detector 1

reaction. Equation (12.4) represents another correlated component caused by periodically pulsed multiple neutrons. Equation (12.5) represents the uncorrelated component caused by the periodically pulsed spallation neutron source.

A numerical example is considered for a one-dimensional slab with infinite height. The thickness of the slab is $H = 55$ cm. The vacuum boundary conditions are imposed on both ends of the slab. The spallation neutron source and neutron detectors are allocated as shown in Fig. 12.1. This chapter considers a one-energy-group problem. The constants used for the numerical example are $\Sigma_t = 0.28$ cm^{-1}, $\Sigma_f = 0.049$ cm^{-1}, $\Sigma_c = 0.05$ cm^{-1}, $v = 2,200$ m/s, $\nu = 2$, and $q = 60$, $T = 0.01$ s (100 Hz). This system is sufficiently subcritical and $k_{\text{eff}} = 0.95865 \pm 0.00002$, which is obtained by a Monte Carlo criticality calculation (it is referred to as "large subcritical system" hereinafter). The Feynman Y function versus counting gate width Δ at the position of the detector 1 in Fig. 12.1 is calculated with a Monte Carlo simulation of the Feynman-α method. The simulation result at detector 1 is shown in Fig. 12.2 as "Monte Carlo." In Fig. 12.2, "Theory (correlated)" shows a

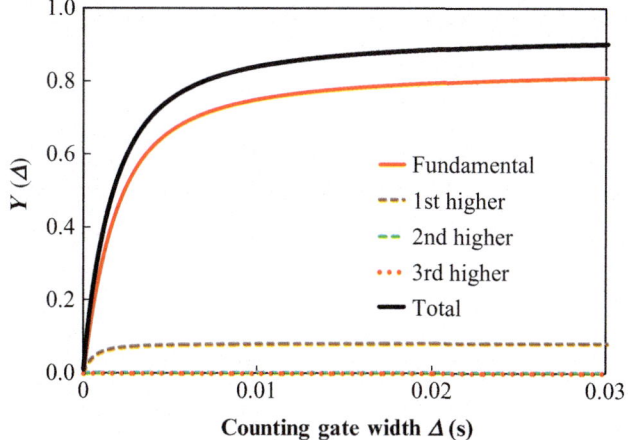

Fig. 12.3 Mode components of the correlated component in the Feynman Y function

theoretical value of the sum of $Y_C(\Delta)$ and $Y_{CS}(\Delta)$ calculated with Eqs. (12.3) and (12.4). "Theory total" shows "Theory (correlated)" plus the theoretical value of the uncorrelated component $Y_{UN}(\Delta)$, calculated with Eq. (12.5). The neutron flux and α_m, which are needed to calculate the theoretical values of Eqs. (12.3), (12.4), and (12.5), are calculated with the Monte Carlo method up to the third-order mode [16]. Beyond the third order, those are approximated with the diffusion theory:

$$\alpha_m = v\left(\Sigma_f + \Sigma_c + DB_m^2 - \nu\Sigma_f\right), m \geq 4 \qquad (12.6)$$

$$\psi_m(x) = \sqrt{\frac{2v}{H+2d}} \sin B_m(x+d), m \geq 4 \qquad (12.7)$$

$$B_m = \frac{(m+1)\pi}{H+2d}. \qquad (12.8)$$

where $d =$ extrapolated length($=0.7104/\Sigma_t$). The summation in Eqs. (12.3), (12.4), and (12.5) is taken up to the 250th mode. As shown in Fig. 12.2, there is good agreement between the Monte Carlo simulation and the theory, which shows verification of the theoretical formula of Eq. (12.2). Using Eqs. (12.3), (12.4), and (12.5), the Feynman Y function is decomposed into mode components. Figures 12.3 and 12.4 show the mode components of the correlated component and of the uncorrelated component, respectively. In these figures, each mode component includes the cross terms with the lower-order mode components. For example, "1st higher" includes the cross terms between the fundamental mode and the first higher-order mode as well as the first higher-order mode itself. Figure 12.1 shows that detector 1 is located at the bottom of the first higher-order mode. Thus, the first higher-order mode has a significant effect on the Feynman Y function of detector 1. Especially, the higher-order mode is more remarkable in the uncorrelated component, as shown in Fig. 12.4.

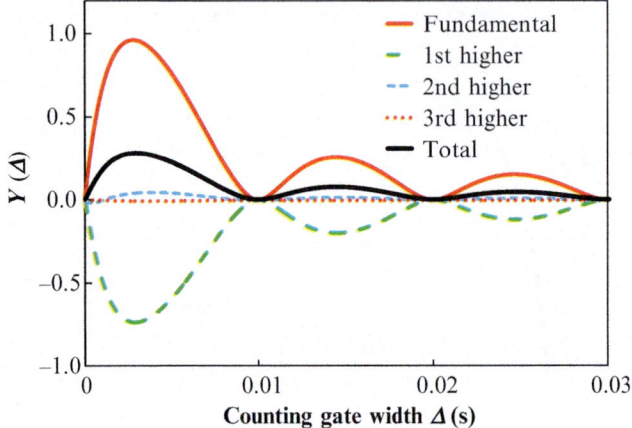

Fig. 12.4 Mode components of the uncorrelated component in the Feynman Y function

Fig. 12.5 Feynman Y function in the nearly critical system

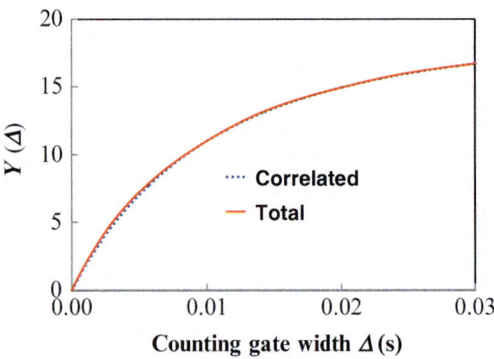

For a "nearly critical system" ($k_{eff} = 0.99242 \pm 0.00002$), the Feynman Y function is calculated using Eq. (12.2). The constants used for the nearly critical system are $\Sigma_t = 0.2834 \, \text{cm}^{-1}$, $\Sigma_f = 0.0524 \, \text{cm}^{-1}$, $\Sigma_c = 0.05 \, \text{cm}^{-1}$, $v = 2,200 \, \text{m/s}$, $\nu = 2$, $q = 60$, and $T = 0.01 \, \text{s}$ (100 Hz). The Feynman Y function versus the counting gate width is shown in Fig. 12.5. "Total" in Fig. 12.5 shows the sum of the correlated and uncorrelated components. As shown in Fig. 12.5, the uncorrelated component is very minor in the nearly critical system. Thus, the Feynman Y function is almost the same as the correlated component. The higher modes are negligibly small in the correlated component in the nearly critical system. Thus, the accurately approximated fundamental mode α can be obtained by fitting the Feynman Y function to the conventional formula, Eq. (12.1). On the other hand, if the subcriticality is not small enough, the uncorrelated component and higher-order modes have significant effects on the Feynman Y function. Therefore, obtaining a fundamental mode α would become difficult by simply fitting the Feynman Y function to Eq. (12.1). The Feynman-α method is not necessarily a suitable method as a subcriticality measurement technique.

12.3 Theory of Power Spectral Density in ADS

Another subcriticality measurement technique is the power spectral density method. This chapter focuses on the cross-power spectral density (CPSD), which is the Fourier transformation of a cross-correlation function between two neutron detector signals. In an infinite and homogeneous subcritical system where the energy and spatial dependence of the neutron is neglected, the CPSD is simply expressed as a function of frequency:

$$\text{CPSD}(\omega) \propto 1/(\omega^2 + \alpha_0^2), \tag{12.9}$$

where ω = angular frequency. The CPSD in an ADS where the energy and spatial dependence is considered, however, is much more involved as

$$\text{CPSD}(\omega) = \text{CPSD}_C(\omega) + \text{CPSD}_{UN}(\omega) + \text{CPSD}_{CS}(\omega), \tag{12.10}$$

$$\text{CPSD}_C(\omega) = \sum_{\ell=0}^{\infty}\sum_{m=0}^{\infty}\sum_{n=0}^{\infty} \frac{S_\ell F_{\ell \to mn} D_{1,m} D_{2,n}}{\alpha_\ell(\alpha_m + \alpha_n)} \frac{1}{(\alpha_n + i\omega)}$$
$$+ \sum_{\ell=0}^{\infty}\sum_{m=0}^{\infty}\sum_{n=0}^{\infty} \frac{S_\ell F_{\ell \to mn} D_{2,m} D_{1,n}}{\alpha_\ell(\alpha_m + \alpha_n)} \frac{1}{(\alpha_n - i\omega)}, \tag{12.11}$$

$$\text{CPSD}_{UN}(\omega) = \frac{2\pi q^2}{T^2} \sum_{n=0}^{\infty}\sum_{\ell=0}^{\infty}\sum_{m=-\infty}^{\infty} D_{1,n} D_{2,\ell} \Psi_n^* \Psi_\ell^* \frac{\delta(\omega - \omega_m)}{(\alpha_n - i\omega_m)(\alpha_\ell + i\omega_m)}, \tag{12.12}$$

$$\text{CPSD}_{CS}(\omega) = -\frac{q}{T} \sum_{n=0}^{\infty}\sum_{\ell=0}^{\infty} D_{1,n} D_{2,\ell} \Psi_n^* \Psi_\ell^* \frac{1}{\alpha_n + \alpha_\ell} \cdot \frac{1}{\alpha_\ell + i\omega}$$
$$- \frac{q}{T} \sum_{n=0}^{\infty}\sum_{\ell=0}^{\infty} D_{2,n} D_{1,\ell} \Psi_n^* \Psi_\ell^* \frac{1}{\alpha_n + \alpha_\ell} \cdot \frac{1}{\alpha_\ell - i\omega}, \tag{12.13}$$

where $i = \sqrt{-1}$, $\omega_m = 2\pi m/T$, and the subscripts C, UN, and CS have the same meanings as in the previous section. For the two systems in the previous section (large subcritical system and nearly critical system), Monte Carlo simulations were performed to obtain CPSDs between detectors 1 and 2 in Fig. 12.1. In the simulations, the pulse period is $T = 0.05$ s (20 Hz). The simulation result for the nearly critical system is compared with the theoretical one in Fig. 12.6. The results of the large subcritical system are shown by Yamamoto [15]. The theoretical results agree well with the Monte Carlo simulations. The uncorrelated component, $\text{CPSD}_{UN}(\omega)$, emerges only at the integer multiples of the pulse frequency as the Delta-function-like peaks. Thus, either of the correlated and uncorrelated components can be easily discriminated from the CPSD. Using Eq. (12.12), the uncorrelated and correlated components of the CPSD in the nearly critical system is decomposed into the mode components, shown in Figs. 12.7 and 12.8, respectively. In the correlated

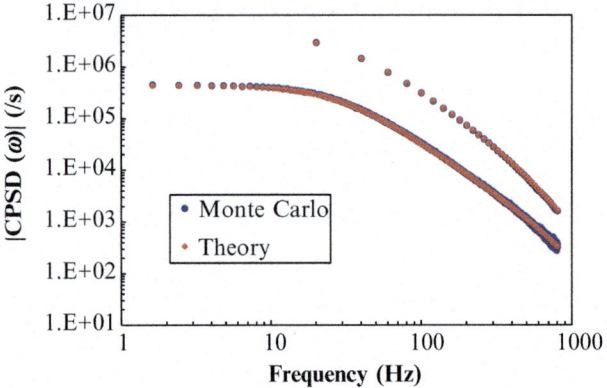

Fig. 12.6 Amplitude of cross-power spectral density (CPSD) in the nearly critical system by Monte Carlo simulation and theoretical value

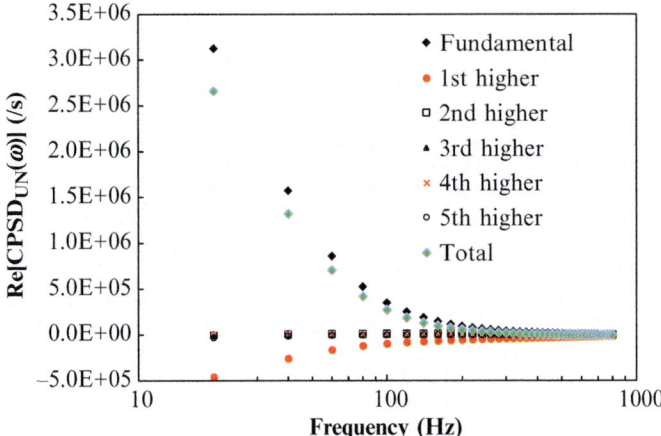

Fig. 12.7 Mode components of the uncorrelated component in the CPSD in the nearly critical system (real part)

component, the higher-order modes are negligibly small, and almost the whole of the CPSD is made up of the fundamental mode. The same condition holds for the large subcritical system. The higher-order mode effect in the correlated component is minor even in the large subcritical system. In the uncorrelated component, the higher-order mode effect is significant even in the nearly critical system. In the large subcritical system, the higher-order mode effect is much more significant. Thus, fitting the uncorrelated component to Eq. (12.9) yields an inaccurate α value unless the system is nearly critical. For example, in the large subcritical system we obtain $\alpha = 789$ (s^{-1}) for the true fundamental mode α value of 940 (s^{-1}) [15] from the uncorrelated component. On the other hand, we obtain $\alpha = 900$ (s^{-1}) from the

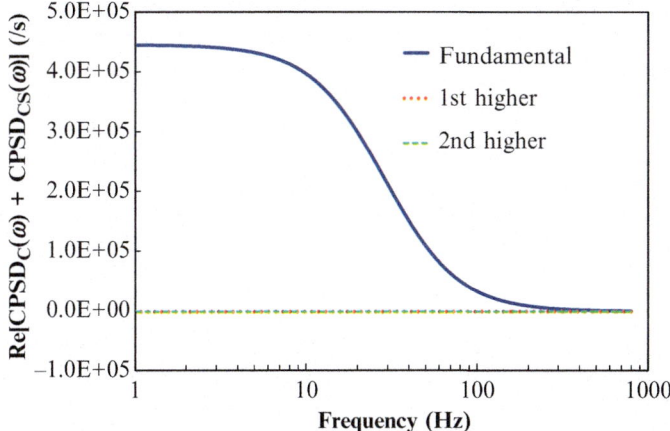

Fig. 12.8 Mode components of the correlated component in the CPSD in the nearly critical system (real part)

correlated component. In the nearly critical system, we obtain $\alpha = 172$ (s^{-1}) and 178 (s^{-1}) from the uncorrelated and correlated component, respectively, for the true fundamental mode α value of 179 (s^{-1}) [15].

12.4 Conclusions

In a subcriticality measurement for an ADS, the Feynman Y function in general appears as the sum of the correlated and uncorrelated components. The higher-mode effect in the correlated component is less significant than in the uncorrelated component. Thus, a relatively good approximation of the true fundamental mode α can be obtained by using the correlated component. However, it is not necessarily easy to separate the correlated component from the measured Feynman Y function. Considering the difficulty of separating the correlated component, the Feynman-α method is not always suitable as a subcriticality measurement technique for ADSs. In an ADS that is nearly critical, the uncorrelated component is very minor. Thus, by fitting the measured Feynman Y function to the correlated component, the fundamental mode α can be accurately estimated.

In a subcriticality measurement using the power spectral density method, the uncorrelated component emerges at the integer multiples of the pulse frequency as delta-function-like peaks. Thus, the uncorrelated component can be easily discriminated from the correlated component. The correlated component is less contaminated by the higher-order modes. An approximate fundamental mode α can be obtained by fitting the Feynman Y function to the correlated component of the power spectral density. The use of the uncorrelated component is not always recommended, because the higher-order modes are more significant in the uncorrelated component.

Open Access This chapter is distributed under the terms of the Creative Commons Attribution Noncommercial License, which permits any noncommercial use, distribution, and reproduction in any medium, provided the original author(s) and source are credited.

References

1. Pázsit I, Ceder M, Kuang Z (2004) Theory and analysis of the Feynman-alpha method for deterministically and randomly pulsed neutron sources. Nucl Sci Eng 148:67–78
2. Pázsit I, Kitamura Y, Wright J, Misawa T (2005) Calculation of the pulsed Feynman-alpha formulae and their experimental verification. Ann Nucl Energy 32:986–1007
3. Kitamura Y, Taguchi K, Misawa T, Pázsit I, Yamamoto A, Yamane Y, Ichihara C, Nakamura H, Oigawa H (2006) Calculation of the stochastic pulsed Rossi-alpha formula and its experimental verification. Prog Nucl Energy 48:37–50
4. Muñoz-Cobo JL, Peña J, González E (2008) Rossi-α and Feynman Y functions for non-Poissonian pulsed sources of neutrons in the stochastic pulsing method: application to subcriticality monitoring in ADS and comparison with the results of Poissonian pulsed neutron sources. Ann Nucl Energy 35:2375–2386
5. Muñoz-Cobo JL, Rugama Y, Valentine TE, Mihalczo JT, Perez RB (2001) Subcritical reactivity monitoring in accelerator-driven systems. Ann Nucl Energy 28:1519–1547
6. Rugama Y, Muñoz-Cobo JL, Valentine TE (2002) Modal influence of the detector location for the noise calculation of the ADS. Ann Nucl Energy 29:215–234
7. Ballester D, Muñoz-Cobo JL (2006) The pulsing CPSD method for subcritical assemblies driven by spontaneous and pulsed sources. Ann Nucl Energy 33:281–288
8. Degweker SB, Rana YS (2007) Reactor noise in accelerator driven systems: II. Ann Nucl Energy 34:463–482
9. Sakon A, Hashimoto K, Sugiyama W, Taninaka H, Pyeon CH, Sano T, Misawa T, Unesaki H, Ohsawa T (2013) Power spectral analyses for a thermal subcritical reactor system driven by a pulsed 14 MeV neutron source. J Nucl Sci Technol 50:481–492
10. Endo T, Yamane Y, Yamamoto A (2006) Space and energy dependent theoretical formula for the third order neutron correlation technique. Ann Nucl Energy 33:521–537
11. Muñoz-Cobo JL, Bergölf C, Peña J, González E, Villamarín D, Bournos V (2011) Feynman-α and Rossi-α formulas with spatial and modal effects. Ann Nucl Energy 38:590–600
12. Yamamoto T (2011) Higher order mode analyses in Feynman-α method. Ann Nucl Energy 38:1231–1237
13. Yamamoto T (2013) Energy-higher order mode analyses in Feynman-α method. Ann Nucl Energy 57:84–91
14. Yamamoto T (2014) Frequency domain Monte Carlo simulation method for cross power spectral density driven by periodically pulsed neutron source using complex-valued weight Monte Carlo. Ann Nucl Energy 64:711–720
15. Yamamoto T (2014) Higher order mode analyses of power spectral density and Feynman-α method in accelerator driven system with periodically pulsed spallation neutron source. Ann Nucl Energy 66:63–73
16. Yamamoto T (2011) Higher order α mode eigenvalue calculation by Monte Carlo power iteration. Prog Nucl Sci Technol 2:826–835

Chapter 13
Study on Neutron Spectrum of Pulsed Neutron Reactor

Takanori Kitada, Thanh Mai Vu, and Noboru Dobuchi

Abstract The neutron spectrum of a pulsed neutron reactor at subcritical state is different from that evaluated by k-eigenvalue mode, because of the time needed in the neutron slowing-down process from fast to thermal energy range. The time needed in slowing down does not depend on the degree of subcriticality, but the decreasing speed of neutron flux becomes fast as the subcriticality becomes deep. Therefore, the neutron spectrum becomes soft as the subcriticality becomes deep. This fact suggests to us that group constants to be used in the design study should change with the degree of subcriticality of the target system, even in the case of the same composition.

Keywords ADSR • Alpha-eigenvalue • k-Eigenvalue mode • Neutron spectrum • Subcriticality • Time-dependent mode

13.1 Introduction

The accelerator-driven subcritical reactor (ADSR) is considered as one of the best candidates to annihilate the radioactivity of nuclear waste and has been investigated in many institutes for many years. The ADSR is operated by the pulsed proton beam as an ignition of spallation reaction to produce many neutrons. Kyoto University Critical Assembly (KUCA) is one of the facilities to demonstrate the ADSR by using accelerated protons for the spallation reaction or deuterons for the deuterium-tritium (DT) reaction.

This study focused on the transient behavior of the neutron spectrum in a subcritical system after the injection of DT neutrons to know and discuss the physical behavior of the neutron spectrum in a subcritical system through the analysis of the experiments performed at KUCA. The subcritical system with pulsed neutrons has been widely analyzed in the steady state, although transient behavior of the neutron spectrum after the injection of DT neutrons can be analyzed in the transient state. This chapter focuses on the neutron spectrum evaluated in

---

T. Kitada (✉) • T.M. Vu • N. Dobuchi
Osaka University, Graduate School of Engineering, 2-1 Yamada-oka, Suita, Osaka 565-0871, Japan
e-mail: kitada@see.eng.osaka-u.ac.jp

steady state, and two kinds of calculation modes in steady state are compared and discussed: the k-eigenvalue mode and the alpha-eigenvalue mode.

Chapter 2 shows a brief explanation of KUCA experiments and the major results. Analyses of the experiments and discussion are described in Chap. 3, and the conclusions are summarized in Chap. 4.

13.2 Experiment at KUCA and Measured Results

This chapter shows a brief explanation of experiments performed at KUCA for the convenience of easy understanding of the analysis described in the next chapter.

Experiments modeled on the ADSR were performed at the A-core with adjacent D+ accelerator. A typical core configuration is shown in Fig. 13.1 for the case of 13 fuel rods. Each square cell is 2 in. $\times$ 2 in. in size, with size in the horizontal direction about 1.5 m, composed of a central 40-cm-thick fuel region and upper and lower polyethylene reflector regions. All control rods are inserted through the experiment. Accelerated D+ ions are hit with tritium target, depicted as T-target in Fig. 13.1, and 14 MeV neutrons produced by D-T reaction at T-target are injected into the core region composed of the fuel rods, polyethylene reflector, etc. Neutrons are injected from outside the core region in this experiment.

The target subcriticality was widely changed by changing the number of fuel rods from 19 to 6 to check the validity of the fiber scintillation counter used in the measurement. The subcriticality of the system was evaluated by the so-called extrapolation area ratio method proposed by Gozani [1]. The counters used in the experiments were set at several positions inside the core, and the measured results summarized in Table 13.1 are the results obtained at the core central area, where the most reliable results are expected. Measured subcriticality is from 2.3 [\$] (19 fuel rods case) to 49 [\$] (6 fuel rods case).

13.3 Analysis and Discussion of Neutron Flux

The analysis results of neutron flux distribution and neutron spectrum are summarized in this chapter with some discussion. Neutron flux distribution and spectrum in the core are shown in Sects. 13.3.1 and 13.3.2, respectively. All analyses are done by a continuous energy Monte Carlo code named MVP-II [2] with JENDL-4.0 [3] library. The code has the function to simulate the experiment not only in eigenvalue mode but also in time-dependent mode, where the necessary time of neutron flight is used to account for the elapsed time after the injection of DT neutrons.

In this chapter, fast and thermal neutrons are in the energy range less than 4 eV and more than 100 keV, respectively.

13 Study on Neutron Spectrum of Pulsed Neutron Reactor

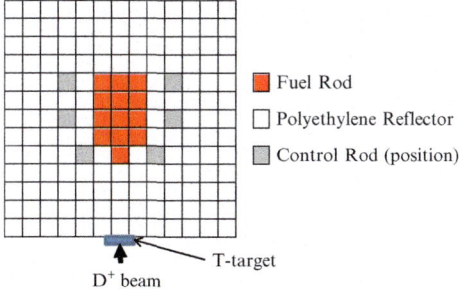

Fig. 13.1 Typical core configuration (13 fuel rods case). *T-target* tritium target

Table 13.1 Measured subcriticality in the Kyoto University Critical Assembly (KUCA) experiment

Number of fuel rods	Measured subcriticality [$] (standard deviation: 1σ)
19	2.32 (0.02)
17	6.40 (0.08)
15	10.9 (0.2)
13	13.4 (0.2)
9	28.2 (1.1)
6	49.4 (1.0)

13.3.1 Neutron Flux Distribution

Neutron flux distribution is drastically changed with the change in subcriticality of the system, because of the change in the number of fuel rods in the core. Figures 13.2, 13.3, and 13.4 show the neutron flux distribution along the central line from T-target to core for 13 fuel rods case in eigenvalue mode and time-dependent mode, respectively. Neutron flux distribution evaluated in time-dependent mode changes as a function of elapsed time after the injection of D-T neutrons, and the shape of neutron flux distribution is almost stable after $1e^{-4}$ s (Figs. 13.3 and 13.4). The comparison of neutron flux between two modes shows the discrepancy (Figs. 13.5 and 13.6). Thermal neutron flux distribution of the time-dependent mode is smaller at fuel region, but higher at the reflector region, than those of eigenvalue mode, although fast neutron flux distribution is almost the same between the two modes. Figures 13.5 and 13.6 also show that the neutron spectrum is different between two modes, and the details are discussed in the next section.

13.3.2 Neutron Spectrum

The neutron spectrum at the fuel region is shown in Figs. 13.7 and 13.8. The neutron spectrum in time-dependent mode changes as a function of elapsed time, although

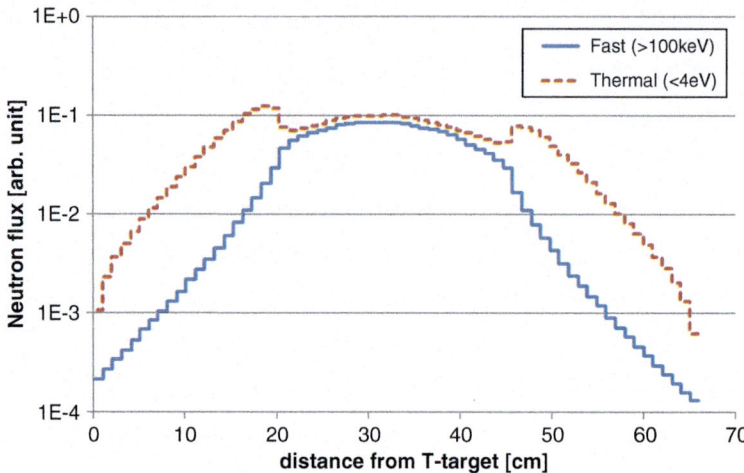

Fig. 13.2 Neutron flux distribution evaluated in eigenvalue mode (13 fuel rods)

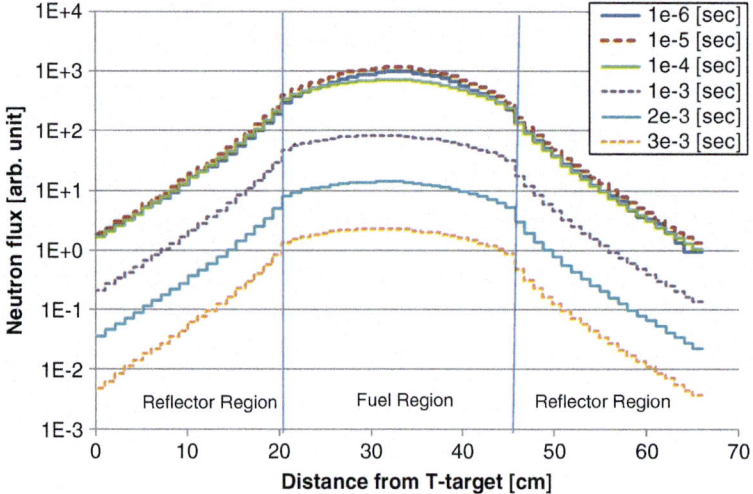

Fig. 13.3 Neutron flux distribution evaluated in time-dependent mode (fast energy range, 13 fuel rods)

the neutron spectrum in eigenvalue mode is singular. In addition to this, the neutron spectrum at the fuel region evaluated in eigenvalue mode is almost the same among different subcritical states, because there is no change in fuel composition through the experiment. Neutron spectrum at the fuel region evaluated in time-dependent mode changes as a function of elapsed time after the injection of D-T neutrons, but the shape of the spectrum becomes stable after around $1e^{-4}$ s, although the

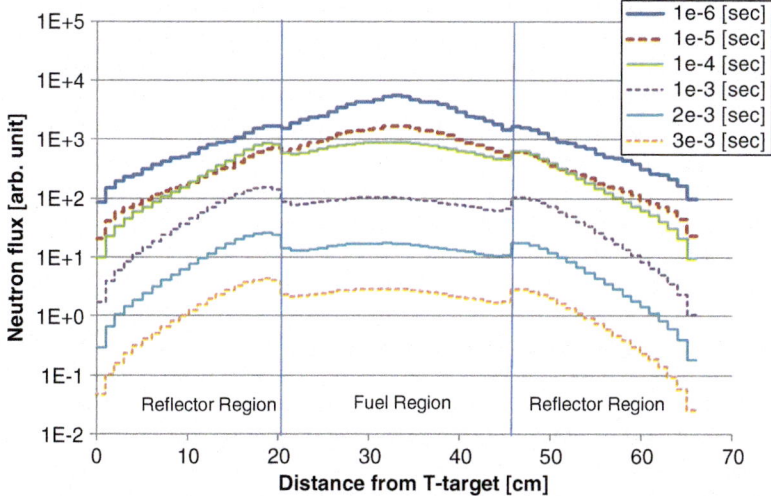

Fig. 13.4 Neutron flux distribution evaluated in time-dependent mode (thermal energy range, 13 fuel rods)

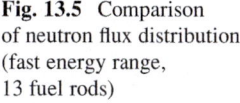

Fig. 13.5 Comparison of neutron flux distribution (fast energy range, 13 fuel rods)

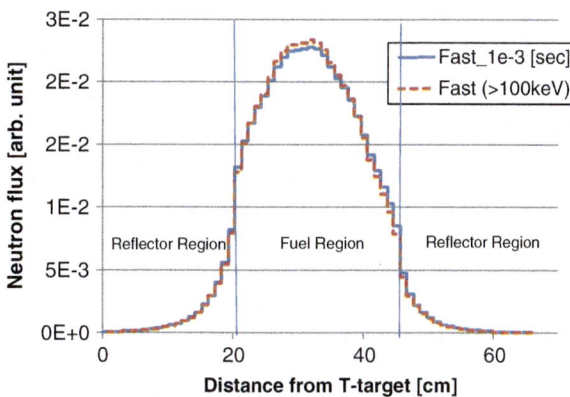

magnitude of neutron flux decreases with elapsed time (Fig. 13.8). The neutron spectrum is compared between two modes, where the shape of the spectrum in time-dependent mode is almost stable (at $1e^{-3}$ s). The comparison of the neutron spectrum is shown in Fig. 13.9. The neutron spectrum in time-dependent mode depends on the subcriticality of the system, and the spectrum becomes soft as the subcriticality becomes deep. This tendency can be understood by considering the following facts. The magnitude of neutron flux decreases as a function of elapsed time in the subcritical system, and the decreasing speed of neutron flux becomes

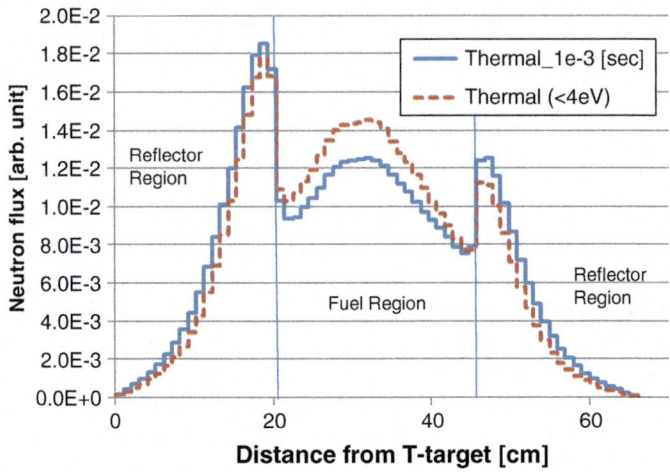

Fig. 13.6 Comparison of neutron flux distribution (thermal energy range, 13 fuel rods)

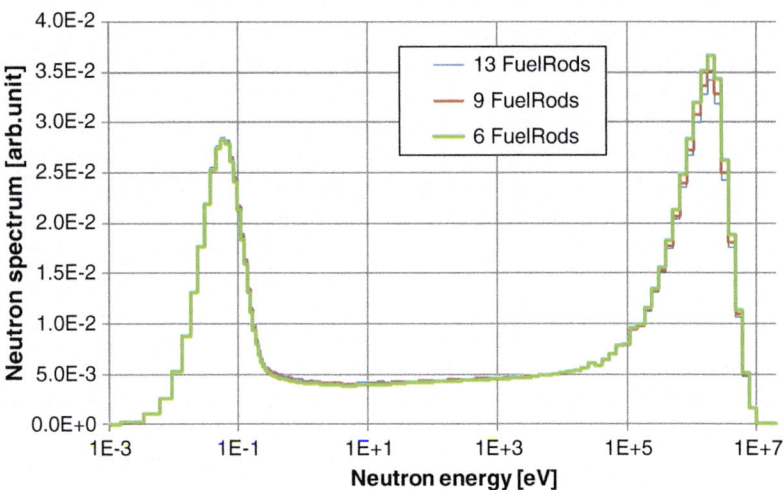

Fig. 13.7 Neutron spectrum at fuel region evaluated in eigenvalue mode

high for the deep subcritical system. Here the neutrons are slowed down by colliding with the medium in the thermal core, and the time needed in the slowing-down process does not depend on the subcriticality of the system. Therefore, the change in the neutron spectrum is caused by the time delay of decrease in thermal energy range where the neutrons are slowed down compared to that in the fast energy range.

The difference in neutron spectrum will cause the difference in collapsed cross sections widely used in design survey calculations, and the degree of the difference

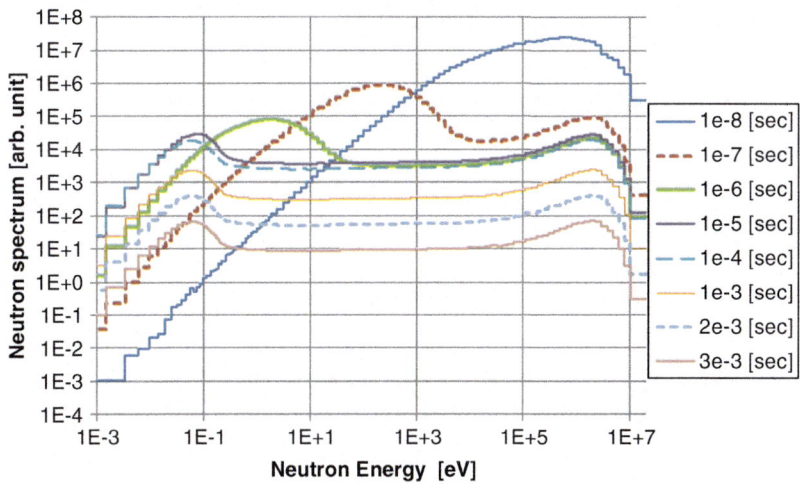

Fig. 13.8 Neutron spectrum at fuel region evaluated in time-dependent mode (13 fuel rods)

becomes remarkable as the subcriticality of the system becomes deep. It should be noted that collapsed cross sections depend on the subcriticality of target system because of the difference in neutron spectrum.

There is one recommendation to evaluate the proper neutron spectrum for collapsing. In eigenvalue mode, the k-eigenvalue mode expressed as Eq. (13.13.1) is usually used because the eigenvalue is an unbiased index to recognize the criticality, but there is another eigenvalue mode, named the alpha-eigenvalue mode, expressed as Eq. (13.13.2):

$$k\text{-eigenvalue mode } L\phi_k = \frac{1}{k}M\phi_k, \qquad (13.1)$$

$$\text{alpha-eigenvalue mode } \left[L + \frac{\alpha}{v}\right]\phi_\alpha = M\phi_\alpha, \qquad (13.2)$$

where L is the destruction operator including leakage and absorption reactions, M is the production operator including fission reactions, k is the k-eigenvalue called the effective multiplication factor, α is the alpha-eigenvalue, v is the neutron speed, and ϕ_k, ϕ_α are the neutron fluxes for each mode. Equation (13.1) is derived from a time-dependent equation by eliminating the term of time derivative, but Eq. (13.13.2) is derived by considering the exponential change of neutron flux in time. Usually a subcritical system such as the ADSR is operated not in stable but in transient conditions.

In the subcritical system, the alpha-eigenvalue is negative, and the impact of the negative alpha-eigenvalue on neutron flux is remarkable at thermal energy range where the neutron speed is small. Therefore, the neutron spectrum evaluated in alpha-eigenvalue mode is softer than that in k-eigenvalue mode. Similar to this consideration, the difference in neutron spectrum could be observed in the

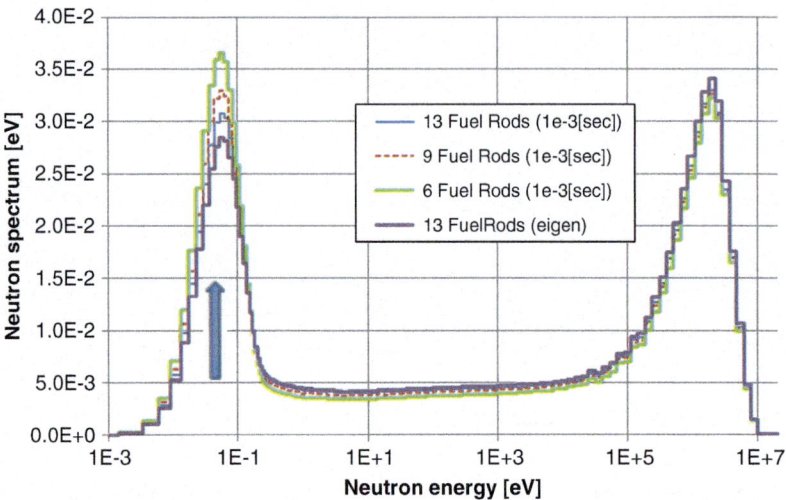

Fig. 13.9 Comparison of neutron spectrum among several cases

supercritical state. However, the difference is expected to be not remarkable compared to subcritical state, because excess reactivity is remarkably small compared to the subcriticality, as readily expected.

13.4 Conclusions

The neutron flux distribution evaluated in time-dependent mode changes with elapsed time after the ignition of neutrons into the subcritical system, and the neutron distribution in energy and space becomes almost stable in about $1\ e^{-4}$ s after the ignition.

There is a remarkable difference in neutron spectrum between two results in k-eigenvalue and time-dependent modes. The neutron spectrum (at 1 μs after the ignition) evaluated in time-dependent mode is softer than that in k-eigenvalue mode, and the difference is more remarkable in a deep subcriticality system. This difference is caused by the fact that additional time is necessary to be moderated before decreasing neutron flux in the thermal energy range, and the time is independent of the subcriticality of the system, depending only on the material composition of the system.

The neutron spectrum of a pulsed neutron reactor is to be evaluated in alpha-eigenvalue mode instead of k-eigenvalue mode to match the neutron spectrum during the decrease with elapsed time after the ignition of pulsed neutrons into the subcritical system.

Open Access This chapter is distributed under the terms of the Creative Commons Attribution Noncommercial License, which permits any noncommercial use, distribution, and reproduction in any medium, provided the original author(s) and source are credited.

References

1. Gozani T (1962) Nukleonik 4:348
2. Nagaya Y, Okumura K, Mori T et al (2005) MVP/GMVP version 2: general purpose Monte Carlo codes for neutron and photon transport calculations based on continuous energy and multigroup methods. JAERI, p 1348
3. Shibata K, Iwamoto O, Nakagawa T, Iwamoto N, Ichihara A, Kunieda S, Chiba S, Furutaka K, Otuka N, Ohsawa T, Murata T, Matsunobu H, Zukeran A, Kamada S, Katakura J (2011) JENDL-4.0: a new library for nuclear science and engineering. J Nucl Sci Technol 48(1):1–30

Part V
Next-Generation Reactor Systems: Development of New Reactor Concepts of LWR or FBR for the Next-Generation Nuclear Fuel Cycle

Chapter 14
Application of the Resource-Renewable Boiling Water Reactor for TRU Management and Long-Term Energy Supply

Tetsushi Hino, Masaya Ohtsuka, Renzo Takeda, Junichi Miwa, and Kumiaki Moriya

Abstract The RBWR (resource-renewable boiling water reactor) is an innovative BWR that has a capability to breed and burn trans-uranium elements (TRUs) using a multi-recycling process. The RBWR can be used as a long-term energy supply, and it reduces the negative environmental impact that TRUs cause as they are otherwise long-lived radioactive wastes. Various design concepts of the RBWR core have been proposed. The RBWR-AC is a break-even reactor and the RBWR-TB and RBWR-TB2 are TRU burners. The RBWR-TB is designed to burn TRUs from the RBWR-TB itself and to burn almost all the TRUs by repeating their recycling. The RBWR-TB is assumed to be applied for a nuclear power phase-out scenario. The RBWR-TB2 is intended to burn TRUs from LWR spent fuels. The RBWR-TB2 is assumed to be applied for reducing the amount of TRUs to be managed in storage facilities. The RBWR cores achieve their TRU multi-recycling capability under the constraint that the void reactivity coefficient must be negative by introducing the parfait core concept. This chapter reviews details of the specific design and core characteristics of the RBWR.

Keywords Break-even • Burner • BWR • Multi-recycle • TRU • Void reactivity coefficient

14.1 Introduction

Nuclear-generated electrical power is one irreplaceable candidate energy source that responds to the needs for energy security and for reduction of greenhouse-gas emissions. However, there has also been growing concern that significant amounts

T. Hino (✉) • M. Ohtsuka • R. Takeda • J. Miwa
Hitachi, Ltd., Hitachi Research Laboratory, 7-1-1, Omika-cho, Hitachi-shi
Ibaraki-ken 319-1292, Japan
e-mail: tetsushi.hino.kd@hitachi.com

K. Moriya
Hitachi-GE Nuclear Energy, Ltd., 3-1-1, Saiwai-cho, Hitachi-shi, Ibaraki-ken 317-0073, Japan

of trans-uranium elements (TRUs) are becoming long-lived radioactive wastes. If TRUs could be recycled as nuclear fuel, the benefits attained from nuclear power would increase as a long-term energy supply and the negative environmental impact of TRUs as radioactive wastes could be greatly reduced. For these purposes many types of innovative reactors, including the sodium-cooled fast reactor (SFR), have been proposed. The resource-renewable BWR (RBWR) has been proposed to achieve the same purposes using concepts based on proven BWR technologies and the BWR capability to control the neutron energy spectrum flexibly [1–3]. A major characteristic of the BWR is "boiling" in the core, which includes water that functions as both a moderator and a coolant. The neutron energy spectrum can be hardened by reducing the hydrogen-to-uranium ratio (H/U) using the two-phase flow and using the hexagonal tight fuel lattice, so that the transmutation of ^{238}U to fissile plutonium is promoted with increasing resonance absorption: this enables the multi-recycling process of both breeding and consuming TRUs. On the other hand, there is a tendency that the harder the neutron spectrum becomes in the TRU-loaded core, the more positive the void reactivity coefficient becomes. The void reactivity coefficient is one of the main safety parameters for light water reactors (LWRs) and must be negative. The RBWR achieves the TRU multi-recycling capability under the constraint of the negative void reactivity coefficient by introducing the parfait core concept [4].

This chapter reviews details of the specific design and core characteristics of the RBWR.

14.2 RBWR System

14.2.1 Overview

Figure 14.1 shows the reactor pressure vessel (RPV) of the RBWR. The common plant specifications of the RBWR and the latest commercial BWR, the ABWR, are listed in Table 14.1. The rated thermal power, electric power, diameter of the RPV, and core pressure are identical for both reactor plants. Figure 14.2 shows a horizontal cross-sectional view of the RBWR core configuration, which is composed of 720 hexagonal fuel bundles and 223 Y-type control rods. The axial configuration uses the parfait core concept in which an internal blanket of depleted uranium oxide is placed between the upper and lower fissile zones of the TRU oxides.

Various design concepts of the RBWR core have been proposed. Recent core designs have focused on TRU management. The RBWR-AC is the break-even reactor that can burn depleted uranium by using TRUs extracted from the spent fuel bundles of LWRs without decreasing the amount of TRUs. The RBWR-TB is the TRU burner that can fission almost all the TRUs, leaving only the minimum critical mass of TRUs, by repeating their recycling and collecting. The RBWR-TB2 is a modified version of the TRU burner. The RBWR-TB2 is designed to be able to burn

Fig. 14.1 Reactor pressure vessel of the resource-renewable boiling water reactor (RBWR) [3]

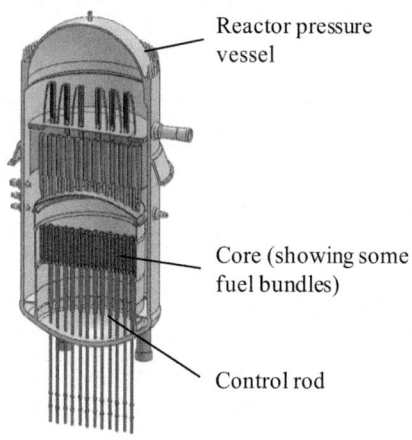

Table 14.1 Plant specifications [3]

Item	RBWR	ABWR
Thermal power (MWt)	3,926	3,926
Electric power (MWe)	1,356	1,356
RPV diameter (m)	7.1	7.1
Core pressure (MPa)	7.2	7.2
Number of fuel bundles	720	872
Fuel lattice type	Hexagonal	Square
Lattice pitch (mm)	199	155
Number of control rods	223	205
Control rod type	Y-type	Cross shape

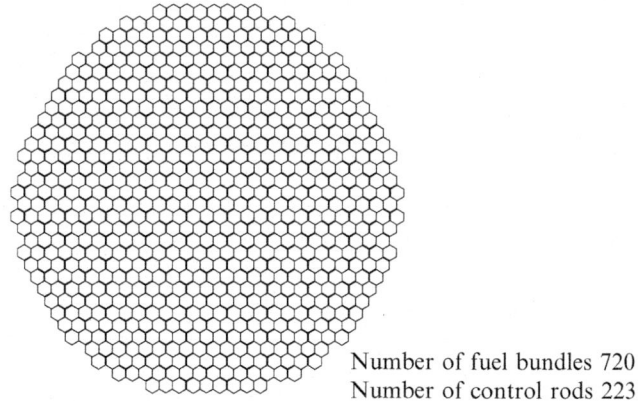

Fig. 14.2 Horizontal cross-sectional view of the RBWR core configuration [3]

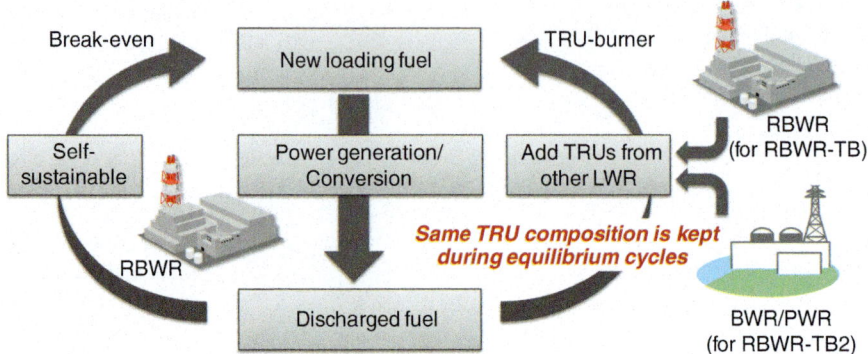

Fig. 14.3 Utilization concept of the RBWR-AC, -TB, and -TB2

TRUs from LWR spent fuels, whereas the RBWR-TB is designed as a burner for the TRUs from the RBWR-TB itself, assuming the RBWR-TB would be utilized when the TRU usefulness is exhausted and almost all should have been fissioned. Figure 14.3 shows the utilization concept of the RBWR-AC, -TB, and -TB2.

In core designs for the RBWR-AC, -TB, and -TB2, keeping charged TRU composition preserved at every operation cycle is mandatory. This criterion ensures the multi-recycling capability, fission, and recycling process of TRUs can be continued while maintaining the criticality and fulfilling the various operation constraints, such as sufficient reactor shutdown margin and negative void reactivity coefficient. As mentioned in the Introduction, the multi-recycling capability is achieved by hardening the neutron energy spectrum and promoting the transmutation of ^{238}U to fissile plutonium using the hexagonal tight fuel lattice, which has a H/U less than that of the conventional BWR square fuel lattice. Figure 14.4 shows the relationship between the volume ratio of water to fuel and the breeding ratio in the RBWR-AC, -TB, -TB2, and the conventional BWR. Because the RBWR-AC and -TB need to continue operation cycles without feeding fissile materials other than those contained in the discharged fuel from themselves, the volume ratios of water to fuel are set lower than those of the RBWR-TB2 and the conventional BWR.

In the following sections, the core calculation method is described first, and then each type of RBWR is described.

14.2.2 Core Calculation Method

An outline of the calculation methods used for the core design is as follows. Group constants of 12 energy groups for the core neutronic calculation were evaluated for the horizontal cross section of the fuel bundle lattice by the Monte Carlo calculation code with 190 energy groups [5]. In the burn-up calculation, 45 actinides from

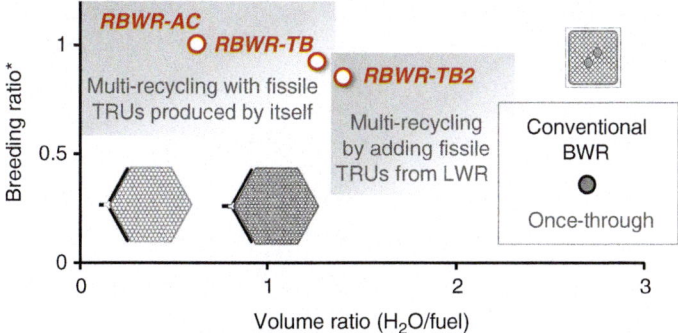

Fig. 14.4 Relationship between water to fuel volume ratio and fissile breeding ratio

^{228}Th to ^{253}Es and 84 fission products (83 nuclides treated explicitly and 1 lumped fission product) were treated. In the core neutronic calculation, the 12-energy group, three-dimensional neutron flux was obtained by solving the diffusion equation with 1 mesh for each fuel bundle in the horizontal direction and 34 meshes in the vertical direction.

In the thermal hydraulic calculation, the in-channel coolant flow rate, the two-phase flow pressure drop, and the axial void fraction distribution were calculated based on the power distribution obtained by the core neutronic calculation, so that the pressure drops between fuel bundles were balanced. The core neutronic calculation and the thermal hydraulic calculation were iterated until the power distribution and in-channel coolant flow distribution converged.

The void reactivity coefficient was evaluated by decreasing the core coolant flow rate to 95 % of the rated flow and dividing the change of the neutron multiplication factor by the change of core averaged void fraction, from the respective values at the rated flow.

14.2.3 RBWR-AC

The axial fuel bundle configuration of the RBWR-AC is shown in Fig. 14.5. The axial configuration is the parfait core, where an internal blanket (520 mm) of depleted uranium oxide is placed between two fissile zones (upper, 280 mm; lower, 193 mm). The upper and lower blankets (70 and 280 mm) are attached above and below the upper and lower fissile zones, respectively.

The neutron absorber zones are placed above and below the fuel zone (fissile and blanket) to increase the margin to maintain the negative void reactivity coefficient. The upper neutron absorber zone is composed of the neutron absorber rods placed between the plenums, which are connected to the fuel rods. The neutron absorber rods are filled with B_4C pellets in a sealed tube with an outside diameter of 7.7 mm.

Fig. 14.5 Axial configuration of the RBWR-AC fuel bundle [3]

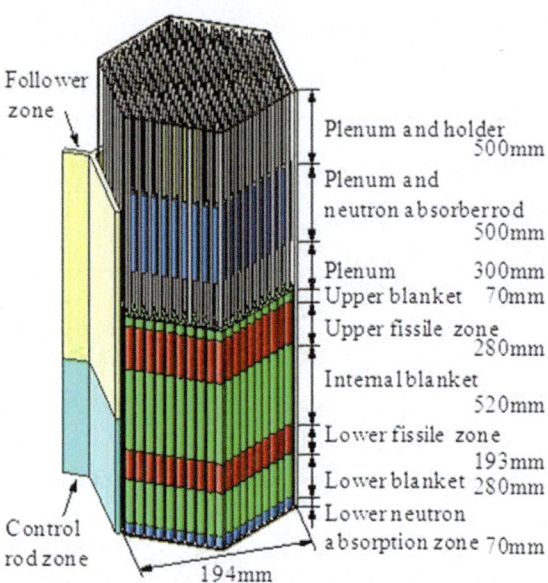

Each neutron absorber rod is attached to support rods fixed with the upper tie-plate of the fuel bundle. The neutron absorber rods are installed in a ratio of one per one fuel rod. Each neutron absorber rod is 500 mm long, and the distance between the upper end of the fuel zone and the lower end of the neutron absorber rod is 300 mm. The lower neutron absorber zone is composed of B_4C pellets filled in the fuel cladding. The length of the lower neutron absorber zone is 70 mm.

Figure 14.6 shows a horizontal cross-sectional view of the configuration of the RBWR-AC fuel bundle and its fissile Pu enrichment distribution. The lattice pitches of the fuel bundles are 199.2 mm on the side with the control rod and 194.7 mm on the side without it. The channel box of the fuel bundle is hexagonal with an inner width of 189.1 mm, and its wall thickness is 2.4 mm. The control rod is 6.5 mm thick, and the gap between the rod outer surface and the channel box is 1.6 mm on each side; the gap between channel boxes on the side without the control rod is 0.8 mm.

The fuel rod gap and pitch are 1.3 and 11.4 mm, respectively. For the equilibrium core of the RBWR-AC, the bundle-averaged fissile plutonium enrichment is 15.7 wt% for the upper fissile zone (Fig. 14.6a) and 20.1 wt% for the lower fissile zone (Fig. 14.6b). Both the upper and lower fissile zones utilize five different fissile Pu enrichments.

The main core specifications and performance values of the RBWR-AC in the equilibrium core are shown in Table 14.2. The core coolant flow is 2.6×10^4 t/h at a subcooling temperature of 5 K at the entrance and has a steam quality of 35 w/o at the core exit. The void fraction of core coolant is about 30 % at the bottom of the lower fissile zone because of heating in the lower blanket; it reaches 80 % at the top of the core. A breeding ratio of 1.01 is achievable under a 45 GWd/t exposure

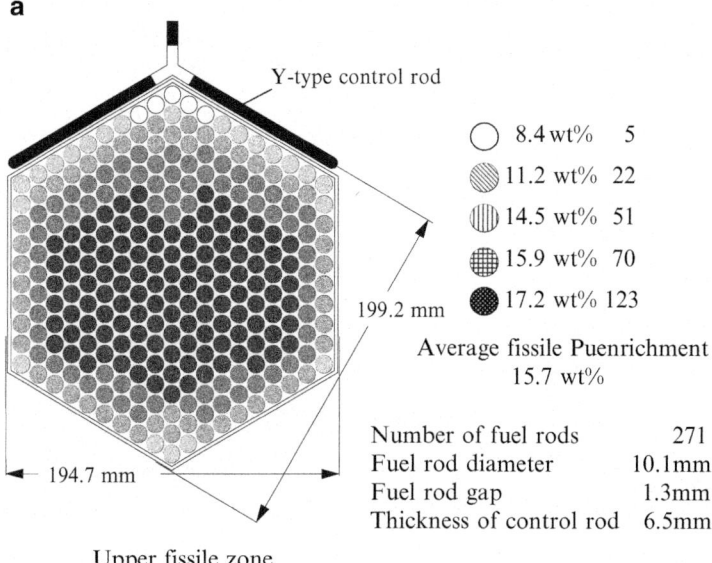

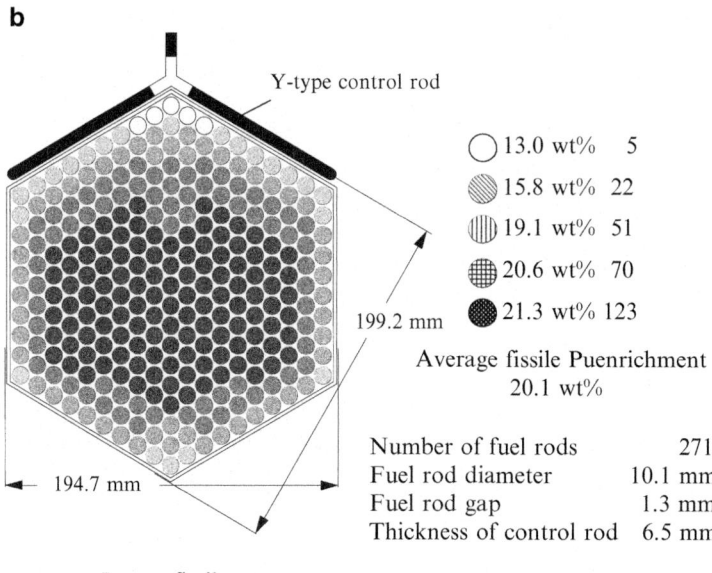

Fig. 14.6 Horizontal cross-sectional view showing configuration of the RBWR-AC fuel bundle and fissile Pu enrichment distribution [3]

Table 14.2 Core specifications and performance values [3]

Item	RBWR-AC	RBWR-TB	RBWR-TB2
Core height (mm)	1,343	993	1,025
Fuel rod diameter (mm)	10.1	7.4	7.2
Fuel rod pitch (mm)	11.4	9.4	9.4
Fuel rod gap (mm)	1.3	2.0	2.2
Pellet diameter (mm)	8.7	6.1	6.0
Number of fuel rods	271	397	397
Coolant flow rate (t/h)	2.6×10^4	3.8×10^4	2.4×10^4
Core exit quality (%)	35	21	36
Void fraction (%)	53	42	56
Pressure drop (MPa)	0.14	0.19	0.06
HM inventory (t)	144	77	76
Puf/HM in fissile zone (wt%)	15.7/20.1	13.9	25
Puf inventory (t)	9.0	4.5	8.3
Burn-up (GWd/t)	45	55	65
MLHGR (kW/m)	47	47	47
MCPR	1.28	1.3	1.28
Void reactivity coefficient (Δk/k/%void)	-2.4×10^{-4}	-2×10^{-4}	-4×10^{-4}
Breeding ratio	1.01	–	–
TRU fission efficiency (%)	–	51	45

averaged with the upper, internal, and lower blankets. Here the breeding ratio is defined as the number of atoms of fissile plutonium left in the discharged fuel bundles per fissile plutonium loaded in the initial charged fuel bundles.

The loading pattern of the fuel bundles in the equilibrium core adopts zone loading with the reflective boundary condition of 60 ° in the azimuthal direction. After the control rod scheduling is done, the radial power peaking factor is about 1.2 and the axial power peaking factor is about 1.8, including the blanket zones, which results in the minimum critical power ratio of 1.3 and the maximum linear heat-generating rate of 47 kW/m.

The RBWR-AC has a void reactivity coefficient of -2.4×10^{-4} Δk/k/%void, which is comparable with that of the current BWR, about -7×10^{-4} Δk/k/%void.

14.2.4 RBWR-TB

The axial fuel bundle configuration of the RBWR-TB is shown in Fig. 14.7. The axial configuration is similar to that of the RBWR-AC, but the RBWR-TB does not have a lower blanket because breeding of fissile plutonium is not needed. Other blanket and fissile zones have different heights from those in the RBWR-AC to enable multi-recycling of TRUs under the different neutron energy spectrum from the RBWR-AC. The upper and internal blanket zones of depleted uranium oxides

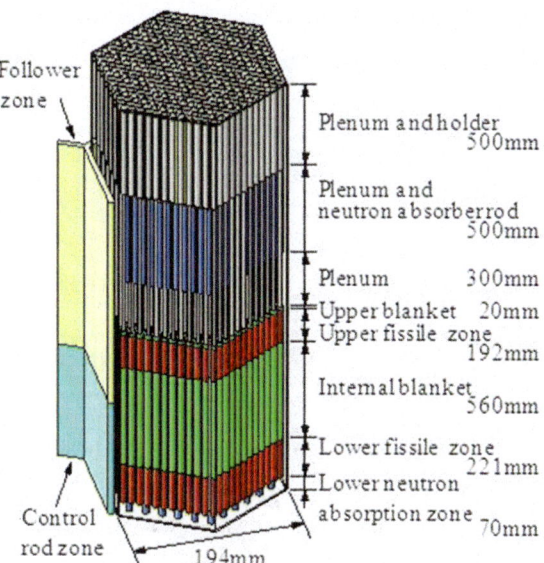

Fig. 14.7 Axial configuration of the RBWR-TB fuel bundle [3]

have heights of 20 and 560 mm, respectively; the upper and lower fissile zones have heights of 192 and 221 mm, respectively.

The RBWR-TB also utilizes the neutron absorber zones above and below the fuel zone. The upper neutron absorber zone has the same structure as that of the RBWR-AC. The number of neutron absorber rods in the lower neutron absorber zone is 91, which was determined so as to keep the void reactivity coefficient negative.

Figure 14.8 shows the horizontal configuration of the RBWR-TB. The fuel bundle of the RBWR-TB is composed of the uniform fissile plutonium enrichment of 13.9 wt%. The lattice pitches of the fuel bundles are 199.3 mm on the side with the control rod and 194.4 mm on the side without it. The channel box of the fuel bundle is hexagonal with an inner width of 189.6 mm and wall thickness of 2 mm. The control rod is 7.5 mm thick, and the gap between the rod outer surface and the channel box is 1.6 mm on each side. The gap between channel boxes on the side without the control rod is 0.8 mm. Geometries of the channel boxes and the control rods are slightly different from those of the RBWR-AC. However, because the center positions of the control rods are the same in the RBWR-AC and -TB and reactor internals fixed to the RPV, such as the core support plate, control rod guide tubes, etc., can be shared, their cores are easily exchanged with each other by changing the fuel bundles, control rods, and some attachments between the core support plate and fuel bundles.

Because the RBWR-TB equilibrium core has a shorter height than that of the RBWR-AC, the number of fuel rods of the RBWR-TB (397) is larger than that of the RBWR-AC (271) to keep the averaged linear heat-generating rate almost the same.

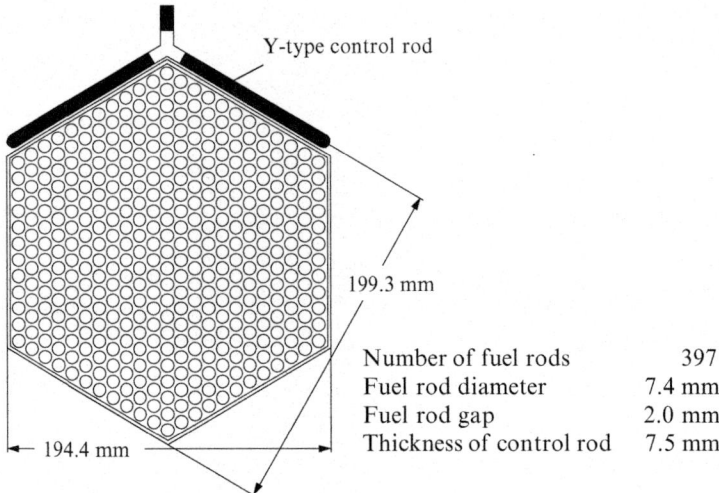

Fig. 14.8 Horizontal cross-sectional view showing configuration of the RBWR-TB fuel bundle [3]

The RBWR-TB aims at burning MAs by transmuting them into fissile isotopes using relatively low energy neutrons as well as by direct fissioning using relatively high energy neutrons. Both capture and fission reactions occur in a broad neutron energy range from thermal to fast. It is expected that the balance of these reactions at different neutron energies enables TRU burning while keeping the isotopic composition almost the same before and after burning, as mentioned in the next subsection.

The main core specifications and performance values of the RBWR-TB in the equilibrium core were shown earlier in Table 14.2. The core coolant flow is 3.8×10^4 t/h at a subcooling of 10 K at the entrance and has a steam quality of 21 % at the core exit. The concept of the loading pattern of fuel bundles in the equilibrium core is the same as that of the RBWR-AC: it adopts zone loading and the reflective boundary condition of 60 ° in the azimuthal direction. A maximum linear heat generation rate of 47 kW/m and an MCPR of 1.3 after the control rod scheduling are achieved. The RBWR-TB has a void reactivity coefficient of -2×10^{-4} Δk/k/%void.

The fission efficiency of TRUs in the RBWR-TB is 51 %. Here the fission efficiency is defined as the net decrease in TRUs divided by the total amount of fissioned actinides through the total fuel residence time in the core. This value indicates what amount of the TRUs can be used as fuel for generating electric power and is related to fissioning cost of the TRUs. As the fission efficiency of TRUs becomes higher, it is expected that the electricity-generating cost needed for burning the same amount of TRUs becomes smaller, if the other costs such as fuel fabrication cost are comparable.

Fig. 14.9 Axial configuration of the RBWR-TB2 fuel bundle [3]

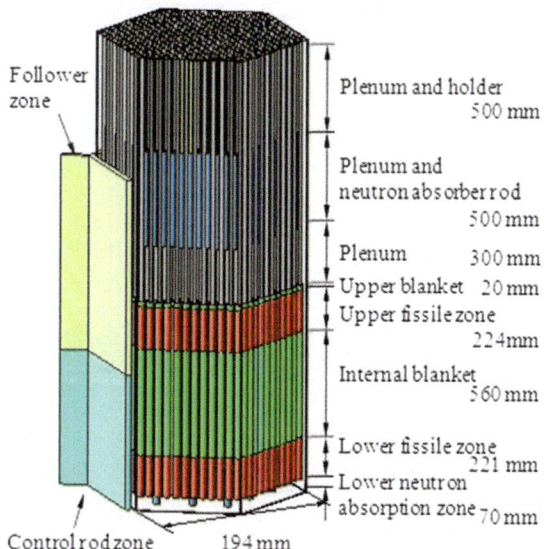

14.2.5 RBWR-TB2

The core concept of the RBWR-TB2 was initiated by an Electric Power Research Institute (ERRI)-organized team of three universities in the United States [6] to compare its core performance values with those of the ABR, which is the SFR having the same purpose [7]. Although the RBWR-TB is assumed to be utilized in the final stage of the nuclear power phase-out scenario, the RBWR-TB2 is assumed to be utilized to control the amount of TRUs during the period while LWRs are being operated as base load power sources.

The axial configuration of the RBWR-TB2 (Fig. 14.9) and it is similar to that of the RBWR-TB. The RBWR-TB2 also does not have a lower blanket because breeding of fissile plutonium is not needed. The upper and internal blanket zones of depleted uranium oxide have heights of 20 and 560 mm, respectively; the upper and lower fissile zones have heights of 224 and 221 mm, respectively. The RBWR-TB2 also uses a lower neutron absorption zone in which the number of neutron absorber rods is 19. This number of the neutron absorber rods is sufficient to keep the void reactivity coefficient negative in the RBWR-TB2.

Figure 14.10 shows the horizontal configuration of the RBWR-TB2. The fuel bundle of the RBWR-TB2 is composed of the uniform fissile plutonium enrichment of 25 wt%. As the RBWR-TB2 fuel includes TRUs from LWRs, the fissile plutonium enrichment becomes higher than that of the RBWR-TB, which uses TRUs from itself and other RBWR-TBs in the equilibrium core. Geometries of the channel box and the control rods are the same as those of the RBWR-TB. The lattice pitches of the fuel bundles are 199.3 mm on the side with the control rod and 194.4 mm on the side without it. The channel box of the fuel bundle is hexagonal

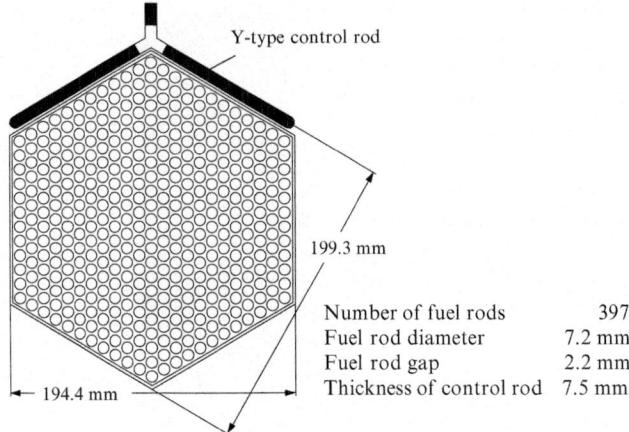

Fig. 14.10 Horizontal cross-sectional view showing configuration of the RBWR-TB2 fuel bundle [3]

with an inner width of 189.6 mm and wall thickness of 2 mm. The control rod is 7.5 mm thick, and the gap between the rod outer surface and the channel box is 1.6 mm on each side. The gap between channel boxes on the side without the control rod is 0.8 mm.

The fuel rod diameter and gap of the RBWR-TB2 are 7.2 and 2.2 mm, respectively; these values result in a larger moderator-to-fuel ratio and a softer neutron energy spectrum than those of the RBWR-TB. Because the fissile composition of the RBWR-TB2 is larger than that of the RBWR-TB, fissile TRUs need to be fissioned with a relatively larger rate to preserve TRU isotopic composition for multi-recycling. The number of fuel rods is the same as that of the RBWR-TB to make the averaged linear heat-generating rate almost the same.

The main core specifications and performance values of the RBWR-TB2 in the equilibrium core were shown earlier in Table 14.2. The core coolant flow is 2.4×10^4 t/h at a subcooling of 10 K at the entrance and has a steam quality of 36 % at the core exit. The concept of the loading pattern of fuel bundles in the equilibrium core is the same as that of the RBWR-TB: it adopts zone loading and the reflective boundary condition of 60 ° in the azimuthal direction. A maximum linear heat generation rate of 47 kW/m and an MCPR of 1.28 after the control rod scheduling are achieved. The RBWR-TB2 has a void reactivity coefficient of -4×10^{-4} Δk/k/%void.

The fission efficiency of TRUs in the RBWR-TB2 is 45 %. This value corresponds to about twice the production efficiency of TRUs, 22 %, in the ABWR. Here, the production efficiency of TRUs is defined with the opposite meaning of the fission efficiency of TRUs, that is, the net increase in TRUs divided by the total amount of fissioned actinides through the total fuel residence time in the core. As the electricity output of the RBWR-TB2 is the same as that of the ABWR, this means accumulation of TRUs would be suppressed by introducing one RBWR-TB2 for two ABWRs.

Table 14.3 Trans-uranium (TRU) composition and Puf/TRU weight per fuel batch [3]

Nuclide	RBWR-AC		RBWR-TB		RBWR-TB2		LWR spent fuel
	Charge	Discharge[a]	Charge	Discharge[a]	Charge	Discharge[a]	
Np-237	0.4	0.4	0.1	0.1	1.9	1.4	6.7
Pu-238	2.9	2.9	4.7	4.7	6.3	6.7	2.8
Pu-239	43.5	43.5	9.5	9.5	27.7	25.5	48.8
Pu-240	36.3	36.3	39.5	39.6	38.5	40.1	23
Pu-241	5.1	5.1	4.4	4.4	5.5	5.4	7
Pu-242	5.1	5.1	25.4	25.4	9.6	10.1	5
Am-241	3.6	3.6	4.7	4.7	5.4	5.4	4.7
Am-242m	0.2	0.2	0.2	0.2	0.2	0.2	0
Am-243	1.3	1.3	4.7	4.7	2.4	2.4	1.5
Cm-244	1.1	1.1	4.1	4	1.8	2	0.5
Cm-245	0.4	0.4	1.2	1.2	0.5	0.6	0
Cm-246	0.1	0.1	1	1	0.2	0.2	0
Cm-247	0	0	0.2	0.2	0	0	0
Cm-248	0	0	0.2	0.2	0	0	0
Cm-249	0	0	0.1	0.1	0	0	0
Puf(t)	1.94	1.96	1.14	1.06	2.06	1.74	0.32
TRU(t)	3.99	4.03	8.18	7.62	6.20	5.63	0.58

[a]Three-year cooling time after discharge was assumed

Table 14.3 summarizes TRU compositions and weights of fissile plutonium and TRU of charged and discharged fuels in the RBWR-AC, -TB, and -TB2. In evaluation of the discharged fuel compositions, a 3-year cooling time after discharge from the core is considered. Because the RBWR-AC and -TB satisfy the multi-recycling criteria under the condition that both reactors charge TRUs that were discharged from themselves, the TRU compositions of the RBWR-AC and -TB are kept the same in the charged and discharged fuels. The weights of fissile plutonium and TRU increase slightly in the discharged fuel in the RBWR-AC, the break-even reactor, whereas they decrease in the discharged fuel in the RBWR-TB, the TRU burner. As TRUs from LWR spent fuels are added to TRUs discharged from the RBWR-TB2 itself, the TRU composition of their mixture is to be the same at every operation cycle with the constant mixing ratio of TRUs discharged from the RBWR-TB2 and LWR. The weight of TRU decreases in the discharged fuel in the RBWR-TB2.

14.3 Conclusion

The specific design and core characteristics of the RBWR were summarized from a review of published studies. The RBWR is categorized as a low moderation LWR. By utilizing a tight lattice fuel and two-phase flow of coolant, the latter of which is a feature of BWRs, the moderator-to-fuel ratio of the RBWR can be reduced to a small enough value as to achieve multi-recycling of TRUs.

Different RBWR cores have been designed for different purposes. The RBWR-AC is a break-even reactor with a Pu breeding ratio more than 1.0. The RBWR-TB and RBWR-TB2 are TRU burners that can fission TRUs at a rate more than twice the rate of TRU production by the ABWR. Each of the reactor types achieves the foregoing performances under the condition requiring negative void reactivity coefficient and multi-recycling capability. With the multi-recycling capability, the RBWR-AC/TB/TB2 can continue to fission or recycle TRUs while maintaining the criticality and fulfilling the various constraints, such as sufficient core shutdown margin and negative reactivity coefficient.

The RBWR appears to be a promising candidate energy source that responds to the needs for energy security, for reducing greenhouse-gas emissions, and for mitigating the negative environmental impact of TRUs.

Open Access This chapter is distributed under the terms of the Creative Commons Attribution Noncommercial License, which permits any noncommercial use, distribution, and reproduction in any medium, provided the original author(s) and source are credited.

References

1. Takeda R, Miwa J, Moriya K (2007) BWRs for long-term energy supply and for fissioning almost all transuraniums. In: Proceedings of the GLOBAL 2007, Boise, ID, September 9–13
2. Takeda R, Miwa J, Moriya K (2012) RBWRs for fissioning almost all uranium and transuraniums. Trans Am Nucl Soc 107:853
3. Hino T, Ohtsuka M, Takeda R, Miwa J, Moriya K (2014) Core designs of RBWR (resource-renewable BWR) for recycling and transmutation of transuranium elements: an overview. In: Proceedings of ICAPP 2014, Charlotte, April 2014
4. Ducat GA et al (1974) Evaluation of the Parfait blanket concept for fast breeder reactors. MITNE-157. http://inis.iaea.org/search/search.aspx?orig_q=RN:5141813
5. Morimoto Y, Maruyama H, Ishii K, Aoyama M (1989) Neutronic analysis code for fuel assembly using a vectorized Monte Carlo method. Nucl Sci Eng 103:351–358
6. Downar T, Hall A, Jabbay D, Ward A, Greenspan E, Ganda F, Bartoloni F, Bergmann R, Varela C, Disanzo C, Kazimi M, Karahan A, Shwageraus E, Feng B, Herman B (2012) Technical evaluation of the Hitachi Resource-Renewable BWR (RBWR) design concept. EPRI Technical Report 1025086
7. Yang WS, Kim TK, Grandy C, Hill RN (2008) Performance characteristics of metal and oxide fuel cores for a 1000 MWt advanced burner reactor. In: Proceedings of ARWIF08, Fukui, February 2008

Chapter 15
Development of Uranium-Free TRU Metallic Fuel Fast Reactor Core

Kyoko Ishii, Mitsuaki Yamaoka, Yasuyuki Moriki, Takashi Oomori, Yasushi Tsuboi, Kazuo Arie, and Masatoshi Kawashima

Abstract A TRU-burning fast reactor cycle associated with a uranium-free transuranium (TRU) metallic fuel core is one of the solutions for radioactive waste management issue. Use of TRU metallic fuel without uranium makes it possible to maximize the TRU transmutation rate in comparison with uranium and plutonium mixed-oxide fuel because it prevents the fuel itself from producing new plutonium and minor actinides, and furthermore because metallic fuel has much smaller capture-to-fission ratios of TRU than those of mixed-oxide fuel. Also, adoption of metallic fuel enables recycling system to be less challenging, even for uranium-free fuel, because a conventional scheme of fuel recycling by electrorefining and injection casting is applicable.

There are some issues, however, associated with a uranium-free TRU metallic fuel core: decrease in negative Doppler reactivity coefficient from the absence of uranium-238, which has the ability to absorb neutrons at elevated temperatures, increase in burn-up swing, because fissile decreases monotonically in uranium-free core, and so on. The purpose of this paper is to evaluate the feasibility of the uranium-free TRU metallic fuel core by investigating the effect of measures taken to enhance Doppler reactivity feedback and to reduce burn-up swing. The results show a TRU-burning fast reactor cycle using uranium-free TRU metallic fuel is viable from the aforementioned points of view because the introduction of diluent Zr alloy, spectrum moderator BeO, and lower core height enables Doppler reactivity coefficient and burn-up reactivity swing of uranium-free TRU metallic fuel to be as practicable as those of conventional fuel containing uranium.

Keywords Burn-up swing • Doppler reactivity feedback • Fast reactor • Metallic fuel • Trans-uranium • Uranium-free

---

K. Ishii • M. Yamaoka • Y. Moriki • T. Oomori • Y. Tsuboi • K. Arie (✉)
Toshiba Corporation, 8 Shinsugita-Cho, Isogo-Ku, Yokohama 235-8523, Japan
e-mail: kazuo.arie@toshiba.co.jp

M. Kawashima
Toshiba Nuclear Engineering Service Corporation, 8 Shinsugita-Cho, Isogo-Ku, Yokohama 235-8523, Japan

15.1 Introduction

For sustainable nuclear power deployment, not only ensuring its enhanced safety but also reduction of the environmental burden associated with radioactive waste management is a challenging issue for the international community. History has shown that obtaining public support is difficult for waste management plans that involve mass disposal of radioactive waste with a half-life of tens of thousands of years. Therefore, as one of the solutions, Toshiba has been developing a system that takes into account that, for the time being, light water reactors (LWRs) have a leading role in commercial nuclear power plants, which enables toxicity and radioactivity of high-level waste to be reduced to those of natural uranium within a few hundred years. This system is mainly characterized by a fast reactor core that does not contain uranium in its fuel, that is, uranium-free TRU fuel. The use of uranium-free TRU fuel makes it possible to maximize the TRU transmutation rate in comparison with fuel containing uranium because it prevents the fuel itself from producing new plutonium and minor actinides.

Although there was much research focused on TRU transmutation with uranium-free fuels, each of these seems to have drawbacks from some aspect. First, for instance, candidates such as Tc-based and W-based oxide fuel, inert matrix fuel such as the rock-like oxide fuel containing mineral-like compounds, and MgO-based oxide fuel provide solutions against issues associated with uranium-free operation, that is, decrease in Doppler reactivity feedback and increase in sodium void reactivity [1–3], but such types of inert matrix fuel may require new technologies for reprocessing. Additionally, many processing phases necessary for fabrication are costly. Second, an accelerator-driven transmutation system coupled with a fast reactor using uranium-free metallic fuel is another candidate that also can relax the issue of the reduced Doppler effect owing to its subcritical system [4–7], but installation of the accelerator facility at a fast reactor site is less cost competitive, especially when the system is not only a TRU burner but also a commercial power plant. Thus, it is worthwhile to develop the TRU transmutation system with uranium-free TRU fuel from the aspect of technological maturity and simplicity, which results in lower cost. Subsequently, the concepts for the TRU burner system with uranium-free TRU are derived from this background: fewer R&D needs and a simple system.

First, by contrast with inert matrix fuels, metallic fuel can be fabricated by the well-known injection casting method [8]. Moreover, metallic fuel is compatible with pyro-process reprocessing that has been developed since the 1960s [9]. Application of an accelerator-driven system for transmutation needs further R&D than that of a fast reactor system. Thus, the metallic fuel fast reactor is preferred for the system.

Second, we aim to develop the TRU-burning system in commercial power reactors while avoiding cost impact. For this reason, a system that can employ the pyro-process for fuel reprocessing would be preferable because it does not need

15 Development of Uranium-Free TRU Metallic Fuel Fast Reactor Core

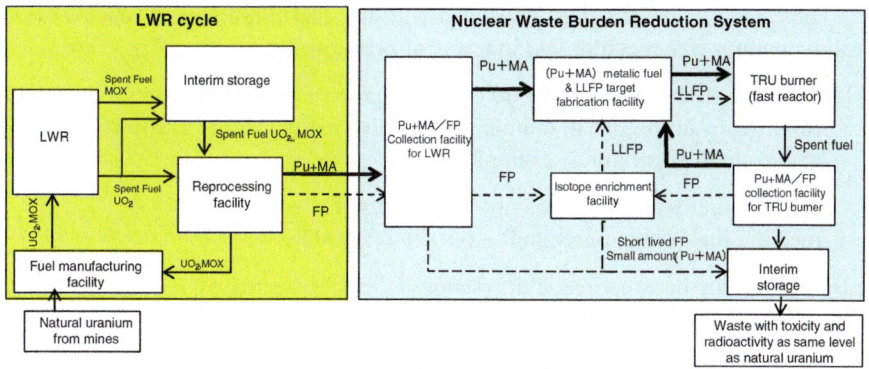

Fig. 15.1 Configuration diagram of the system to reduce nuclear waste burden

complex processes. Therefore, we introduce a metal fuel alloy that can be simply fabricated by injection casting and reprocessed by pyro-processing.

Additionally, in terms of reduction of nuclear waste burden, a metallic fuel fast reactor cycle has the great potential to transmute long-lived fission products (LLFPs) because of its excellent neutron economy [10, 11]. Moreover, it has an advantage for long-term energy security because the basic technology of the metallic fuel fast reactor cycle is also applicable to the future sustainable nuclear energy supply system.

For these reasons, Toshiba is developing a system to reduce nuclear waste burden using a TRU burner as shown in Fig. 15.1. The system is characterized by a closed fuel cycle that encompasses the following main facilities: fuel manufacturing plant to fabricate uranium-free TRU metallic fuel and LLFPs target from TRU and LLFPs extracted from LWR spent fuel, a fast reactor to burn those fuels, and recycling facilities to reprocess and refabricate the spent fuel from the fast reactor by pyro-processing. Although substances remain after reprocessing that must finally be disposed outside the cycle, their toxicity and radioactivity are diminished to the same level as those of natural uranium by enhancing burning and processing rates and storing them for a few hundred years within the system. Among the aforementioned facilities in the system, this study focuses on the TRU-burning fast reactor and investigates the practicability of the uranium-free TRU metallic fuel core.

15.2 Issues and Measures Against the Uranium-Free TRU Metallic Fast Reactor Core

This chapter presents issues and measures against the uranium-free TRU metallic fast reactor core. Also, the targets and constraints in parametric survey and selection of core and fuel specification are briefly described.

There are two main issues associated with the TRU burning fast reactor cycle using uranium-free metallic fuel in terms of practicability:

(1) Decrease in the absolute value of the negative Doppler reactivity coefficient resulting from absence of uranium-238, which has the ability to absorb neutrons at elevated temperatures. example,

metallic fuel with uranium: -1×10^{-3} Tdk/dT
metallic fuel without uranium: -6×10^{-4} Tdk/dT

(2) Increase in burn-up reactivity swing as fissile decreases monotonically in uranium-free core. example,

metallic fuel with uranium: ~1 %dk/kk'/150 days
metallic fuel without uranium: ~6 %dk/kk'/150 days

To solve these issues, there are several candidates, as follows:

(1) Enhance Doppler feedback

- Introduce diluent material in the metallic fuel
- Introduce spectrum moderator

(2) Reduce burn-up reactivity swing

- Reduce the core height
- Introduce neutron absorber outside the core
- Increase the number of refueling batches

Generally, if it is conventional fast reactors with U-Pu fuel, the burn-up reactivity swing depends mainly on decrease of fissile amount and increase of neutron parasitic capture of fission products and actinides from burn-up. Therefore, the typical ways to reduce burn-up reactivity swing are to increase conversion ratio via fissile enrichment reduction and to reduce neutron parasitic capture. Here, the conversion ratio is defined as the amount of fissile materials production divided by the amount of neutron absorption, that is, fission and capture, and natural decay of fissile materials. It is difficult, however, for a uranium-free core to increase the conversion ratio because fissile enrichment cannot be controlled in the absence of uranium. Although the reduction of neutron parasitic capture by neutron spectrum hardening improves burn-up reactivity swing, it also harms the Doppler effect. For these reasons, when it comes to uranium-free core, increase of the fissile amount at the beginning of the cycle makes sense because it reduces the ratio of the fissile consumption to the fissile amount at the beginning of the cycle.

These candidates were parametrically surveyed to evaluate the feasibility of the uranium-free TRU metallic fuel fast reactor core in light of aforementioned issues. The targets assumed were the core performances with the Doppler reactivity coefficient equivalent to a conventional U-Pu metallic fuel core. Furthermore, constrains associated with fuel fabrication such as melting temperature was taken into consideration because, in this evaluation, diluent material was assumed to be used as a fuel slug alloy, not cladding material. Hence, the slug was assumed be

15 Development of Uranium-Free TRU Metallic Fuel Fast Reactor Core

Table 15.1 Assumed condition of the 300 MWe fast reactor core for the parametric survey

Items	Value
Reactor thermal power	714 MW
Operation cycle length	150 days
Fuel type	TRU 10 wt% Zr alloy
Number of fuel pins per S/A	169
Core diameter	180 cm
Fuel pin diameter	0.65 cm
Core height	93 cm
TRU composition	LWR discharged
	10 years cooled

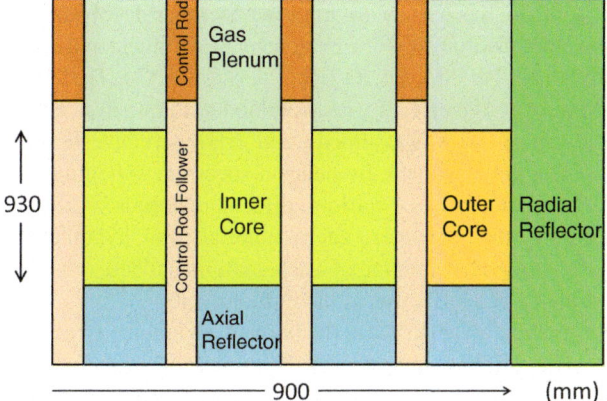

Fig. 15.2 RZ geometry for parametric survey

fabricated by injection casting as the same as the conventional metallic fuel. This step makes the allowable maximum melting temperature of the fuel alloy less than 1,200 °C to prevent Am volatilization during injection casting [12].

15.3 Parametric Analysis on the Effect of Measures

This chapter describes parametric analysis methodology and analysis results for Doppler feedback enhancement and burn-up reactivity swing reduction.

15.3.1 Parametric Analysis Methodology

A hypothetical 300 MWe fast reactor core was used for the parametric survey to enhance Doppler feedback and burn-up reactivity swing. Table 15.1 and Fig. 15.2

show the assumed core conditions and RZ geometry for parametric survey, respectively. The calculation methods were as follows. Core burn-up characteristics were analyzed with the burnup calculation code STANBRE [13]. Reactivity coefficients were analyzed using the diffusion calculation code DIF3D [14]. The effective cross sections used in these calculations were obtained by the cell calculation code SLAROM-UF [15], based upon 70 group cross sections from JENDL-4.0 [16] with a self-shielding factor table as a function of background cross section. This method for the production of the effective cross sections is considered to be adequate to take into account the influence of each diluting material upon the self-shielding effect of heavy isotopes for the parametric study. Concerning material compositions, a homogeneous model of fuel, diluent, and spectrum moderator was used.

To begin with, in the survey to improve Doppler feedback, 21 elements to enhance resonance absorption were evaluated as a diluent material for the TRU alloy: Cr, Mn, Fe, Ni, Nb, Mo, Tc, Ru, Rh, Pd, Nd, Sm, Gd, Tb, Dy, Er, Tm, Ta, W, Os, and Au. Moreover, the effect by neutron moderators such as BeO, $^{7}Li_2O$, $^{11}B_4C$ (100 % enrichment of ^{11}B was assumed), and ZrH_2 were investigated to clarify the impact against Doppler feedback by neutron spectrum softening. To compare the Doppler effect enhancement of various diluent materials and neutron spectrum moderators in a simple manner, each material was hypothetically added to TRU-10wt%Zr alloy. The amount of each material added was adjusted case by case to maintain 1.0 of k-effective at the end of cycle.

Next, in the evaluation to decrease the burn-up swing, the effects of the measures taken to increase the fissile amount at the beginning of the cycle were studied. The effects on burn-up reactivity swing were evaluated by reducing the core height, installing B_4C shield at core peripheral, and increasing the number of refueling batches, which all lead to increase of the fissile amount at the beginning of the cycle.

Last, reflecting the results obtained by the parameter surveys, an optimal uranium-free TRU metallic fuel core was specified, and its feasibility in light of Doppler feedback and burn-up swing was evaluated by core performance analysis.

15.3.2 Analysis Results for Doppler Feedback Enhancement

The effects of measures taken to enhance Doppler feedback, that is, diluent and spectrum moderator, are evaluated in this section.

As shown in Fig. 15.3, 6 among 21 diluent materials are found to enhance Doppler feedback more than Zr, the typical metallic fuel alloy. Although Nb, Ni, W, Mo, Fe, and Cr have greater potential to enhance Doppler feedback than Zr, there are some deficiencies that cannot be ignored. First, the melting points of Pu-Ni alloy and Pu-Fe alloy are below 500 °C, which is too low for nuclear fuel [17]. Second, the melting point of Pu-W alloy is too high to fabricate fuel by injection casting because the melting temperature of W itself is above 3,000 °C.

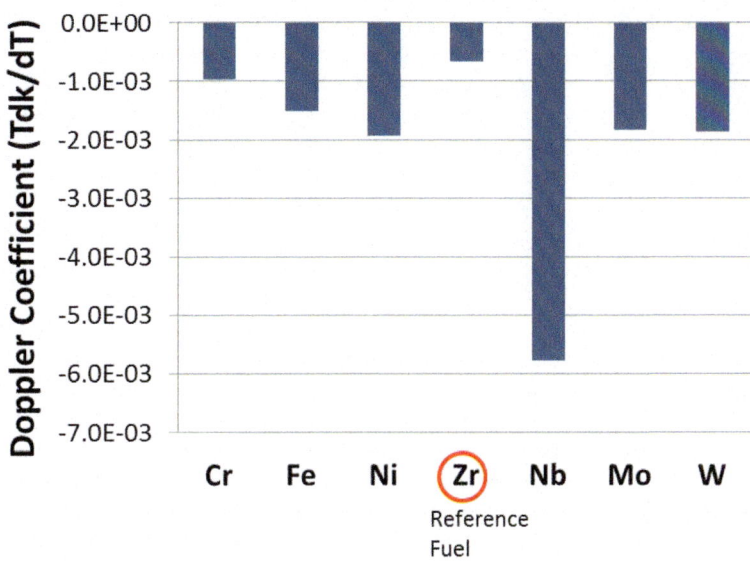

Fig. 15.3 Doppler coefficients associated with major diluent elements

Third, the allowable contents of Mo and Nb in the metal fuel alloy are too small to enhance the Doppler coefficients, which are 5 wt% and 3 wt%, respectively, under the condition to maintain their melting temperatures below 1,200 °C to prevent Am vaporization during injection casting [18]. Finally, the number of past experiences with Cr, for example, irradiation testing of Pu-Cr alloy, is less than enough to employ it as a diluent material for uranium-free fuel. Consequently, Zr was chosen as the fuel diluent material.

Then, as shown in Fig. 15.4, the absolute value of the negative Doppler coefficient remarkably increased by introducing a spectrum moderator such as BeO, $^{11}B_4C$, or ZrH_2. The adoption of ZrH_2, however, may cause dissociation of hydrogen upon accident. Besides, the usage of $^{11}B_4C$ is costly because almost 100 % enrichment of ^{11}B is necessary to enhance Doppler feedback significantly. Therefore, BeO was selected as a moderator material for the uranium-free core.

15.3.3 Analysis Results for Burnup Reactivity Swing Reduction

This section evaluates the effects of measures taken to reduce burnup reactivity swing of the uranium-free TRU metallic fuel core. In the parameter surveys, the operation cycle length, that is, 150 days, the core volume, and the core power density were kept constant to compare the effect of each countermeasure. The average fuel burnup was also kept constant, save for the survey of the number

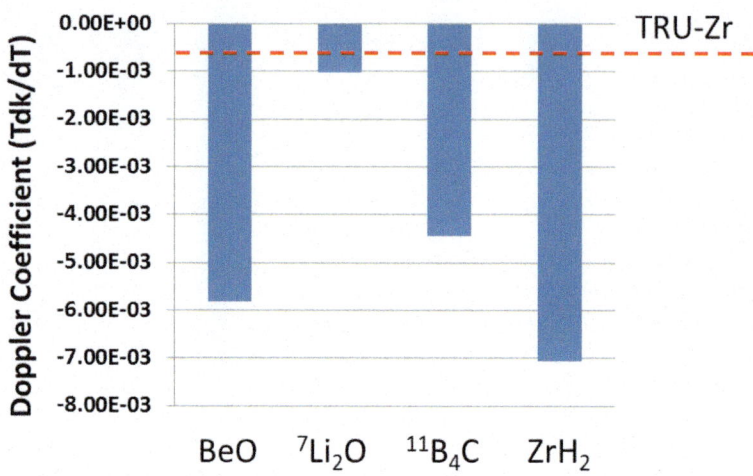

Fig. 15.4 Doppler coefficients associated with neutron spectrum moderator

Table 15.2 Results of burn-up reactivity swing reduction

Items	Reduction (%)
Core height changed from 93 to 65 cm	12
Peripheral S/A reflector changed to B_4C absorber	5
Number of refueling batches changed from 5 to 7	5

of refueling batches. The adjusting parameter to increase the fissile amount was the zirconium content in TRU-Zr alloy fuel to keep k-effective = 1.0 at the end of the cycle.

Table 15.2 shows the summary of the analysis results. The reduction of the core height from 93 cm to 65 cm resulted in a 12 % decrease of burn-up reactivity swing. The introduction of a B_4C shield, where natural boron was assumed, at the core periphery region resulted in only about a 5 % decrease in burn-up reactivity swing. On the other hand, the penalty of this countermeasure is the increase of core power peaking because the leakage of neutrons from the core surface increases. Hence, this measure was not adopted in the subsequent core design. Regarding the effect of the number of refueling batches, the larger is the number of refueling batches, the smaller the burn-up reactivity swing becomes. The effect was approximately a 5 % decrease in burnup reactivity swing for a 40 % increase in the number of refueling batches. This measure was not adopted in the subsequent core design because its effect on the burn-up reactivity swing is small and it leads to significant increase of core power peaking because of the increased difference of burn-up between most burnt fuel and fresh fuel.

Table 15.3 Specification of the uranium-free TRU metallic core

Items	Value
Reactor thermal power	714 MW
Operation cycle length	150 days
Fuel type	TRU-Zr alloy
Number of fuel pins per S/A	135
Fuel pin diameter	0.48 cm
Core diameter	250 cm
Core height	65 cm
Spectrum moderator	BeO pins in Fuel S/A (number of pins, 196)
TRU composition	LWR discharged
	10 years cooled

15.4 Developed Uranium-Free TRU Metallic Core

This chapter describes specifications for selection of a uranium-free TRU metallic core and performance of the uranium-free TRU metallic core. Then, the core and fuel are developed on the basis of those results and the feasibility of the developed core is evaluated.

15.4.1 Specification Selected for Uranium-Free TRU Metallic Core

On the basis of the results of the parametric surveys, the uranium-free TRU burning core was specified as shown in Table 15.3 and Fig. 15.5. TRU-Zr alloy fuel pins and BeO pins were employed to enhance the Doppler coefficient. The reason for adopting the TRU-Zr alloy fuel is to use a simpler fuel fabrication method, that is, injection casting, in contrast to a TRU-Zr particle fuel in a zirconium metal matrix. Then, the zirconium content in TRU-Zr alloy was assumed to be limited below 35 wt% to keep the melting point of the TRU-Zr alloy below 1,200 °C to prevent Am vaporization during injection casting [19]. The fuel pins and the BeO pins were separately located in the fuel subassemblies (Fig. 15.6). The diameter of fuel pins was reduced from 0.65 to 0.48 cm to compensate for the increase of the average linear heat rate caused by employment of the BeO pins. Core height is 65 cm to reduce burn-up reactivity swing, whereas the core diameter was increased from 180 to 250 cm to keep the linear heat rate of the fuel pin similar to the 93-cm-height core. The operation cycle length is 150 days, which can be controlled by conventional control rods and fixed neutron absorbers.

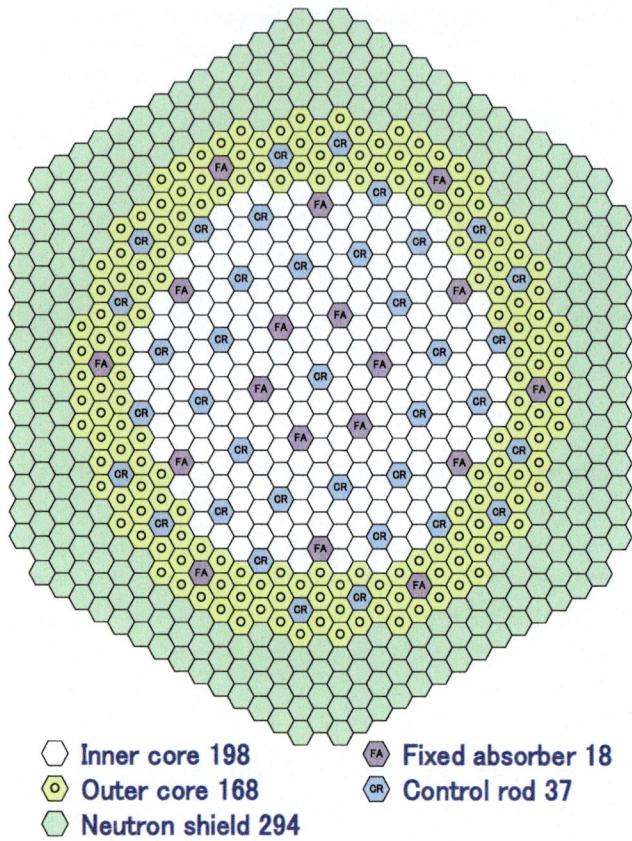

○ Inner core 198   (FA) Fixed absorber 18
(○) Outer core 168   (CR) Control rod 37
○ Neutron shield 294

Fig. 15.5 Uranium-free core layout

15.4.2 Performance of the Uranium-Free TRU Metallic Core

The core performance of the developed uranium-free core was evaluated as shown in Table 15.4. The Zr content in the fuel alloy was determined to maintain criticality during the operation cycle under the conditions of the upper limit of the melting point, 1,200 °C. According to the results, the uranium-free TRU metallic core is viable in terms of core performance, safety performance, fuel fabrication, and TRU burner.

The Doppler coefficient is similar to that of the conventional metallic fuel fast reactor cores, and the burn-up reactivity swing is considered to be controllable by conventional control rods and fixed absorbers. Moreover, core sodium void reactivity including the upper plenum region is negative because of neutron leakage at the upper plenum region and neutron spectrum moderation from the presence of BeO during sodium voiding. Although the restriction for sodium void reactivity

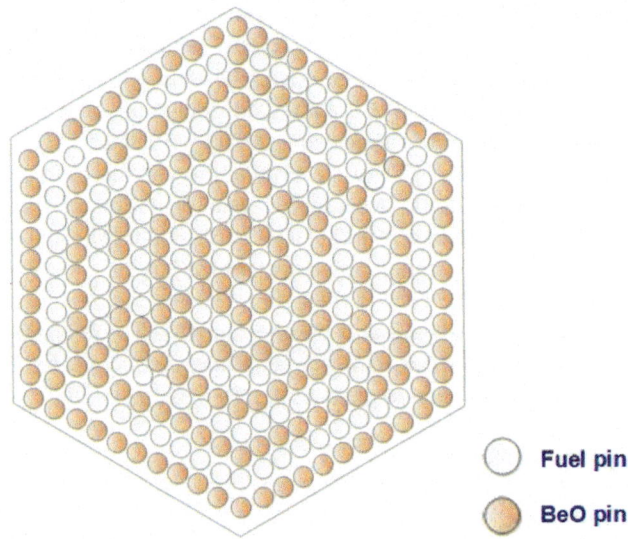

Fig. 15.6 Fuel subassembly cross section

Table 15.4 Performance of the uranium-free TRU metallic core

Items	Value
Fuel composition	TRU-35%Zr/TRU-19%Zr
Inner core/outer core	
TRU inventory (Pu/MA)	2.17 t at BOEC (1.89/0.28 t)
Burn-up reactivity swing	5.1 % dk/kk'
Power density (average)	260 W/cc
Linear heat rate (average)	220 W/cm
TRU burning rate (Pu/MA)	260 kg/EFPY (230/30 kg/EFPY)
Doppler coefficient at EOEC	-3×10^{-3} Tdk/dT
Na void reactivity at EOEC	<0 %dk/kk'

EOEC end of equilibrium cycle, *EFPY* effective full-power year

was not assumed for the core in this study, low sodium void reactivity is a significant factor for sodium-cooled fast reactors.

Furthermore, the developed core design has the potential to achieve passive safety features against unprotected events such as unprotected loss of flow (ULOF) and unprotected transient overpower (UTOP) similar to a conventional metallic fuel core because the basic core safety parameters, that is, average and peak linear heat rates for lower fuel temperatures, the enhanced Doppler coefficient, and low sodium void coefficient (negative sodium coefficient in whole core), were maintained within the similar ranges of a conventional metallic fuel core design [20].

Feasibility in the light of decay heat is also confirmed to be practicable, as the decay heat of the fresh fuel material is 32 W/kgHM, which is less than 10 % of that of the minor actinide (MA)-only fuel. Also, the decay heat of the fresh fuel subassembly is approximately 240 W. Taking advantage of some cooling scheme such as air flow, this fuel can be fabricated as a fuel pin bundle [21].

Moreover, the results also shows the profitability of the uranium-free TRU metallic fuel fast reactor itself, because a 1-year operation of this 300 MWe TRU-burning fast reactor burns 260 kg TRU, corresponding to the amount produced by a 1.2 GWe-year operation of a conventional LWR.

For all these reasons, the TRU-burning fast reactor using uranium-free TRU metallic fuel is considered to be feasible. Further study such as reduction of burn-up reactivity swing and trade-off of various countermeasures considering economic aspect helps improve and optimize the core design in the next phase.

15.5 Conclusions

A TRU transmutation system associated with the uranium-free metallic fuel fast reactor is a practical way to burn TRU with sustainability, fewer R&D needs, and a simple system, because it can be used as both a TRU burner and a power supply plant. Employment of pyro-processing for recycling reduces the burden of R&D requirements, and introduction of a conventional fuel fabrication method and pyro-processing allows less complex facilities.

In this study, two main issues related to the uranium-free core were investigated and discussed to clarify the feasibility of a TRU-burning fast reactor cycle using such a core: Doppler coefficient for reactor safety, and burn-up reactivity swing for acceptable reactor operating cycle length.

The results show that the uranium-free fast TRU fast reactor core is viable because those issues can be solved by TRU-Zr alloy fuel, BeO neutron moderator, and reduced core height. Thanks to the BeO pins that function not only as a neutron moderator but also as a diluent material, the 35 %Zr alloy fuel can be fabricated without Am vaporization because its melting point is maintained below 1,200 °C, the temperature that causes Am vaporization during injection casting fuel fabrication. Moreover, the decay heat of the fresh fuel is considered to be an acceptable level for the fuel fabrication. Also, a 1-year operation of this 300 MWe core burns the TRU that is produced by 1.2 GWe-year operation of a conventional LWR.

In conclusion, the prospect of a TRU-burning fast reactor cycle using uranium-free metallic fuel was confirmed. Further study, not only to improve core performances but also to develop a recycling process associated with this uranium-free system, which is currently under way, promotes realization of the system.

Open Access This chapter is distributed under the terms of the Creative Commons Attribution Noncommercial License, which permits any noncommercial use, distribution, and reproduction in any medium, provided the original author(s) and source are credited.

References

1. Yamashita T, Kuramoto K-I, Akie H, Nakano Y, Nitani N, Nakamura T, Kusagaya K, Ohmichi T (2002) Rock-like oxide fuels and their burning in LWRs. J Nucl Sci Technol 39(8):865–871
2. Miwa S, Osaka M, Usuki T, Sato I, Tanaka K, Hirosawa T, Yoshimachi H, Onose S (2011) MgO-based inert matrix fuel for a minor actinide recycling in a fast reactor cycle. In: Proceedings of Global 2011, Makuhari, Japan, Dec 11–16, 2011
3. Messaoudi N, Tommasi J (2002) Fast burner reactor devoted to minor actinide incineration. Nucl Technol 137:84–96
4. Hill RH, Khalil HS (2000) Physics studies for sodium cooled ATW blanket. In: Proceedings of the IAEA technical committee meeting on core physics and engineering aspects of emerging nuclear energy systems for energy generation and transmutation. Argonne National Laboratory
5. Beller DE, Van Tuyle GJ, Bennett D, Lawrence G, Thomas K, Pasamehtoglu K, Li N, Hill D, Laidler J, Finck P (2001) The U.S. accelerator transmutation of waste program. Nucl Instrum Methods Phys Res A 463:468–486
6. Heidet F, Kim TK, Taiwo TA (2013) Multiple-stage fuel cycle options based on subcritical systems. In: Proceedings of the 2013 ANS Winter Meeting, Washington, DC, November 10–14, 2013
7. Meyer MK et al (2001) Fuel design for the U.S. accelerator driven transmutation system. Nuclear Application in the New Millennium (accApp-ADITTA '01), Reno, Nov 2001
8. Carmack WJ et al (2009) Metallic fuels for advanced reactor. J Nucl Mater 392:139–150
9. Stevenson CE (1987) The EBR-II fuel cycle story. American Nuclear Society, La Grange Park, Illinois, USA
10. Arie K, Kawashima M, Araki Y, Sato M, Mori K, Nakayama Y, Ishiguma K, Fujiie Y (2007) The sustainable system for global nuclear energy utilization. In: GLOBAL 2007, Boise, ID, September 9–13, 2007
11. Arie K, Kawashima M, Oomori T, Okita T, Kotake S, Fujiie Y (2013) Role of fast reactor and its cycle to reduce nuclear waste burden, In: GLOBAL 2013, Salt Lake City, UT, September 29–October 3, 2013
12. Hayes S et al (2009) Status of transuranic bearing metallic fuel development. In: Global 2009, Paris, September 6–11, 2009
13. OECD-NEA (2000) Nuclear Energy Agency, Organisation for Economic and Co-operation Development (NEA, OECD), Paris, France
14. Derstine KL (1984) DIF3D: a code to solve one-, two- and three-dimensional finite difference diffusion theory problems. ANL-82-64, Argonne National Laboratory
15. Hazama T et al (2009) SLAROM-UF: ultra fine group cell calculation code for fast reactor: version 20090113. JAEA-Review 2009-003
16. Shibata K, Iwamoto O, Nakagawa T, Iwamoto N, Ichihara A, Kunieda S, Chiba S, Furutaka K, Otuka N, Ohsawa T, Murata T, Matsunobu H, Zukeran A, Kamada S, Katakura J (2011) JENDL-4.0: a new library for nuclear science and engineering. J Nucl Sci Technol 48(1):1–30
17. Kittel JH et al (1971) Plutonium and plutonium alloys as nuclear fuel materials. Nucl Eng Des 15:373–440
18. ASM Alloy Phase Diagram Center (2007) ASM International, Materials Park. (http://www.asminternational.org/AsmEnterprise/APD)
19. Hecker SS, Stan M (2008) Properties of plutonium and its alloys for use as fast reactor fuels. J Nucl Mater 383:112–118
20. Wade DC, Fujita EK (1989) Trends versus reactor size of passive reactivity shutdown and control performance. Nucl Sci Eng 103:182
21. Kawaguchi K et al (2007) Conceptual study of measures again heat generation for TRU fuel fabrication system. In: Global 2007, Boise, ID, September 2007

Chapter 16
Enhancement of Transmutation of Minor Actinides by Hydride Target

Kenji Konashi and Tsugio Yokoyama

Abstract A hydride target including minor actinides (MA) is able to enhance the transmutation rate in a fast breeding reactor (FBR) without degradation of core safety characters. Fast neutrons generated in the core region of the FBR are moderated in the MA-hydride target assemblies and then efficiently absorbed by MA. The MA-hydride target pin has been designed in the light of recent research of hydride materials. This chapter shows the feasibility of MA transmutation by an existing reactor, Monju.

Keywords Fast breeder reactor • Hydride • Minor actinide

16.1 Introduction

High-level wastes generated after reprocessing spent nuclear fuels include long-lived radioactive nuclides of minor actinides (MA), such as ^{237}Np, ^{241}Am, ^{243}Am, and ^{244}Cm. A currently available method for the final disposal of the high-level wastes is to vitrify them under rigid control, to store them in monitored locations until the radiation decays to allowable levels, and then to dispose them underground.

The transmutation of MA by the fast breeder reactor (FBR) has been intensively studied to reduce radioactivity of the wastes [1]. Transmutation rate, which is one of the most important factors for transmutation methods, is determined by values of neutron flux and nuclear reaction cross section as follows:

$$\lambda = \int \Phi_n(E) \sigma_n(E) \mathrm{d}E$$

K. Konashi (✉)
Institute for Materials Research, Tohoku University, Oarai, Ibaraki-ken 311-1313, Japan
e-mail: konashi@imr.tohoku.ac.jp

T. Yokoyama
Toshiba Nuclear Engineering Services Corporation, 8 Shinsugita, Isogo-ku, Yokohama 235-8523, Japan

FBRs provide high fast neutron flux, wherein the neutron reaction cross sections are small compared with those in the thermal energy region. Moderation of fast neutrons by hydride materials was considered to increase the transmutation rate [2–5]. In this chapter, enhancement of transmutation of MA by an MA-hydride target is studied. Target assemblies containing MA-hydrides are placed in the radial blanket region. Fast neutrons generated in the core region are moderated in the hydride target assembly and then produce high flux of thermal neutrons, which have large nuclear reaction cross sections to actinides. The MA-hydride target also has another advantage to load MA to limited space. The target of $(MA, Zr)H_x$ increases mass of MA and hydrogen density in the blanket region compared with MA and $ZrH_{1.6}$ loaded separately [4, 5].

Hydride fuels have been used in TRIGA reactors of General Atomics (GA) for many years [6]. On the other hand, hydride materials do not have much history of use in FBRs. Recently, a control rod of FBR with hafnium (Hf)-hydride has been studied [7]. In this chapter, the MA-hydride target pin was designed using experimental data of Hf-hydride.

16.2 Design of MA-Hydride Target

The TRIGA fuel consists of a U-metal phase and a Zr-hydride phase at high temperature in the reactor. The MA-hydrides are stable at high temperature [8]. The phase relationship of the U-Th-Zr hydride has been studied, considering Th as a surrogate of MA. Figure 16.1 shows the microstructure of $UTh_4Zr_{10}H_{24}$: black areas are Zr hydride, gray region is $ThZr_2H_x$, and white areas are uranium metal. The thermodynamic analysis shows that the MA-hydride consists of MA-hydride, MA-Zr-hydride, and Zr-hydride (Fig. 16.1).

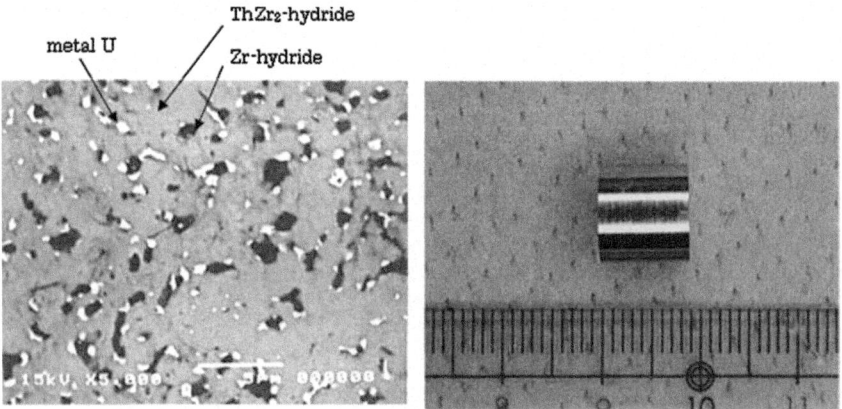

Fig. 16.1 Pellet of $(U,Th,Zr)H_x$ and microstructure

Fig. 16.2 MA-hydride target pin

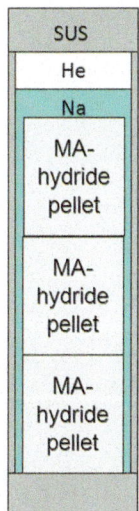

For the application of Hf-hydride to neutron absorber material in FBR [3], the Hf-hydride pin has been developed. The fabrication of sodium-bonded Hf-hydride pins has been demonstrated. The pins were successfully irradiated in BOR-60 for 1 year [7]. The MA-target pin was designed based on the foregoing experiences. Figure 16.2 shows a target pin that includes MA-hydride ((MA, Zr)$H_{1.6}$) pellets. The gap of the MA hydride pellet-stainless steel cladding was filled with liquid sodium to keep the temperature of the pellets low. The results of irradiation experiments show that the sodium also reduces loss of hydrogen from the hydride pin.

16.3 Design of Core with MA-Hydride Target

Table 16.1 and Fig. 16.3 show the core specification and layout with hydride targets. The core layout is based on the Japanese prototype fast reactor Monju. The thermal power is 714 MWt, the diameter of the active core is about 1,800 mm, and the height is 930 mm. The 54 hydride MA-hydride target assemblies are located at the inner most row in the three radial blanket rows. Each assembly contains 61 MA-hydride target pins, where the diameter of the pellets is set at 10.4 mm and the stack length is 930 mm. The ratio of H/M (M=MA+Zr) is considered to be 1.6. The composition of MA is assumed that derived from the typical large LWR discharged fuel, that is, ^{237}Np/^{241}Am/^{242m}Am/^{243}Am/^{243}Cm/^{244}Cm/^{245}Cm/^{246}Cm = 0.5200/0.2493/0.0010/0.1663/0.0006/0.0592/0.0031/0.0006.

Table 16.1 Major core specifications for minor actinides (MA) transmutation

Reactor type	Fast breeder reactor
Cooling system	Sodium cooled (loop-type)
Thermal output	714 MW
Electrical output	280 MW
Fuel	Mixed oxide
Plutonium enrichment	Inner/outer 16/21(% Pu fission)
Average burn-up	80,000 MWd/t
Cladding material	SS 316

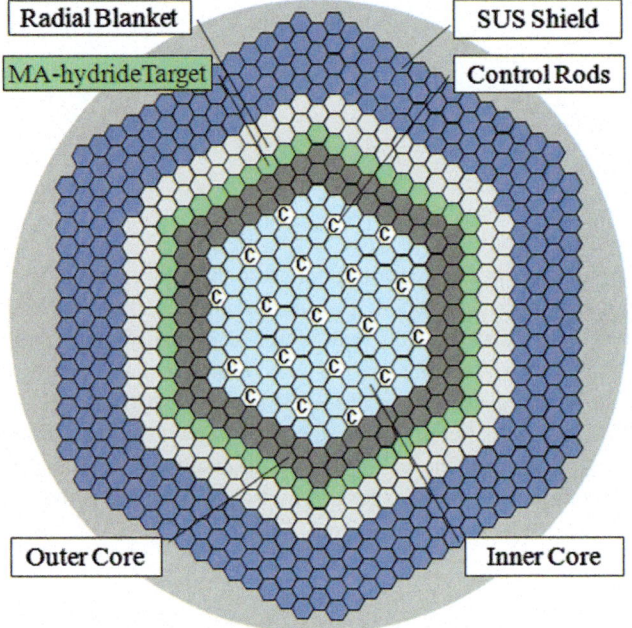

Fig. 16.3 Core layout with hydride targets

16.4 Transmutation Calculation

Transmutation performances of the hydride MA target and related core features in FBRs have been evaluated with the method shown in Table 16.2. A three-dimensional continuous energy Monte Carlo Code MVP [9] and MVP-BURN [10] are used as burn-up calculations for evaluating the transmutation of MAs. The cross-section library applied in the calculations is JENDL-4.0, which is processed to be adjusted to the MVP code. In the burn-up calculation, the prediction-correction method is employed to improve accuracy with millions of neutron histories for the criticality calculation, where the accuracy of Eigen value is about 0.04 %.

Table 16.2 Calculation method for MA transmutation

Items	Methods	Notes
Computation method	Three-dimensional continuation energy Monte Carlo analysis code; MVP (burn-up routine is MVP-BURN)	1,200,000 neutron histories with 120 batches. Initial 20 batches are run to establish the initial neutron source distribution
Nuclear data	JENDL-4.0 library	
Calculation model	Pin heterogeneous model	

Table 16.3 Comparison of reduction ratio of MAs

Target	Loading mass (kg)	Reduction mass (kg/year)	Reduction ratio after 1 year	Effective half life (year)
Case1: MA-hydride	335	91.1 (33.0)[a]	0.272	2.19
Case2: MA-metal	335	27.6 (9.4)[a]	0.082	8.07
Ratio: Case1/Case2	1.00	3.30	3.30	0.27

[a]Values in parentheses are reduction masses by fissions

Calculations have been done for two kinds of transmutation target. In case 1, the transmutation target was the MA hydride of $(MA_{0.1}, Zr_{0.9})H_{1.6}$. Calculation with metal $MA_{0.1}Zr_{0.9}$ target without H was done in case 2. The results of calculations are summarized in Table 16.3, where effective half-life is defined as the time such that the residual amount of MA is decreased to half of the MA loaded during the burn-up. The effective half-life is calculated to be 2.19 years in case 1 and 8.07 years in case 2, mainly because of the softened spectrum effect induced by the MA-hydride. The transmutation rate of the MA-hydride target is about three times higher than that of the MA-metal target. Figure 16.4 shows the change of total MA and each element of MA in the MA-hydride target with increase of time. Major elements in MA, that is, Np and Am are decreased simultaneously during the burn-up. The contribution of long-lived Cm (^{245}Cm and ^{246}Cm) is much smaller than that of Np and Am. The change of total MA in the MA-metal target is also shown in Fig. 16.4 for comparison.

The major mode of the transmutation in the present method is not fission but neutron capture (see Table 16.3). As shown in Fig. 16.5, Am and Np are mainly transmuted to Pu because of neutron capture, beta decay, and alpha decay. Recycled Pu is used as a driver fuel in this reactor.

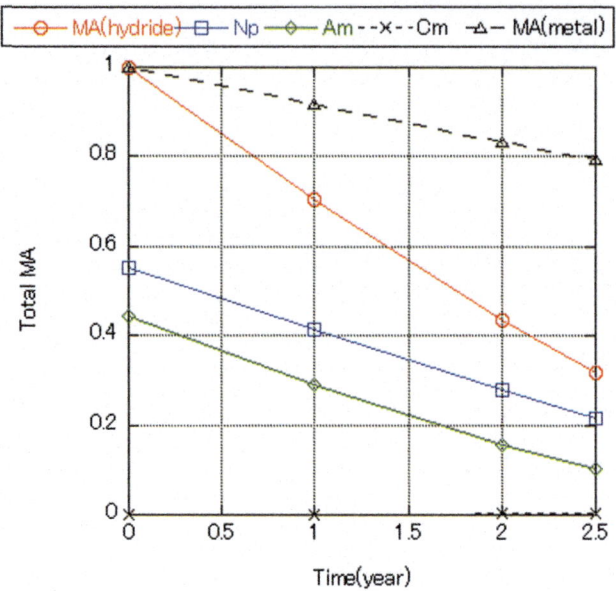

Fig. 16.4 Change of each element in MA assemblies

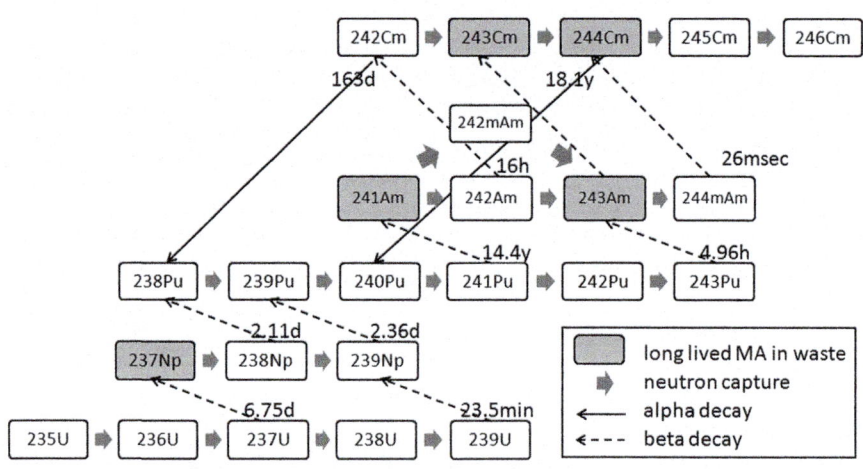

Fig. 16.5 Chain transmutations for actinide nuclides

16.5 Discussion

One of the problems for hydride used in a fast spectrum core is the thermal spike wherein a large power peak occurs in the fuel pins near the hydride used zones. Figure 16.6 shows the intra-assembly radial power distribution in the outermost assembly of the core, that is, the fuel assembly adjacent to the MA-hydride assembly. If a Zr-hydride assembly without MA was used instead of the MA-hydride assembly, a large power peak appeared at the No. 15 pins adjacent to the hydride assembly (shown by pink line of $ZrH_{1.6}$ case in Fig. 16.6). In our proposed case, MA-hydride works as an absorber of thermal neutrons, and thus a thermal spike is suppressed. As a result, the radial power distribution of the core has an ordinary profile in the core zones (Fig. 16.7). The power of the first row in the blanket region is, however, a little larger than that of ordinary fast reactors because of the fission reactions of MAs or daughter nuclides, although this power increase is considered to be controllable by adjusting the assembly flow distribution.

Figure 16.8 shows the mass balance of MA for the system of about three 1GWe-class LWRs and one FBR with MA-hydride target as previously described. The LWR annually produces spent fuel with burn-up of 45 GWd/t containing 23 kg MA.. The mass of transmuted MA per year is almost equivalent to that produced annually in about three LWRs, which means that most of the produced MA is

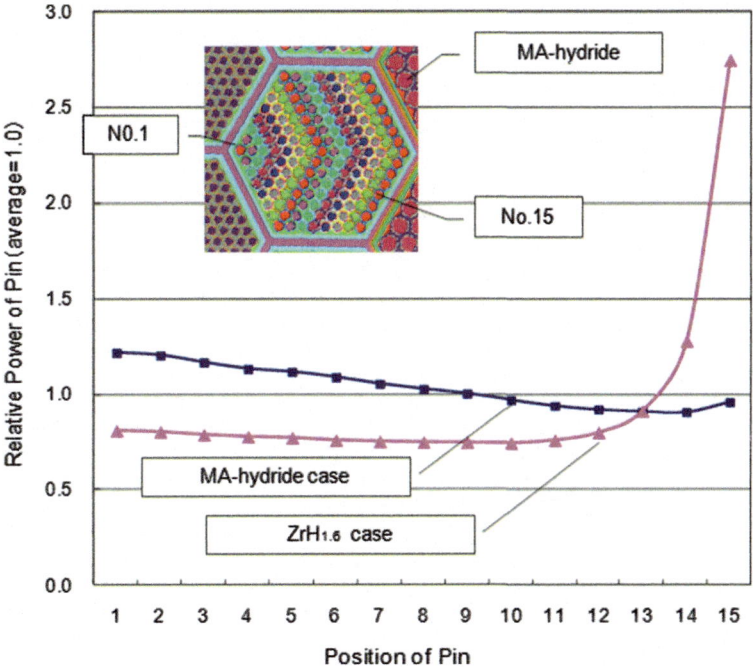

Fig. 16.6 Intra-assembly power distribution of the assembly adjacent to hydride assemblies

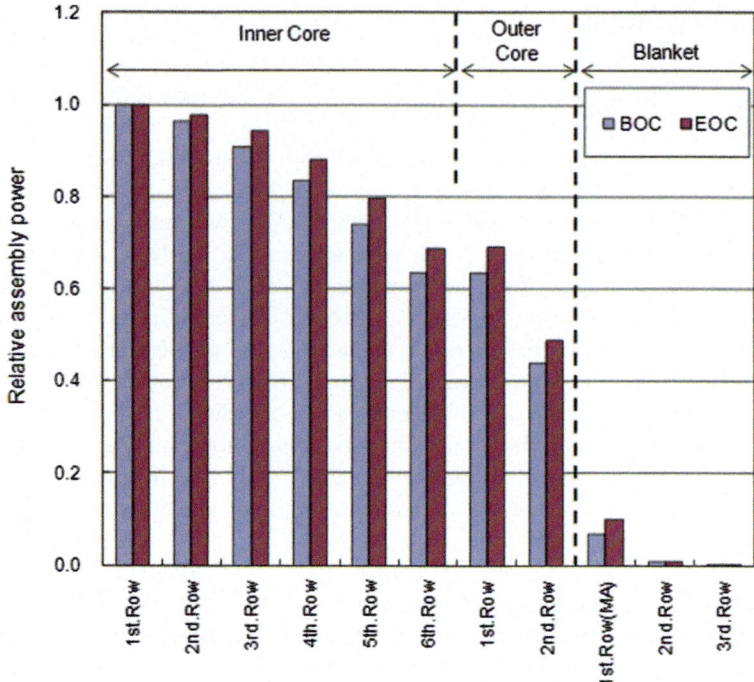

Fig. 16.7 Radial power distribution of the core

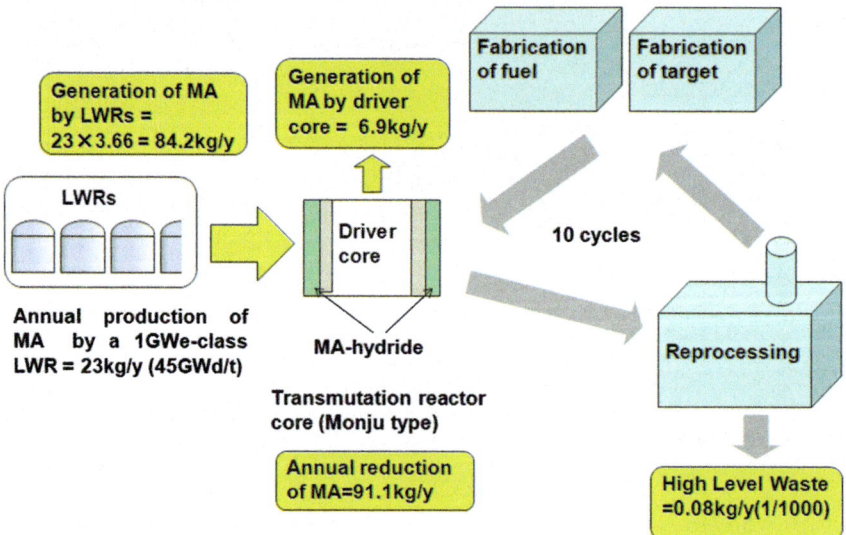

Fig. 16.8 Mass balance of MA in the system of LWRs and FBR with MA-hydride targets

transmuted in the system and only a small amount of MA is transferred to the waste stream. As seen in Fig. 16.8, MA recycling is necessary for higher transmutation efficiency. It takes ten times the effective half-life to reduce the mass of MA to 1/1,000 of the initial mass. When the irradiation time of one cycle is 2.19 year, ten fuel cycles are necessary to reduce the mass of MA to 1/1,000.

16.6 Conclusions

Transmutation by MA-hydride targets has been studied when the hydride target assemblies are loaded in the radial blanket region of the typical fast reactor. The MA-target pin has been designed based on experience accumulated in the development of the Hf-hydride control rod. The following conclusions have been obtained.

1. The MA-hydride target shows an excellent performance for the transmutation of wastes. The effective half life is 2.19 years, which is one third of that of the MA-metal target case.
2. The MA-hydride target has advantages in lowering of peak power and increase of loading mass in the radial blanket region.
3. The transmutation mass of MA is almost equivalent to the annual production of MA from about three 1GWe-class LWRs.
4. After MA is irradiated for 21.9 year, a mass of MA reduces to 1/1,000 of that of initially loaded MA. Thus, only 1/1,000 of the MA generated in a LWR should be transferred to the waste stream.

Open Access This chapter is distributed under the terms of the Creative Commons Attribution Noncommercial License, which permits any noncommercial use, distribution, and reproduction in any medium, provided the original author(s) and source are credited.

References

1. Homogeneous versus heterogeneous recycling of transuranics in fast nuclear reactors. OECD/NEA, Paris
2. Rome M et al (1996) Use of fast reactors to burn long-life actinides, especially Am, produced by current reactors. In: Proceedings of the Physor'96, Mito, September 16–20, 1996
3. Konashi K et al (1999) Transmutation of actinide nuclear wastes using hydride fuel target. Trans Am Nucl Soc 81:124–125
4. Konashi K et al (2001) Development of actinide-hydride target for transmutation of nuclear waste. In: Proceedings of the international conference on future nuclear systems (GLOBAL '01), Paris, September 9–13, 2001
5. Bates S et al (2009) Minor actinide recycle in sodium cooled fast reactors using heterogeneous targets. In: Proceedings of the meeting on advances in nuclear fuel management IV (ANFM 2009), Hilton Head Island, 12–15 April, 2009

6. Simnad MT (1981) The U-ZrHx alloy: its properties and use in TRIGA fuel. Nucl Eng Des 64:4033–4422
7. Konashi K, Itoh K, Kido T, Kosaka Y, Seino S (2013) Development of hydride neutron absorber for fast reactor: irradiation experiment on hydride neutron absorber in BOR-60. In: Proceedings of ICAPP 2013, Jeju Island, April 14–18, 2013, paper FF232
8. Olander D, Konashi K, Yamawaki M (2012) 3.12 Uranium-zirconium hydride fuel. In: Comprehensive nuclear materials, Elsevier, pp 314–357
9. Nagaya Y, Mori T, Okumura K, Nakagawa M (2005) MVP/GMVP version 2: general purpose Monte Carlo Codes for neutron and photon transport calculations based on continuous energy and multigroup methods. JAERI 1348
10. Okumura K, Nakagawa M, Kaneko K (1997) Development of burn-up calculation code system MVP-BURN based on continuous energy Monte Carlo method and its validation (M&C and SNA 97), vol 1. Saratoga Springs

Chapter 17
Method Development for Calculating Minor Actinide Transmutation in a Fast Reactor

Toshikazu Takeda, Koji Fujimura, and Ryota Yamada

Abstract To effectively transmute minor actinides (MAs), which have long-lived radioactivity and high decay heat, fast reactors are very promising because many minor actinides can be loaded and transmutation rates are high compared to light water reactors. With the increase of loaded minor actinides, the neutron spectrum becomes hard and core safety parameters will deteriorate. Especially, the sodium void reactivity increases with MA addition to cores. To overcome the difficulty, we propose MA transmutation fast reactors using core concepts with a sodium plenum and internal blanket region in reactor cores. Therefore, cores become complex, and calculation accuracy becomes poor. To accurately evaluate the neutronic properties such as MA transmutation rate and sodium void reactivity, we improved calculation methods. In this chapter we show new methods for calculating MA transmutation rates for each MA nuclide, for calculating the uncertainty of MA transmutation using sensitivities. A new sensitivity is derived that is defined as a relative change of core parameters relative to infinite-dilution cross sections, not effective cross sections. To eliminate bias factors in estimating core parameter uncertainties, a new method is proposed. This method is used to reduce the calculation uncertainty through the use of adjusted cross sections.

Keywords Calculation methods • Fast reactors • Minor actinide • Sensitivity • Sodium void reactivity • Transmutation

17.1 Introduction

The importance of nuclear energy, as a realistic option to solve the issues of the depletion of energy resources and the global environment, has been acknowledged worldwide. However, acceptance of large-scale contributions would depend on satisfaction of key drivers to enhance sustainability in terms of economics, safety,

---

T. Takeda (✉) • R. Yamada
Research Institute of Nuclear Engineering, University of Fukui, Fukui, Japan
e-mail: t_takeda@u-fukui.ac.jp

K. Fujimura
Hitachi Works, Hitachi-GE Nuclear Energy, Ltd., Ibaraki, Japan

adequacy of natural resources, waste reduction, nonproliferation, and public acceptance. Fast reactors with fuel recycle enhance the sustainability indices significantly, leading to the focus on sodium-cooled fast reactors (SFR) in the Generation IV International Forum (GIF) and the International Project on Innovative Nuclear Reactors and Fuel Cycles (INPRO) initiative of the International Atomic Energy Agency (IAEA).

The necessary condition for successful fast reactor deployment is the understanding and assessment of innovative technological and design options, based on both past knowledge and experience, as well as on ongoing research and technology development efforts. The severe accident at Tokyo Electric Power Company's Fukushima Dai-Ichi Nuclear Power Station caused by the Great East Japan Earthquake and tsunami on March 11, 2011 prompted all countries to redefine their fast reactor programs. To achieve the successful deployment of fast reactors, drastic safety enhancement is the most important issue to be established, especially in Japan, where the restart of nuclear power plants once these have been stopped is a serious matter of argument.

The safety aspects of fast reactors (FRs) have been reviewed [1–4] in representative countries that have developed or have a plan to develop fast reactors in the near future, especially after the Fukushima accident. These countries are improving the safety of SFRs by considering the DiD (defense in depth). The designs of SFRs should have tolerance to DBA (design basis accidents) and BDBA (beyond design basis accident) caused by internal and external events. The inherent safety and passive safety should be effectively utilized for reactor shutdown and reactor cooling. For the case of severe accidents, it is indispensable first to shut down the reactors. Furthermore, decay heat removal is also indispensable even in the case of SBO (station black out). For SFRs, natural circulation can be expected in the sodium heat transport systems and the decay heat can be removal to atmosphere by the air cooling system.

In Japan, the Ministry of Education, Culture, Sports, Science and Technology has launched a national project entitled "Technology development for the environmental burden reduction" in 2013. The present study is one of the studies adopted as the national project. The objective of the study is the efficient and safe transmutation and volume reduction of MAs with long-lived radioactivity and high decay heat contained in HLW in sodium-cooled fast reactors. We are aiming to develop MA transmutation core concepts harmonizing MA transmutation performance with core safety. The core concept is shown in Chap. 2. Also, we are aiming to improve design accuracy related to MA transmutation performance. To validate and improve design accuracy of the high safety and high MA transmutation performance of SFR cores, we developed methods for calculating transmutation rates of individual MA nuclides and estimating the uncertainty of MA transmutation.

A new definition of transmutation rates of individual MA nuclides is derived in Chap. 3. Using the definition, one can understand the physical meanings of transmutation for individual MA nuclides. Sensitivities are required to estimate the uncertainty of MA transmutation rates from cross-section errors. In Chap. 4, sensitivity calculation methods are derived. First, the sensitivity calculation method relative to infinite-dilution cross sections is introduced. The MA transmutation rates are burn-up properties. Thus, the sensitivity calculation method for burn-up-dependent properties is derived. Finally, we investigate how many energy groups are required in sensitivity calculations. Calculated MA transmutation rates have large uncertainties resulting from the large uncertainties in MA cross sections. To reduce these uncertainties in MA transmutation rates, we introduce a new method to reduce prediction uncertainties of MA transmutation rates in Chap. 5. In this method, we eliminate bias factors included in experiments and calculations by using ratios of the calculation to the experiment of core performance parameters. After removing the bias factors, the cross section is adjusted using measured data. The conclusions are shown in Chap. 6.

17.2 MA Transmutation Core Concept

MA transmutation core concepts are developed by considering the amount of MA loading and the safety-related core parameters. Increase of MA loading in the core of a SFR makes the amount of MA transmutation large, which may decrease long-term radiotoxicity and decay heat of MA. On the other hand, loading a large amount of MA into the core of a SFR increases the sodium void reactivity. Therefore, harmonization of MA transmutation and sodium void reactivity is a key issue in designing the core concepts. As an example, Fig. 17.1 shows the relationship of MA content and sodium void reactivity; when the MA content is about 10 %, the sodium void reactivity increases by about 1$.

A homogeneous MA-loaded core of 750 MWe was designed in the FaCT project [1, 2] (Fig. 17.2). The configuration of this core is a conventional homogeneous core and homogeneous MA loading into the core fuel increases sodium void reactivity. Therefore, MA content in the core fuel assembly is limited to less than about 5 wt%. On the other hand, the safety issue has become more and more important since the Fukushima Daiichi NPP accident. Further, low void reactivity SFR designing has been pursued in Russia and France [5]. In this study, the coexistence of enhanced MA transmutation and zero void reactivity, that is, the harmonization of MA transmutation and core safety, is set as an objective.

Hitachi proposed an axially heterogeneous core (AHC) concept with sodium plenum [6, 7]. It was clarified that an increase of flux level at the top of the core fuel caused by the presence of the internal blanket and decrease of the height of the inner core fuel greatly decreased sodium void reactivity. In the core concept, sodium void reactivity can be extremely reduced without disrupting core performance for normal operation. The difference in core configurations between the Hitachi AHC with sodium plenum proposed in FR '91 [6] and the ASTRID ACV [5], which has been recently studied in France, is that absorber material is loaded in the upper shield for the ASTRID ACV.

We are going to optimize the specifications of the core shown in Fig. 17.3 to realize the high MA transmutation and zero sodium void reactivity. Figure 17.4

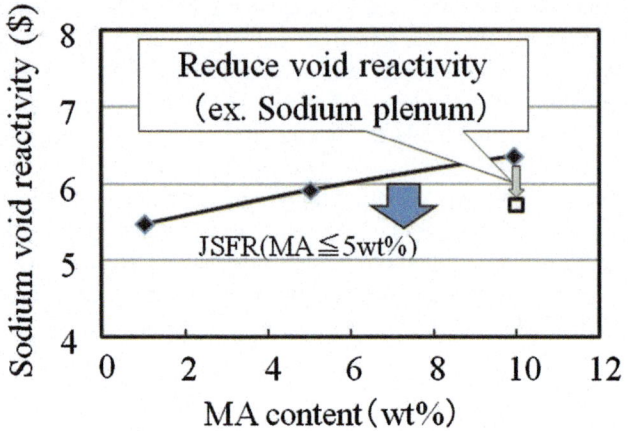

Fig. 17.1 Dependency of sodium void reactivity on minor actinide (*MA*) content

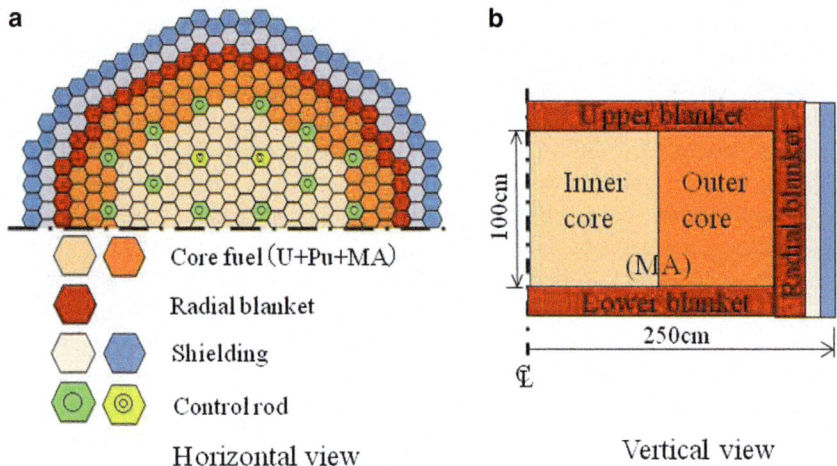

Fig. 17.2 Core configuration of the homogeneous MA-loaded core

shows the axial distribution of coolant density and the density coefficient of that core. The sodium density coefficient ($\%\Delta k/kk'/\Delta\rho$) in the sodium plenum becomes positive. On the other hand, sodium density change ($\Delta\rho$) in the sodium plenum becomes negative because of the increase of coolant temperature for an accident such as ULOF (unprotected loss of flow accident). Therefore, net sodium void reactivity becomes negative.

The mechanism for reducing sodium void reactivity of the core is that the axial neutron leakage is largely enhanced with coolant voiding in the sodium plenum. It is known that the evaluated leakage component of sodium void reactivity with diffusion theory might be overestimated by about 50 %. Therefore, calculation

Fig. 17.3 Vertical view of axially heterogeneous core with sodium plenum

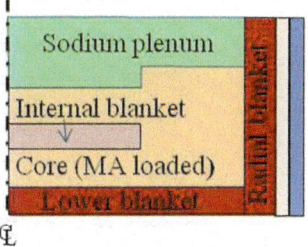

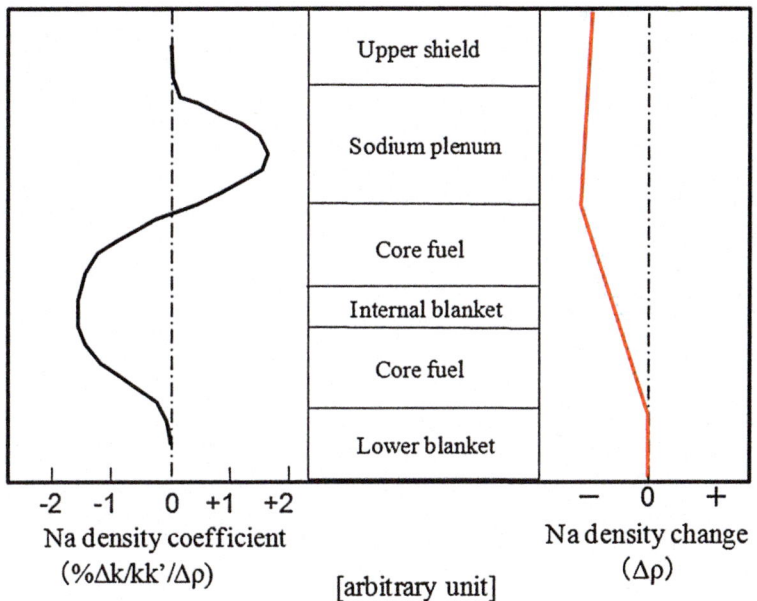

Fig. 17.4 Axial distribution of coolant density and density coefficient

accuracy for sodium void reactivity of the core with a sodium plenum might be poor. Thus, we should consider change of the neutron spectrum in the heterogeneous MA loaded core. Figure 17.5 shows the neutron spectra for MOX driver fuel without MA and 10 % MA-mixed fuels in transmutation target with and without Zr-Hx. The spectrum of 10 % MA-mixed fuel in transmutation target without Zr-Hx is slightly softer than that of the MOX driver fuel without MA, because the 10 % MA-mixed fuel in the transmutation target has no fissile plutonium but the MOX driver fuel includes ^{239}Pu and ^{241}Pu. The neutron spectrum of the moderator mixture target fuel is clearly softer than other fuels.

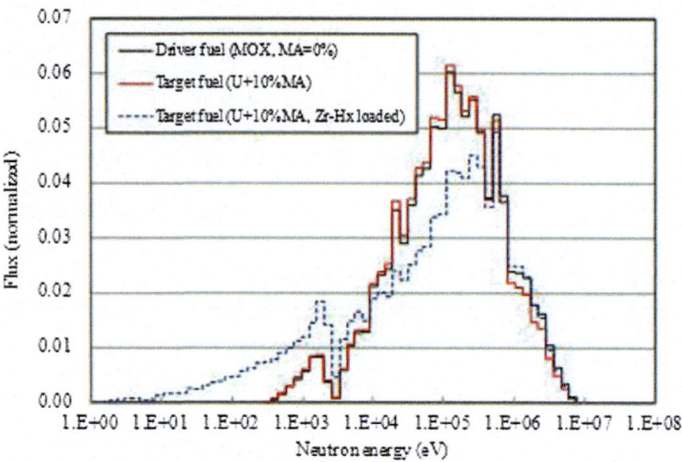

Fig. 17.5 Neutron spectra for MOX fuel and MA transmutation target fuels

17.3 MA Transmutation Rate

Let us first introduce a definition of transmutation rates of individual MA nuclides fueled in a reactor core. The calculation method of the transmutation rate relative to the nuclides is as follows. First, conventional burn-up calculations are carried out and burn-up-dependent flux in each region is calculated, which is used in the second step calculation. In the second step, we consider only the relevant MA in each region and perform burn-up calculations using the flux obtained in the first step.

In this second step of calculation, nuclide k is produced from the original nuclide l. There are many passes of reactions of transmutation of the initial nuclide (shown by N) and the production of N as is shown in Fig. 17.6. We can calculate the production rate of nuclide k at time T from the initial nuclide l as

$$P_{lk} = \widetilde{N}_k(T)/\widetilde{N}_l(0) \tag{17.1}$$

where $\widetilde{N}_l(0)$ is number density of nuclide l at time 0 and $\widetilde{N}_k(T)$ is number density of nuclide k at time T, assuming nuclide l is present alone at $t=0$. Using $\widetilde{N}_k(T)$, the overall fission (see Fig. 17.6) relative to the initial nuclide l is calculated as

$$\mathrm{OF}^l = \sum_k \int_0^T \sigma_f^k(t)\widetilde{N}_k(t)\phi(t)\mathrm{d}t \tag{17.2}$$

where $\sigma_f^k(t)$ is fission cross section of nuclide k at time t, $\phi(t)$ is neutron flux at time t, and Σ is summation over all nuclides k resulting from initial nuclide l; this

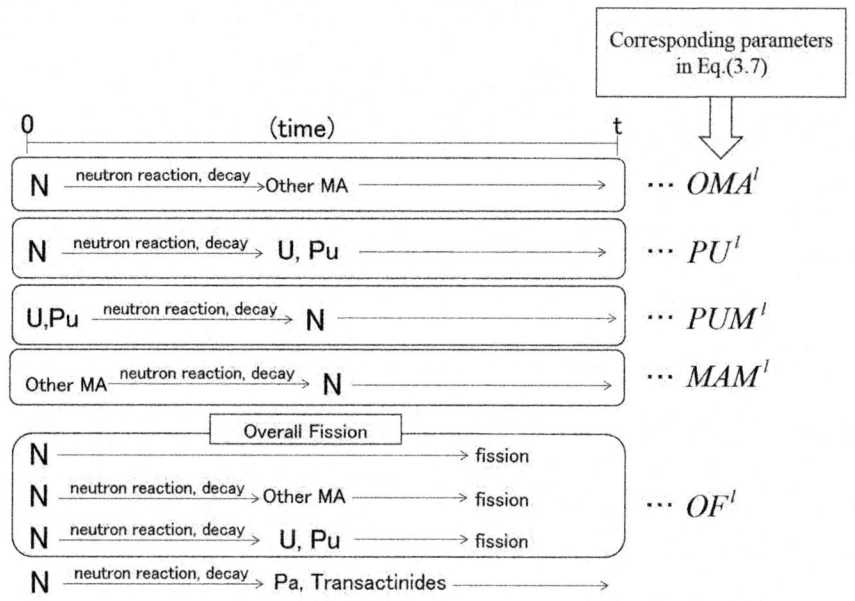

Fig. 17.6 Transmutation of initial MA nuclides and production of the MA nuclide

includes all the fissions from the initial nuclide l. Furthermore, the production of other MA nuclides except the initial nuclide l can be calculated by

$$\text{OMA}^l = \widetilde{N}_l(0) \sum_{k \in \text{MA}, k \neq l} P_{lk} \qquad (17.3)$$

The Pu and U production from nuclide l is given by

$$\text{PU}^l = \widetilde{N}_l(0) \sum_{k \in \text{U, Pu}} P_{lk} \qquad (17.4)$$

The production of MA nuclide l from Pu and U is given by

$$\text{PUM}^l = \sum_{k \in \text{U, Pu}} \widetilde{N}_k(0) P_{kl} \qquad (17.5)$$

The production of MA nuclide l from other MA is given by

$$\text{MAM}^l = \sum_{k \in \text{MA}, k \neq l} \widetilde{N}_k(0) P_{kl} \qquad (17.6)$$

Using Eqs. (17.2), (17.3), (17.4), (17.5), and (17.6), the net transmutation of nuclide l is calculated by

$$TR^l = OF^l + OMA^l + PU^l - PUM^l - MAM^l \qquad (17.7)$$

In Fig. 17.6, the individual parameters OMA^l, PU^l, PUM^l, MAM^l, and OF^l are shown. PUM^l and MAM^l denote the productions of the relevant MA nuclide from fuel (Pu, U) and other MA nuclides, respectively. Thus, there are minus signs in these parameters in Eq. (17.7), whereas other parameters show the elimination of the relevant MA nuclide, so the signs are positive.

When we consider the total MA transmutation for all MA nuclides, the second and the fifth terms cancel each other, so the whole transmutation is given by

$$TR = \sum_{l \in MA} TR^l = \sum_{l \in MA} \left(OF^l + PU^l - PUM^l \right) \qquad (17.8)$$

Therefore, we can define the MA transmutation of MA nuclide l by

$$TR^l = OF^l + PU^l - PUM^l \qquad (17.9)$$

Thus, the transmutation rate is composed of two terms: the first is the amount of incineration rate by fission and the second is the net transmutation rate to fuel (U and Pu). The first fission rates of individual nuclides contain the direct fission of the relevant nuclide plus the fission of other nuclides transmuted by decays or neutron reactions as "overall fission" [OF^l in Eq. (17.9)] (Fig. 17.7). It was found that the indirect fission contribution by ^{238}Pu and ^{239}Pu is remarkably large for nuclides ^{239}Np and ^{241}Am. The net production rates of U and Pu are calculated from the difference between the production rates of U and Pu from the relevant MA nuclide and the MA production from the initial U and Pu. Figure 17.7 shows the overall fission rate of ^{237}Np in a thermal advanced pressurized water reactor (APWR) and two fast reactors, a MOX-fueled sodium-cooled fast reactor and a metal-fueled lead-cooled fast reactor [8]. In the thermal reactor, the overall fission rate is about 5 % in one cycle and is very small compared with the fast reactors. In fast reactors, the direct fission of ^{237}Np is rather large, and the ^{238}Pu fission contribution is also large. The ^{239}Pu fission contribution is small for fast reactors.

We are developing a calculation code system based on the foregoing method and are planning to apply the system to MA transmutation core design. In the core design we consider a homogeneous MA loading core, and a heterogeneous MA loading core, in which MA is loaded in special assemblies with moderators.

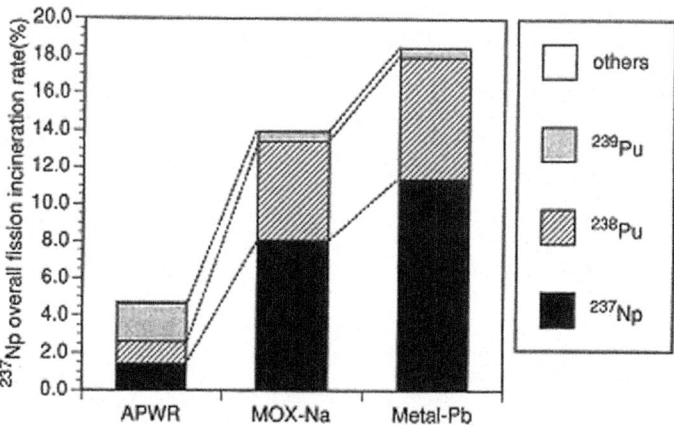

Fig. 17.7 Calculation example of overall fission rate of ^{237}Np by developed calculation. *APWR* advanced pressurized water reactor

17.4 Sensitivity Calculation Method

17.4.1 Sensitivity to Infinite-Dilution Cross Section

Core characteristics such as k_{eff}, power distribution, and control rod worth are calculated by using effective cross section in deterministic methods. Sensitivities are usually calculated by using sensitivity calculation codes such as SAGEP [9], SAGEP-T [10], and SAINT [11]. However, the sensitivities are for relative changes of effective cross sections. Here we derive a calculation method of sensitivities relative to infinite-dilution cross sections for fast reactor analysis. Usually the effective cross sections are calculated by using the Bondarenko self-shielding factor method, the subgroup method. In that method the effective cross sections are expressed by the infinite-dilution cross section and the self-shielding factors f

$$\tilde{\sigma} = f \cdot \sigma \qquad (17.10)$$

The self-shielding factors depend on the background cross section and temperature. The background cross section for nuclide i' in a homogeneous medium is calculated by the formula

$$\sigma_{b^{i'}} = \frac{1}{N_{i'}} \sum_{k \neq i'} N_k \cdot \sigma_t^k \qquad (17.11)$$

where N_k is the atomic number density of light nuclide k and σ_t^k is the microscopic total cross section. The sensitivity coefficient is defined by the relative changes of the core characteristics caused by the relative changes of the cross sections. Here we consider the following two sensitivities, the sensitivity S, which results from the relative change of the infinite-dilution cross sections, and the approximate sensitivity $\widetilde{S}$, which is the result of the relative change of the effective cross sections. From Eq. (17.10), the change of the effective cross section can be expressed by

$$\frac{d\widetilde{\sigma}}{\widetilde{\sigma}} = \frac{df}{f} + \frac{d\sigma}{\sigma} \qquad (17.12)$$

Therefore, the improved sensitivity is expressed by using the approximate sensitivity as follows:

$$S \equiv \frac{dR/R}{d\sigma/\sigma} = \frac{dR/R}{d\widetilde{\sigma}/\widetilde{\sigma}} \cdot \left(1 + \frac{df/f}{d\sigma/\sigma}\right) \qquad (17.13)$$

Sensitivities and cross sections are dependent on nuclides, reaction types (such as fission, capture, and scattering), and energy groups. Here we consider the case where there is a perturbation in σ of nuclide i, reaction type j, in energy group g. This perturbation causes a change in the self-shielding factor f of nuclide i', reaction j', in energy group g'. The second term of the right-hand side of Eq. (17.14) has to cover the contributions for all nuclides i', reaction types j', in energy groups g'; therefore, we have to take the summation over i', j', and g'. The sensitivity for the nuclide i, reaction type j, in energy group g is given by

$$S_{i,j,g} = \widetilde{S}_{i,j,g} + \sum_{i'}\sum_{j'} \widetilde{S}_{i',j',g'} \cdot \frac{df_{j'}^{i'}/f_{j'}^{i'}}{d\sigma_j^i/\sigma_j^i} \qquad (17.14)$$

The first term is the direct contribution to S; it can be calculated using the conventional tools evaluating sensitivity coefficients such as SAGEP, SAGEP-T, and SAINT. The second term represents the indirect contribution through the change of self-shielding factor. These coefficients can be calculated as follows: here we apply the resonance approximation for heavy nuclides, which is suitable for treating fast reactors with hard neutron spectra rather than light water reactors. The self-shielding effect depends on the neutron spectrum; where the neutron spectrum for the heavy nuclide i' is written as

$$\phi(E) \propto \frac{1}{\sigma_t^{i'}(E) + \sigma_b^{i'}} \cdot \frac{1}{E} \qquad (17.15)$$

Equation (17.15) indicates that when $\sigma_t^{i'}(E)$ and $\sigma_b^{i'}$ change by the same factor, the neutron spectrum remains the same; this shows that the ratio h has an effect on the neutron spectrum, and also on the self-shielding factor. Following a similar method [12], the coefficient in the second term of the right-hand side of Eq. (17.14) is called TERM and can be written as

$$\frac{df_j^{i'}(h^{i'})/f_j^{i'}(h^{i'})}{d\sigma_j^i(E)/\sigma_j^i(E)} = \frac{\sigma_j^i(E)}{f_j^{i'}(h^{i'})} \cdot \left\{ \frac{N_i}{N_{i'}}(1-\delta_{ii'}) - \frac{\sigma_b^{i'}}{\sigma_t^i(E)}\delta_{ii'} \right\} \cdot \frac{\partial f_j^{i'}(h^{i'})}{\partial \sigma_b^{i'}} \quad (17.16)$$

where the derivative of f to σ is calculated by

$$\frac{\partial f_j^{i'}(h^{i'})}{\partial \sigma_b^{i'}} = \frac{f_j^{i'}(\sigma_b^{i'} + \Delta\sigma_b^{i'}) - f_j^{i'}(\sigma_b^{i'})}{\Delta\sigma_b^{i'}} \quad (17.17)$$

17.4.2 Burn-up Sensitivity

To calculate reliable MA transmutation rates, it is important to evaluate the uncertainty of calculated MA transmutation rates. The uncertainty can be calculated when the sensitivity of the MA transmutation rates to the cross sections called burn-up sensitivity is known. Therefore, we developed a calculation code of burn-up sensitivity based on the generalized perturbation theory [9]. The burn-up sensitivity $\widetilde{S}$ relative to effective cross sections is calculated by

$$\begin{aligned}\widetilde{S} &= \frac{dR/R}{d\widetilde{\sigma}/\widetilde{\sigma}} \\ &= \frac{\sigma}{R}\left[\sum_{i=1}^{I}\left\langle\int_{t_i}^{t_{i+1}} dt \frac{\partial R}{\partial \widetilde{\sigma}}\right\rangle + \sum_{i=1}^{I}\left\langle\int_{t_i}^{t_{i+1}} dt N^* \frac{\partial M}{\partial \widetilde{\sigma}} N\right\rangle \right. \\ &\left. + \sum_{i=1}^{I+1}\left\langle \Gamma_i^* \frac{\partial B_i}{\partial \widetilde{\sigma}} \phi_i \right\rangle + \sum_{i=1}^{I+1}\left\langle \Gamma_i \frac{\partial B_i^*}{\partial \widetilde{\sigma}} \phi_i^* \right\rangle - \sum_{i=1}^{I+1} P_i^* \left\langle KN\phi \frac{\partial \sigma_f}{\partial \widetilde{\sigma}} \right\rangle \right]\end{aligned} \quad (17.18)$$

In Eq. (17.18), the term containing $\frac{\partial R}{\partial \sigma}$ is called the direct term; the second term is the number density term, which represents the effect of the change of nuclide number densities caused by cross-section changes; the third term shows the effect of the change of flux from to cross-section changes; the fourth term shows the effect of the change of adjoint flux caused by cross-section changes, and the last term shows the effect of constant power production even when there are cross-section changes. The adjoint number density N^* is calculated from the end of a burn-up period to the beginning of the period, and the generalized flux and generalized adjoint flux are calculated at each burn-up step. The adjoint number density N^* is not continuous but has a discontinuity at each burn-up step. To calculate true burn-up sensitivities S relative to infinite-dilution cross sections, we introduce $\widetilde{S}$ to Eq. (17.14) to obtain S.

17.4.3 Dependence of Sensitivities on Numbers of Energy Groups

In sensitivity and uncertainty analysis, multi-group sensitivities are usually used, but there is no theoretical basis for the effect of number of energy groups to sensitivities. Here we derive a relationship between sensitivities calculated with different numbers of energy groups by considering the case where multi-groups are collapsed to a few groups. The sensitivity of core parameter R to the microscopic cross section of nuclide i and reaction j in group g in multi-groups is denoted by S and is defined by

$$S = \frac{dR/R}{d\sigma_{i,j}^g / \sigma_{i,j}^g} \tag{17.19}$$

The sensitivity of R to microscopic cross section in few groups is given by

$$S = \frac{dR/R}{d\sigma_{i,j}^G / \sigma_{i,j}^G} \tag{17.20}$$

Cross sections from few groups are calculated from a multi-group cross section by using neutron flux ϕ^g in group g:

$$\sigma_{i,j}^G = \sum_{g \in G} \sigma_{i,j}^g \phi^g \Big/ \sum_{g \in G} \phi^g \tag{17.21}$$

where the summation about g is performed over energy groups g included in few groups G. Let us consider the case where multi-group cross sections change as follows:

$$\sigma_{i,j}^g \to \sigma_{i,j}^g + \delta\sigma_{i,j}^g \tag{17.22}$$

With the cross-section change, the neutron flux also changes:

$$\phi^g \to \phi^g + \delta\phi^g \tag{17.23}$$

The few group cross sections change as follows:

$$\sigma_{i,j}^G \to \sigma_{i,j}^G + \delta\sigma_{i,j}^G = \sigma_{i,j}^G \left\{ 1 + \frac{\sum_{g \in G} \left(\delta\phi^g \sigma_{i,j}^g + \delta\sigma_{i,j}^g \phi^g \right)}{\sum_{g \in G} \phi^g \sigma_{i,j}^g} - \frac{\sum_{g \in G} \delta\phi^g}{\sum_{g \in G} \phi^g} \right\} \tag{17.24}$$

Therefore we obtain

17 Method Development for Calculating Minor Actinide Transmutation in a Fast...

$$\frac{\delta\sigma_{i,j}^G}{\sigma_{i,j}^G} = \frac{\sum_{g\in G}\left(\delta\phi^g \sigma_{i,j}^g + \delta\sigma_{i,j}^g \phi^g\right)}{\sum_{g\in G} \phi^g \sigma_{i,j}^g} - \frac{\sum_{g\in G} \delta\phi^g}{\sum_{g\in G} \phi^g} \quad (17.25)$$

Here we apply the narrow resonance approximation to express the flux perturbation caused by cross-section change.

$$\phi^g = \frac{C}{N_i\left(\sigma_{i,j}^g + \sigma_0^g\right)} \quad (17.26)$$

where C is a constant, N_i is number density of nuclide i, $\sigma_{i,j}^g$ is microscopic total cross section of nuclide i, and σ_0^g is background cross section. When using only the j reaction cross section of nuclide i, $\sigma_{i,j}^g$, the flux perturbation is expressed by

$$\frac{\delta\phi^g}{\phi^g} = -\frac{\delta\sigma_{i,j}^g}{\sigma_{i,j}^g + \sigma_0^{g'}} \quad (17.27)$$

where $\sigma_0^{g'} = \sigma_0^g + \sum_{j\neq j} \sigma_{i,j}^g$ in the first-order approximation. Introducing the preceding equation to Eq. (17.25) leads to

$$\frac{\delta\sigma_{i,j}^G}{\sigma_{i,j}^G} = \sum_{g\in G} \frac{\delta\sigma_{i,j}^g}{\sigma_{i,j}^G} \cdot \frac{\phi^g \sigma_{i,j}^g}{\sum_{g\in G} \phi^g \sigma_{i,j}^g}$$

$$- \sum_{g\in G} \frac{\delta\sigma_{i,j}^g}{\sigma_{i,j}^G} \cdot \frac{\sigma_{i,j}^g}{\sigma_{i,j}^g + \sigma_0^{g'}} \left\{ \frac{\phi^g \sigma_{i,j}^g}{\sum_{g\in G} \phi^g \sigma_{i,j}^g} - \frac{\phi^g}{\sum_{g\in G} \phi^g} \right\} \quad (17.28)$$

We change the multi-group cross sections $\sigma_{i,j}^g$ at constant rate α (for example, 1 %) within few groups G:

$$\frac{\delta\sigma_{i,j}^g}{\sigma_{i,j}^g} = \alpha \quad (17.29)$$

In this case, the few-groups cross-section change is expressed by the multi-group sensitivity as follows

$$\frac{\delta\sigma_{i,j}^{G}}{\sigma_{i,j}^{G}} = \alpha(1 - X^{G}) \quad (17.30)$$

where X is given by

$$X^{G} = \sum_{g \in G} \frac{\sigma_{i,j}^{g}}{\sigma_{i,j}^{g} + \sigma_{0}^{g'}} \left\{ \frac{\phi^{g}\sigma_{i,j}^{g}}{\sum_{g \in G} \phi^{g}\sigma_{i,j}^{g}} - \frac{\phi^{g}}{\sum_{g \in G} \phi^{g}} \right\} \quad (17.31)$$

Therefore, the few groups sensitivity is given by

$$S^{G} \equiv \frac{dR/R}{d\sigma_{i,j}^{G}/\sigma_{i,j}^{G}} = \sum_{g \in G} S^{g}(1 + X^{G}) \quad (17.32)$$

Thus, in general,

$$S^{G} \neq \sum_{g \in G} S^{g} \quad (17.33)$$

We use this relationship to choose energy groups N ($G = 1$–N) such that

$$S^{G} \approx \sum_{g \in G} S^{g} \quad (17.34)$$

As an example, we calculated k_{eff} sensitivities in 7, 33, and 70 energy groups, and compared sensitivities. In 7 groups, the sensitivities to ^{235}U capture cross section are different from the corresponding integrated sensitivities calculated from 70 groups by 10–20 % above 100 eV. However, in 33 groups, the sensitivities are different from the 70 groups result by at most 5 %. This result convinced us that calculations of sensitivities for 33 groups or 70 groups are sufficient.

17.5 Reduction of Prediction Uncertainty

To accurately calculate neutronics parameters, we have to use reliable calculation methods and nuclear data. For this purpose, we can use valuable measured data obtained from fast critical assemblies and fast reactors by applying the bias factor method [13] and the cross-section adjustment method [14]. In these two methods, it is necessary to consider that there are two kinds of errors, systematic and statistical errors, in measured and calculation errors. Here we propose a method to remove the systematic errors to improve prediction accuracy. Measured data R_e have a systematic error R_{eb} and a statistical error R_{es} and are expressed by

$$R_e = R_{e0}(1 + R_{eb} + R_{es}) \tag{17.35}$$

where R_{e0} is the true value. Also, calculated neutronics parameters R_c are expressed by

$$R_c = R_{c0}(1 + R_{cb} + R_{cs} + S\Delta\sigma) \tag{17.36}$$

where R_{c0} is the true value, R_{cb} is systematic error, R_{cs} is statistical error from calculation methods, and $S\Delta\sigma$ is the error from cross-section error. To eliminate the systematic errors in measurements and calculations, we consider the ratio of measurement to calculation, called bias factors:

$$f = \frac{R_e}{R_c} = \frac{1 + R_{eb} + R_{es}}{1 + R_{cb} + R_{cs} + S\Delta\sigma} \tag{17.37}$$

Because the average of statistical errors becomes zero, the variance of f becomes

$$V(f) = V(R_{es}) + V(R_{cs}) + SWS^T \tag{17.38}$$

where W is the variance of nuclear data used. In deriving Eq. (17.38) it was assumed that all the systematic and statistical errors are smaller than unity and that there is no correlation between statistical errors of measurements and calculations. From Eq. (17.38) we can say that if there is no statistical error, the bias factor f is within the range of

$$1 - c\sigma < f < 1 + c\sigma \quad \sigma = \sqrt{V(f)} \tag{17.39}$$

with the confidence level of 65 %($c = 1$), 95 % ($c = 2$), or 99 %($c = 3$). Therefore, if f is outside the range, we can say in the foregoing confidence level, there is a systematic error of

$$R_{eb} - R_{cb} = |1 - f| - c\sigma \tag{17.40}$$

For sodium void calculations, calculated values are the sum of positive nonleakage components and negative leakage components. The negative leakage components are difficult to estimate because the transport effect has to be considered in calculating the neutron steaming. Therefore, there may be a nonnegligible systematic error in the leakage term R_{cb}^L when the void pattern is leaky. By considering such a void pattern, we can discard the leakage term in systematic errors. Thus, we can determine the systematic errors. After the removal of the systematic errors, we can apply the cross-section adjustment method or the bias factor method to improve the calculation accuracy. In the cross-section adjustment method [13], the adjusted cross section is determined so as to minimize the functional J

$$J = (T - T_0)W^{-1}(T - T_0)^t + [R_e - R_c(T)][V_e + V_m]^{-1}[R_e - R_c(T)]^t \quad (17.41)$$

where W is the cross-section covariance data, and V_e and V_m are the variance of measured data and calculation method, respectively. In Eq. (17.41), we replace R_e and $R_c(T)$ by

$$\begin{aligned} R_e &\rightarrow R_e - R_{e0}R_{eb} \\ R_c(T) &\rightarrow R_c(T) - R_{c0}R_{cb} \end{aligned} \quad (17.42)$$

because the true values are unknown, they are approximated by R_e and R_c. The adjusted cross section is given by

$$T = T_0 + WG^t[GWG^t + V_e + V_m]\left[R_e(1 - R_{eb}) - R'_c(T_0)(1 - R_{cb})\right] \quad (17.43)$$

The covariance of the adjusted cross section is expressed by

$$W' = W - WG[GWG^t + V_e + V_m]GW \quad (17.44)$$

This expression is the same as in [13], but we have to use the adjusted cross section shown in Eq. (17.43).

Using the adjusted cross section, the MA transmutation rate can be estimated by

$$R_c^{(2)}(T) = R_c^{(2)}(T_0) + G^{(2)}(T - T_0) \quad (17.45)$$

where the superscript (2) indicates the MA transmutation rate of the target reactor.

The variance, the uncertainty, of the MA transmutation rate is given by

$$V\left[R_c^{(2)}(T)\right] = G^{(2)}W'G^{(2)t} + V_m^{(2)} - NV_m^{(12)} - V_m^{(12)t}N^t \quad (17.46)$$

where $V_m^{(2)}$ is the covariance of the calculation method used for the MA transmutation rate in the target reactor core, $V_m^{(12)}$ is the correlation between the calculation method errors for the critical assemblies and the target core, and N is defined by

$$N = G^{(2)}WG^{(1)t}\left\{G^{(1)}WG^{(1)t} + V_e^{(1)} + V_m^{(1)}\right\}^{-1} \quad (17.47)$$

We will estimate the MA transmutation amount and the uncertainties by using these methods.

17.6 Conclusion

To realize the harmonization of MA transmutation and sodium void reactivity, the MA transmutation fast reactor core concept, with an internal blanket between the MA-loaded core fuel region and the sodium plenum above the core fuel, was proposed. The feature of this core concept is that sodium void reactivity can be greatly reduced without spoiling core performance for normal operation.

To accurately evaluate neutronics parameters in a MA transmutation fast reactor, we improved the calculation methods for estimating MA transmutation rates and safety-related parameters such as sodium void reactivity. For the MA transmutation rate, we introduced a definition of MA transmutation for individual MA nuclides and a method for calculating the MA transmutation rates. To evaluate the prediction accuracy of neutronics parameters, we proposed a new method that can eliminate systematic errors of measurements and calculations, and introduced a method to reduce the prediction uncertainty based on the cross-section adjustment method or the bias factor method. Furthermore, we improved the sensitivity, which is necessary to evaluate the uncertainty, by considering the effect of self-shielding.

Acknowledgments A part of the present study is the result of "Study on minor actinide transmutation using Monju data" entrusted to University of Fukui by the Ministry of Education, Culture, Sports, Science and Technology of Japan (MEXT).

Open Access This chapter is distributed under the terms of the Creative Commons Attribution Noncommercial License, which permits any noncommercial use, distribution, and reproduction in any medium, provided the original author(s) and source are credited.

References

1. Nakai R (2009) Design and assessment approach on advanced SFR safety with emphasis on the core disruptive accident issue. In: Proceedings of the international conference on fast reactor and related fuel cycles (FR09), Kyoto, 7–11 December, 2009
2. Aoto K, Uto N, Sakamoto Y, Ito T, Toda M, Kotake S (2011) Design study and R & D progress on Japan sodium-cooled fast reactor. J Nucl Sci Technol 48:463–471
3. Beils S, Carluec B, Devictor N, Fiorini GL, Sauvage JF (2011) Safety approach and R & D program for future French sodium-cooled fast reactors. J Nucl Sci Technol 48:510–515
4. Carluec B, Lo Point P, Mariteau P, Capelle S (2012) Severe accident countermeasure of SFR. In: Proceedings of the JAEA-IAEA international workshop on prevention and mitigation of severe accidents in sodium-cooled fast reactors, Tsuruga, 11–13 June, 2012
5. Grouiller JP et al. Transmutation in ASTRID. In: Proceedings of international conference on fast reactors and related fuel cycles: safe technologies and sustainable scenarios. FR13, Paris, March 4–7, paper IAEA-CN-199-140
6. Kawashima K et al (1991) Study of the advanced design for axially heterogeneous LMFBR cores. In: Proceedings of international conference on fast reactors and related fuel cycles, FR91, vol 1, Kyoto, October 28–November 1
7. Kawashima K et al (1992) Conceptual core design to enhance safety characteristics in MOX fueled large LMFBRs (I). Neutronics and transient safety performance characteristics. In:

Proceedings of the international conference on design and safety of advanced nuclear power plants, ANP-92, vol 2, Tokyo, October 25–29
8. Takeda T, Narabayashi H, Hirokawa N (1998) Interpretation of transmutation rates of minor actinides in thermal and fast reactors. Ann Nucl Energy 25(9):653–665
9. Takeda T, Umano T (1985) Burnup sensitivity analysis in a fast breeder reactor. Part I: Sensitivity calculation method with generalized perturbation theory. Nucl Sci Eng 91:1–10
10. Takeda T, Asano K, Kitada T (2006) Sensitivity analysis based on transport theory. J Nucl Sci Technol 43(7):743–749
11. Nakano M, Takeda T, Takano H (1987) Sensitivity analysis of cell neutronics parameters in high-conversion light-water reactors. J Nucl Sci Technol 24(8):610–620
12. Foad B, Takeda T (2014) Importance of self-shielding for improving sensitivity coefficients in light water reactors. Ann Nucl Energy 63:417–426
13. Takeda T, Yoshimura A, Kamei T, Shirakata K (1989) Prediction uncertainty evaluation method of core performance parameters in large liquid-metal fast breeder reactors. Nucl Sci Eng 103:157–165
14. Takeda T, Kamei T (1991) Uncertainty evaluation of burnup properties of large fast reactors using data adjustment method. Nucl Sci Eng 28(4):275–284

Chapter 18
Overview of European Experience with Thorium Fuels

Didier Haas, M. Hugon, and M. Verwerft

Abstract Since the early 1970s, studies and experimental projects have been undertaken in Europe to examine the potential of thorium-based fuels in a variety of reactor types. The first trials were mainly devoted to the use of thorium in high-temperature reactors. These projects can be seen as scientific successes but were not pursued on a commercial basis because of the priority given in Europe to the development of light water reactors. Later on, thorium oxide was considered as a potential matrix for burning plutonium (possibly also minor actinides), and several core design studies, as well as experiments, were undertaken. The most recent such concern the BR2 and HFR Material Test Reactor (MTR) irradiations in Belgium and in the Netherlands, respectively, as well as the KWO PWR in Obrigheim in Germany, in which thorium-plutonium oxide fuel (Th-MOX) was successfully irradiated up to 38 GWd/tHM. The results of these experiments have shown that Th-MOX behaves in a comparable way as conventional uranium-plutonium oxide fuel (U-MOX). More work is still needed before Th-MOX will reach sufficient maturity to implement it on a large scale in power reactors, but all currently available results indicate that licensing Th-MOX for LWRs should be feasible. Finally, European research projects are still devoted to the study of thorium salts in molten salt reactors, a design that incorporates on-line reprocessing and needs no specific thorium fabrication, adding therefore the benefits of thorium without its main challenges.

Keywords Fuel • HTR • LWR • MSR • Plutonium • Thorium

International Symposium on Nuclear Back-end Issues and the Role of Nuclear Transmutation Technology after the accident of TEPCO's Fukushima Daiichi Nuclear Power Stations

D. Haas (✉)
Consultant, 101 rue de la Station, 1457 Walhain, Belgium
e-mail: Didier.haas@hotmail.be

M. Hugon
European Commission, Brussels, 1049 Brussels, Belgium

M. Verwerft
SCK•CEN, Boeretang 200 B-2400 Mol, Belgium

18.1 Introduction

Natural thorium (Th) has only one isotope, ^{232}Th, which is fertile. In a thermal reactor, Th can absorb neutrons and, following nuclear reactions, produces ^{233}U, which is fissile. Under optimized breeding conditions, a sustainable Th-^{233}U cycle can be reached, but the thorium cycle needs a seed or driver fuel, which can be based on ^{235}U or on Pu.

^{233}U as a fissile nuclide features high neutron production in a thermal and epithermal neutron spectrum. This ability offers improved neutron economy for reactors fueled with ^{233}U rather than ^{235}U or ^{239}Pu, particularly at thermal energies in light water reactors (LWRs). In theory, breeding (formation of fissile nuclides) is achievable at thermal energies with a Th/233U fuel, which is not the case with U-MOX fuel. However, even though breeding can be demonstrated at an experimental level, optimal breeding is not achieved in the current fleet of LWRs. In today's context, U-MOX fuels are not reprocessed, and here Th-MOX offers perhaps its best advantage over U-MOX. The excellent chemical stability of the thorium oxide matrix makes it an excellent candidate for direct disposal, and thus also for once-through fuels allowing burning excess Pu without production of higher actinides. Another alternative would be to use Th-MOX fuels in LWRs as a means to initiate the breeding of ^{233}U for future use in other reactor types, as an option to save natural U and to further improve the U-Pu fuel cycle.

In addition to the LWR/FR scenario, two reactor types have been considered for a breeding Th fuel cycle in the future: high-temperature reactors (HTRs) and molten salt reactors (MSRs). HTRs represent the fastest route to implement a closed breeding Th fuel cycle. The technology exists conceptually but needs to be developed before commercialization (which is pending). Also, supporting technologies associated with fuel manufacturing, reprocessing, transport, waste management, and final disposal need to be developed. MSRs represent a longer-term development option for Th fuel cycles. In MSRs loaded with Th-based fuels, breeding may be achieved over a wide range of neutron energies. On-line reprocessing is an important feature of MSRs, which enables continuous re-use of the nuclear fuel by extracting the fission products.

The potential development of a closed Th fuel cycle faces some obstacles. Reprocessing is one of these, as Th oxide is more stable than U oxide. In contrast to the Purex process, which has been industrially operational in the U-Pu fuel cycle for more than 30 years, the Thorex process, which has been investigated for many years in laboratories, faces some difficulties: it requires stronger acids (and therefore more advanced corrosion-free materials for process vessels) and longer dissolution times. Remote-controlled fuel manufacturing represents another challenge as Th-based fuels have high-energy gamma radiation from the presence of ^{232}U after irradiation, which requires remote fabrication and handling in heavily shielded facilities. Thus, this fuel fabrication, transport, and reprocessing are more complex than the present practice for U oxide fuel, for instance.

This presentation summarizes the history and status of the main European research programs (cordis.europa.eu) with Th use. These programs concerned HTRs, LWRs, and MSRs. Emphasis is given here on the latest two developments.

18.2 Thorium European Research Programme History

During the early years of nuclear energy R&D in Europe, between 1960 and 1980, the main experimental projects involving Th fuels were related to the HTRs (DRAGON OECD international project in the UK, ATR and THTR reactors in Germany) and also to an irradiation of Th-MOX fuel in the Lingen BWR in Germany. These projects can be seen as scientific successes, but they were not pursued on a commercial basis because of the priority given in Europe to the development of LWRs (except in the UK, where low-temperature gas-cooled reactors were developed), with UO_2 as reference fuel, and, for countries having selected the reprocessing cycle strategy, the recycling of the recovered Pu as MOX fuel.

Afterward, several studies were undertaken to examine worldwide interest in Th. In 1997, M. Lung wrote a report entitled "A present review of the thorium fuel cycle" [1] at the request of the European Commission. Then, in the 4th EURATOM Framework Programme, a review of the benefits of the Th cycle as a waste management option was carried out [2].

As a result of these studies, it was recognized that this option presented major advantages in term of actinides management through the "burning" of excess Pu in a non-U matrix (Th oxide), at least for those countries in Europe that considered Pu as a waste and not a source of energy for future utilization in fast reactors. These assessments opened the door to several European irradiation experiments during the 5th EURATOM Framework Programme using Th-MOX, namely in the KWO PWR in Obrigheim (Germany), in the HFR MTR in the Netherland (operated by NRG), and in the BR2 MTR in Mol (SCK•CEN) ("THORIUM CYCLE [3]" and "OMICO [4]" projects). These efforts were pursued and completed within the 6th EURATOM Framework Programme, with the demonstration at laboratory scale that this fuel would behave in a comparable way as current MOX fuel (see Sect. 18.3). In the 6th EURATOM Framework Programme, the fuels irradiated in the programs THORIUM CYCLE and OMICO were further investigated (postirradiation examination, radiochemical analysis, and leaching tests) in the "LWR-DEPUTY" project [5] and a strategy study on the "Impact of Partitioning, Transmutation and Waste Reduction Technologies on the Final Nuclear Waste Disposal" ("RED-IMPACT") was performed [6].

In parallel, efforts at the European level started in early 2000 and are still under way concerning the development of the MSR, using a Th-^{233}U cycle in liquid Th fluoride fuel. Between the 5th and the 7th EURATOM Framework Programmes, several projects (MOST, ALISIA, EVOL) were funded (see Sect. 18.4).

Within the European nuclear research community, a Technology Platform named SNETP (Sustainable Nuclear Energy Technology Platform: www.snetp. eu) gathers most of the stakeholders involved in reactor research. SNETP issued a "Strategic Research Agenda" in May 2009 (revised in 2013 following the Fukushima accident) with an Annex (in January 2011) devoted to Th. In the annex, Th systems are noted as having significant long-term potentialities but also significant challenges before reaching industrial implementation. The two aspects (Pu management, molten salts) mentioned in this chapter were specifically recognized in the Th Annex to the Strategic Research Agenda.

18.3 Th-MOX Fuels Irradiated in LWR Conditions

Within the European Framework Programmes, the study of Th fuels behavior in LWRs was first aimed at comparing the behavior and the applicability of various matrices to be used for the transmutation of Pu and minor actinides (projects THORIUM CYCLE, LWR-DEPUTY, OMICO). Comparisons were made with standard fuels (UO_2, MOX), and also with so-called inert matrices fuels (using, for example, Mo or MgO as matrix in CERMET and CERCER fuel types, respectively). As explained earlier, irradiation experiments were performed in three facilities, namely, the KWO PWR, HFR, and BR2 Material Test Reactors.

The THORIUM CYCLE project was a 4-year project with the following participants: the coordinator NRG (NL), BNFL (UK), CEA (F), FZK and KWO (D), and JRC-IE and JRC-ITU (EU). The goals of this project, which started on 1 October 2000, were to supply key data for application of the Th cycle in LWRs. In particular, it included the study of

- The behavior of Th-based fuel at extended burn-up through an irradiation experiment of four short fuel pins [UO_2, $(U,Pu)O_2$, ThO_2, and $(Th,Pu)O_2$] up to 55 GWd/tHM in HFR, and an irradiation experiment of one short fuel pin [$(Th,Pu)O_2$] to 38 GWd/tHM in a PWR (KWO); it should be noted that a previous irradiation of $(Th,Pu)O_2$ in Germany (Lingen) achieved a burn-up of 20 GWd/tHM [7];
- The core calculations for Th-based fuel, including code-to-code validation, sensitivity check for significant isotopes ^{232}Th and ^{233}U, and the calculation up to 80–100 GWd/tHM for Th-MOX fuel.

The irradiation test in KWO enabled the investigation of the operational safety of Th-MOX rod behavior under realistic pressurized water reactor (PWR) conditions. The short test rod was inserted in a MOX assembly to provide the most realistic boundary conditions possible. The foreseen MOX carrier assembly had already been irradiated for one cycle. The cladding appeared in good condition after irradiation, and its creep-down, measured at the reactor site during the shut-down periods, as well as its general behavior, were well within the bounds of experience for UO_2 fuels. The fission gas (Xe and Kr) release was about 0.5 % [8], which is about half that for equivalent MOX fuels at the same burn-up, but the linear power was lower than in equivalent U-MOX studies. Taking into account experimental uncertainties, the fuel behavior seems to be at least as good as U-MOX.

The THORIUM CYCLE project was completed in 2006, but the postirradiation experiments were performed under a subsequent experiment called LWR-DEPUTY (coordinator, SCK.CEN). In this program, the main tests on Th-MOX consisted of additional fuels studies (microscopy, radial distributions of elements and isotopes) and radiochemical analyses. The objective of these analyses was to obtain a reliable experimental database for burn-up analysis and to evaluate changes in the heavy nuclide content:

- To optimize the dissolution and analysis strategies
- To establish the first dataset on heavy nuclide and fission product content in irradiated Th-MOX to assess the overall uncertainties
- To use this dataset in a benchmark analysis program

The OMICO Project [4] was conducted from 2001 to 2007. Its scope included the study and modeling of the influence of microstructure and matrix composition on Th-MOX fuel in-pile behavior in normal PWR conditions. The following tasks were undertaken:

- Fabrication of the Th-MOX fuels at the JRC-ITU
- Irradiation in the "CALLISTO" PWR loop in BR2, representing real PWR conditions; the burn-up achieved at the end of this project was about 13 GWd/tHM
- Nondestructive examinations (gamma-spectrometry, visual examinations) and microstructure studies

It should be noted that the pins were instrumented for pressure and fuel temperature determination. The test matrix was such that the Th-MOX could be compared with U-MOX and UO_2 fuels. Another test parameter consisted of the fabrication process (homogeneous versus heterogeneous powder mixtures). The results of the temperature/pressure readings were primarily used to benchmark computer code models for Th-MOX fuels behavior in the first stage of their life.

Besides the irradiation, fuel characterization was performed, including thermal diffusivity measurements, and the results were published [4, 9]. The results show a similar thermal conductivity for (nonirradiated) Th-MOX as compared to U-MOX.

In the LWR-DEPUTY [5] project, selected samples of the OMICO and THORIUM CYCLE programs were extensively studied to provide experimental datasets suitable for evaluating their in-pile performance. The experimental data were the basis of a benchmark exercise on the Th-MOX fuel pin irradiated at the NPP KWO to investigate the qualification of the numerical tools and software packages. A scoping study of the leaching behavior was also conducted. In addition to the experimental work, steady-state and transient analyses were performed for different PWR designs fueled completely or partially with Th-MOX fuel. An assessment of steady-state parameters (reactivity, shutdown margin, and reactivity feedback coefficients) has been performed in comparison with UO_2. All feedback coefficients are favorable for a safe operation under steady-state conditions. A comparative analysis of control rod ejection scenarios has also been performed, and it was found that the maximum values obtained for fuel and clad temperature and maximum fuel enthalpy are in line with the acceptance criteria for the current generation PWRs.

After 10 years of research sponsored through the EURATOM programs, the following conclusions can be drawn regarding the behavior of Th-MOX fuel in LWR conditions:

- Th-MOX has great potential and its fabrication as an oxide fuel is feasible
- Even at a laboratory-scale production route, Th-MOX shows a good in-pile performance
- Know-how on Th-MOX has increased, but
- Fuel performance obviously needs to be further improved before code calculations can predict specific Th-MOX behavior

As a general conclusion, the results of these experiments have shown that Th-MOX behaves in a comparable way (even better in some aspects) to MOX, and that licensing Th-MOX in a LWR should not be problematic, although more experimental data on fuels representative of the future commercial fuels would be needed. Experimental data also demonstrate that Th fuels will be more resistant to corrosion than U fuels in the case of spent fuel geological disposal.

18.4 The Molten Salt Reactor

The MSR, which incorporates the reprocessing on line and needs no specific Th fabrication, adds the benefits of Th without its main challenges. In particular, breeding may be achieved over a wide range of neutron energies, which is not the case for the U-Pu cycle.

Under the European Framework Programs, conceptual developments on fast neutron spectrum molten salt reactors (MSFRs) using fluoride salts open promising possibilities to exploit the ^{232}Th-^{233}U cycle. In addition, they can also contribute to significantly diminishing the radiotoxic inventory from present reactor spent fuels, in particular by lowering the masses of transuranic elements. Finally, if required because of expansion of nuclear electricity generation breeding beyond the iso-generation could be achieved. With the Th-U cycle, doubling times values are only slightly higher than those predicted for solid-fuel fast reactors working in the U/Pu cycle (in the range 40–60 years). The characteristics of different launching modes of the MSFR with a thorium fuel cycle have been studied, in terms of the safety, proliferation, breeding, and deployment capacities of these reactor configurations [10].

Between Framework Programmes 5 and 7, several projects ("MOST", "ALISIA", "EVOL") were conducted, and promising developments and results were obtained in particular in the following areas:

- Conceptual design studies
- Safety developments, in particular, to study the residual heat extraction; tests with liquid salts have been undertaken to prove the ability of the cold plug system to act as a security valve on the loop circuit
- Fabrication of the salt mixture (LiF-NaF-KF) to be used in the French molten salt loop (FFFER project) has been achieved
- Experimental investigation of physicochemical properties of fluoride salts
- Experimental tests of the metallic-phase extraction process;
- Corrosion studies and experiments (this remains one of the main challenges for the development of the reactor system)

Finally, it should be noted that the MSR with its Th cycle is one of the six reference systems selected for R&D collaboration in the framework of the Generation IV International forum. The main contributors are the European partners, supported by Russia as observer.

18.5 Conclusions

Since the early 1970s, studies and experimental projects have been undertaken in Europe to examine the potential of Th-based fuels in a variety of reactor types. These projects have all been successful from a scientific point of view, but not all were followed up relative to the overall development of nuclear industry in Europe. High-temperature reactors (HTRs), although very well suited for Th use, have not been deployed to the benefit of LWRs. Results on the use of Th matrices in Th-MOX fuels in LWRs are encouraging, but still need demonstration at a larger scale in commercial conditions. Finally, the probably most efficient use of Th would be in a salt, to feed MSRs. Conceptual studies and related experimental programs are under way.

Open Access This chapter is distributed under the terms of the Creative Commons Attribution Noncommercial License, which permits any noncommercial use, distribution, and reproduction in any medium, provided the original author(s) and source are credited.

References

1. Lung M (1997) A present review of the thorium fuel cycle. EURATOM Report EUR 17771
2. Gruppelaar H, Shapira JP (2000) Thorium as a waste management option. In: 4th EURATOM framework programme, EURATOM, EUR 19142 EN
3. Thorium cycle: development steps for PWR and ADS application. 5th EURATOM Framework Programme Project, FIKI-CT-2000-00042
4. Verwerft M et al (2007) Oxide fuels: microstructure and composition variations (OMICO). Final report EUR 23104, SCK•CEN, Mol

5. Verwerft M et al (2011) LWR-DEPUTY: light water reactor fuels for deep burning of Pu in thermal systems. Final Report (FI6W-036421). Report Nr: ER-200, SCK•CEN, Mol
6. Grenèche D et al. RED-IMPACT (Impact of Partitioning, Transmutation and Waste Reduction Technologies on the Final Nuclear Waste Disposal). In: 6th EURATOM Framework Programme Project, FI6W-CT-2004-002408, Jülich Forschungszentrum
7. Welhum P (1981) Isotopic analysis on PuO_2-ThO_2 fuel irradiated in Lingen BWR power plant. European Applied Research Reports, vol 2, no. 6
8. Somers et al (2013) Safety assessment of plutonium mixed oxide fuel irradiated up to 37.7 GWd/tonne. J Nucl Mater 437(1-3):303–309C
9. Cozzo C et al (2011) Thermal diffusivity and conductivity of thorium–plutonium mixed oxides. J Nucl Mater 416(1-2):135–141
10. Merle-Lucotte E et al (2011) Launching the thorium fuel cycle with the molten salt fast reactor. In: International congress on advances in nuclear power plants (ICAPP), Nice

Part VI
Reactor Physics Studies for Post-Fukushima Accident Nuclear Energy: Studies from the Reactor Physics Aspect for Back-End Issues Such as Treatment of Debris from the Fukushima Accident

Chapter 19
Transmutation Scenarios after Closing Nuclear Power Plants

Kenji Nishihara, Kazufumi Tsujimoto, and Hiroyuki Oigawa

Abstract With consideration of the phase-out option from nuclear power (NP) utilization in Japan, an accelerator-driven system (ADS) for Pu transmutation has been designed and scenario analysis performed. The ADS is designed based on the existing ADS design for MA transmutation, and the six-batch ADS was selected as a reference design for scenario analysis. In the scenario analysis, the once-through scenario of light water reactor (LWR) spent fuel is referred to as a conventional scenario with a LWR-MOX utilization scenario. As the transmutation scenario, three cases of transmuters that are only-FR, only-ADS, and both-FR +ADS are analyzed. The numbers of necessary transmuters are obtained as 15 to 32 units, and the necessary period for transmutation as 180–240 years. The benefit on repository by reduction of Pu and MA is reduction of repository area by a factor of five and of decay time of toxicity by one order of magnitude. The FR+ADS scenario would be a modest solution, although the ADS scenario is preferable if rapid transmutation is required.

Keywords ADS • Phase-out scenario • Scenario study • Transmutation

19.1 Introduction

After the Fukushima-Daiichi accident, Japan started a discussion of nuclear power (NP) utilization including a "phase-out" option in addition to the usual scenario of utilizing plutonium by deploying fast breeder reactors (FBRs). In the phase-out option, construction of new plants is limited and dependency on NP will be gradually reduced. One of the reasons supporting the phase-out scenario is an ambiguous prospect of conducting underground disposal of radioactive wastes. Increase of wastes can be limited or even stopped in the phase-out scenario, but spent fuels (SFs) containing plutonium (Pu) and minor actinides (MAs) will remain as a legacy of NP. "Direct disposal" to the underground of SFs confined in canisters is considered as a strong option to treat this legacy, but Pu and MAs that exist in the

---

K. Nishihara (✉) • K. Tsujimoto • H. Oigawa
Japan Atomic Energy Agency, 2-4 Shirane, Shirakata, Tokai-mura, Naka-gun, Ibaraki 319-1195, Japan
e-mail: nishihara.kenji@jaea.go.jp

underground can be utilized for nuclear weapons and can cause public dose in the very far future over several tens of thousands of years. Instead of direct disposal, transmutation of Pu and MAs (TRU, trans-uranic) has been studied in many countries for the purpose of eliminating them from the waste.

Transmutation can be performed by a "transmuter" that is dedicated for transmutation with the lesser role of electricity generation. It contains a fast reactor (FR) and an accelerator-driven system (ADS), which are fast neutron systems with metal coolant. FRs have been mainly developed as breeder reactors but they act as a burner reactor in the phase-out scenario. The burner reactor has no blanket region for breeding, larger Pu content, and shorter operation-cycle length [1]. The ADS has been designed as an MA transmuter with a smaller amount of Pu but changes to a Pu transmuter in this scenario.

In the present study, an ADS for Pu transmutation (Pu-ADS) is designed by neutronics calculation based on the ADS for MA transmutation (MA-ADS). In the original design for MA transmutation, drop of criticality during depletion is very small, and a long operation cycle is achieved because MAs behave as fertile material. In the Pu-ADS, criticality decreases much more rapidly and design modification is necessary.

After the design of the Pu-ADS, a scenario study is performed by a nuclear material balance (NMB) code that was developed by the authors. The following items are revealed by the study: accumulation of TRU in the LWR SFs, necessary number of transmuters, reduction of TRU, reduction of repository footprint, and radiotoxicity by transmuters.

In Sect. 19.2, calculation methods for neutronics design and scenario code are introduced. Section 19.3 provides the neutronics design and resulting ADS. Section 19.4 discusses assumptions and results of the scenario study. The results are concluded in Sect. 19.5.

19.2 Methodology

19.2.1 Neutronics Calculation

Several codes were combined for ADS design (Fig. 19.1) containing proton transport, neutron transport, cross-section preparation, and depletion. The PHITS code [2] was used for transportation of protons and neutrons above the energy boundary of 20 MeV. Transportation of neutrons slowing down less than 20 MeV is interrupted, and position, direction, and energy are stored in a cutoff file. This file is processed as to be readable by the PARTISN code [3], which is a neutron transport code with multi-group theory. A 73-group cross section is prepared by SLAROM [4] code with the JENDL4.0 [5] nuclear data library. A 1-group micro-cross section is calculated by multiplying the 73-group cross section to the 73-group flux from PARTISN. One-group micro-cross section and total flux is used in the ORIGEN2

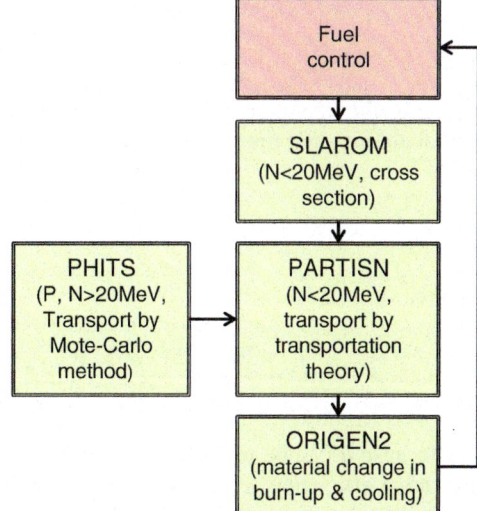

Fig. 19.1 Calculation codes

code [6] to obtain material change after depletion. The material composition from ORIGEN2 is processed by a fuel control program that simulates reprocessing and fuel fabrication with adjustment of MA content ratio.

19.2.2 Scenario Analysis

The NMB code [7] was employed for the scenario analysis. The code calculates material balance of 26 actinides (through Th to Cm, $T_{1/2} >$ several days) in spent fuels with an accuracy comparable to the ORIGEN2 code. LWR, CANDU, gas-cooled reactor, several sodium-cooled FRs, and lead-bismuth-cooled ADS are available. Each reactor can be coupled with appropriate fuel such as UO_2, MOX, ROX, Pu-nitride (PuN), and MA-nitride (MAN). Fission products are estimated by dividing them into several groups (iodine, rare gas, technetium and platinum group metals, strontium, cesium, and others). The number of waste packages and repository size are determined by temperature analysis based on several repository layouts. Potential radiotoxicity that is defined as dose by direct ingestion can be also estimated.

19.2.3 Transmutation Half-Life

In this section we define the effective transmutation rate and transmutation half-life that represent performance of a transmuter in the case of a phase-out scenario. A transmutation amount after an in-core period of T_{in} years is

$$w_{tr} = \frac{P\, T_{in} \cdot 3600 \cdot 24 \cdot 365 \cdot \varepsilon_o \cdot \overline{A}}{E_{fiss}\, N_A}. \tag{19.1}$$

Here, the effective transmutation rate, λ_{tr}, is transmuted amount divided by initial amount and time needed for transmutation including out-core period.

$$\lambda_{tr} = \frac{w_{tr}}{w_i}\frac{1}{T_i + T_o} = a\varepsilon_o\varepsilon_c h, \tag{19.2}$$

where,

$$a = \frac{3600 \cdot 24 \cdot 365 \cdot \overline{A}}{E_{fiss}\, N_A} \cong 3.8 \cdot 10^{-4}.\,(\mathrm{t/MW/year}) \tag{19.3}$$

Because a can be regarded as constant for Pu-transmuters, λ_{tr} is determined by operation efficiency, ε_o, cycle efficiency, ε_c, and specific heat, h. A time evolution of amount of heavy metal after introducing transmuters is expressed as

$$\frac{dw}{dt} = -\lambda_{tr}w,\ w = w_0 e^{-\lambda_{tr} t}, T_{tr} = \frac{\ln(2)}{\lambda_{tr}}, \tag{19.4}$$

where T_{tr} is a transmutation half-life. In the phase-out scenario, there is heavy metal of $w_0\,t$ when transmuters are employed in full scale. T_{tr} means a period needed to transmute half of w_0 in the case that the maximum number of transmuters are introduced. Another fact is that λ_{tr} and T_{tr} depend on two parameters relating to operation time efficiency and one fundamental core parameter, h. The thermal output of core affects a number of transmuters, but not transmutation behavior in the mass-flow analysis.

19.3 ADS Design for Pu Transmutation

19.3.1 Reference ADS (MA-ADS)

An ADS core dedicated for MA transmutation (MA-ADS) [8] bases the present design for Pu transmutation. As listed in Table 19.1, the MA-ADS is a medium-size core loaded with nitride fuel cooled by lead-bismuth eutectic (LBE). Nitride fuel is diluted by zirconium-nitride with a weight fraction of around 50 % (volume fraction around 65 %) that can be changed so that criticality becomes an appropriate level ($k_{eff} = 0.97$). The ADS is operated for 600 effective full-power days (EFPDs) without any fuel reloading while the interval for maintenance of an accelerator can occur. In other words, the ADS is a one-batch core that implies large reactivity drop after depletion in the case of Pu transmutation. Because control rods for criticality are not equipped in the ADS, the drop must be supplemented by

Table 19.1 Parameters of minor actinides-accelerator-driven system (MA-ADS) [8]

Thermal power	800 MWt
Electricity generation	260 MWe
Proton energy	1.5 GeV
Transmutation rate	250 kg/300 EFPDs
Coolant	LBE
Upper limitation of k_{eff}	0.97
Operation period	600 EFPDs
Batch number	1
Fuel composition	(Pu+MA)N+ZrN
Pin outer diameter	7.65 mm
Pin pitch	11.48 mm

Table 19.2 MA composition fed to MA-ADS

	MA-ADS (%)
Np-237	49.5
Am-241	32.2
Am-243	13.4
Cm-244	4.4
Cm-245	0.4

Table 19.3 Pu composition fed to Pu-ADS and Pu+U-ADS

ADS case	Pu (%)	Pu+U (%)
U-235	–	0.1
U-238	–	49.9
Pu-238	2.4	1.2
Pu-239	54.5	27.2
Pu-240	24.2	12.1
Pu-241	11.9	6.0
Pu-242	7.0	3.5

increasing proton beam current. Fortunately, MAs are fertile material that becomes fissile after capturing one neutron, and the drop is very small, as shown in the following section. Table 19.3 provides weight composition of MA fed to MA-ADS. Figure 19.2 illustrates the R-Z model for calculation.

19.3.2 Assumption of Pu Feed

Two cases of Pu feed were assumed in the present design for Pu transmutation: Pu-ADS, to which only Pu is provided from reprocessing process of LWR SF, and Pu+U-ADS, to which Pu accompanied by U with 50 % weight ratio is provided from the process. Treatment of pure Pu raises proliferation concern in a country without nuclear weapons, and the reprocessing plant is designed to add depleted

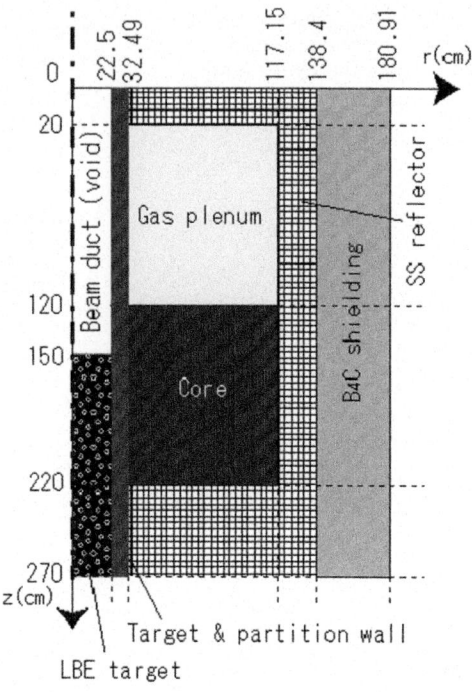

Fig. 19.2 R-Z model of accelerator-driven system (ADS)

Table 19.4 Efficiencies of ADSs

ADS case	MA (Ref.)	Pu	Pu+U
T_i (years)	2	1	2
T_o (years)	3	3	3
ε_o (%)	82.1	82.1	82.1
ε_c (%)	40.0	25.0	40.0

U to Pu just after separation. Addition of ^{238}U results in ^{239}Pu production and generally is undesirable for Pu transmutation. Table 19.3 lists two compositions. Other conditions are similar to the MA-ADS.

19.3.3 Result of One-Batch Core

The reference ADS for MA transmutation is designed with a one-batch core, which means that all the fuel is loaded and unloaded simultaneously. At first, this one-batch design was adopted to Pu transmuters. Table 19.4 lists the in-core and out-core time with operation efficiency. Out-core time of the ADSs is 3 years, which allows decay of ^{244}Cm. In-core time of the reference MA-ADS is 2 years, although that of Pu-ADS is reduced to 1 year because the decrease of criticality is

Table 19.5 ADS inventories and transmutation half-life for one-batch design (equilibrium core)

ADS case		MA (Ref.)	Pu	Pu+U
Volume fraction of inert matrix (%)		69.8	87.1	68.1
Core inventory at BOC (t)	U	0.19	0.02	2.62
	Pu	1.83	1.84	1.88
	MA	2.37	0.25	0.18
Core inventory at EOC (t)	U	0.18	0.02	2.39
	Pu	1.79	1.60	1.63
	MA	1.92	0.25	0.18
Transmutation, BOC-EOC (t)	U	0.00	0.00	0.23
	Pu	0.04	0.24	0.26
	MA	0.45	0.00	0.00
Specific heat, h (MW/tHM)		182	380	171
λ_{tr} (/year)		2.28E-02	2.96E-02	2.14E-02
T_{tr} (year)		30.5	23.4	32.5

too rapid for this ADS. The operation efficiency, ε_o, is 82.1 % assuming 300 days operation annually.

Design and transmutation performance are summarized in Table 19.5. Volume fraction of the inert matrix, ZrN, of the MA-ADS core is 69.8 %, adjusted so that k-effective at the beginning of the cycle (BOC) of the equilibrium core becomes 0.97. The equilibrium core is obtained after calculating ten cycles of burning, cooling, and recycling. Volume fraction of the Pu-ADS is more and that of the Pu+U-ADS is almost the same. The inventory at BOC of the heavy metal in Table 19.5 is proportional to a one-volume fraction of ZrN. An interesting observation is that the amounts of Pu at BOC are equal among three ADSs, which means U and MA contribute very little to the criticality before depletion. However, impacts on the criticality drop after depletion is significant (Fig. 19.3). k_{eff} drop of the MA-ADS is as small as 1.5 %dk, although others lose 14 %dk even at the equilibrium cycle around 6,000 days, which means MA is a better fertile than ^{238}U. The Pu-ADS has a steeper decrease than Pu+U-ADS because of the absence of ^{238}U. The huge drop of the Pu- and Pu+U-ADS is not acceptable in the current design of accelerator and target for the MA-ADS; the acceptable drop is about 3 %dk in the MA-ADS.

The effective transmutation rate and transmutation half-life are listed at the bottom of Table 19.5. The half-life of the Pu-ADS is shortest because its specific heat is twofold larger than others although its cycle efficiency, ε_c, is much smaller than others.

19.3.4 Result of six-Batch Core

In the one-batch design in the previous section, k_{eff} drop of Pu ADSs is 14 %dk, which is too large to be compensated by burnable poison or control rods. As the first step of design improvement, a multi-batch design is introduced. Theoretically, an

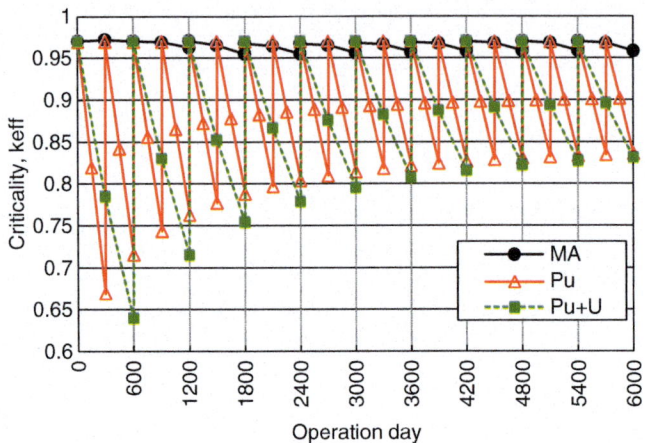

Fig. 19.3 Time evolution of criticality for one-batch design

N-batch core can reduce k_{eff} drop by $1/N$, that is, a drop of 14 %dk can be reduced to 2.3 %dk by six-batch design. Figure 19.4 illustrates the criticality change with an expansion for operation date of 0–1,200 days. The criticality drop for the early operation date is larger than the limit of 3 %dk, that is, 0.94 of k_{eff} at end of burn-up; the drop decreases in the equilibrium cycle. The maximum drop is 5 %dk for Pu-ADS and 6.5 %dk for Pu+U ADS, which can be compensated by control rods or burnable poison or mitigated by shorter operation in the early cycle in future improvements. The drop in the equilibrium cycle is approximately 2 %dk, which is comparable to that of the reference MA-ADS.

Volume fractions and inventories are listed in Table 19.6. Six-batch cores generally require more inventory than a one-batch core because an averaged k_{eff} during operation of multi-batch cores is higher than that of the one-batch core. Transmutation amounts of Pu- and Pu+U-ADSs are much smaller than that of the MA-ADS because the operation period is, respectively, only 50 and 100 days.

To evaluate transmutation half-life, operation and cycle efficiencies must be determined. The short operation period of 50 or 100 days implies frequent fuel exchange and low operation efficiency. There are two kinds of interval: fuel exchange and plant maintenance. We assumed that fuel exchange of a 1/6 core requires 15, 30, or 60 days for the Pu- and Pu+U-ADS and that plant maintenance including accelerator needs 60 days. Because fuel exchange for 15 days is very short, considering shutdown and startup of the ADS plant is included, tentative storage inside a core vessel should be applied for such a short interval. In the case of Pu-ADS, the 50-day operation and 15-, 30-, or 60-day interval are repeated five times, then 50-day operation and 60-day maintenance are done. In the case of Pu+U-ADS, 100-day operation and 15-, 30-, or 60-day interval are repeated two times, then 100-day operation and 60-day maintenance are done. The total operation period before a long plant maintenance of 60 days in both ADSs is 300 days.

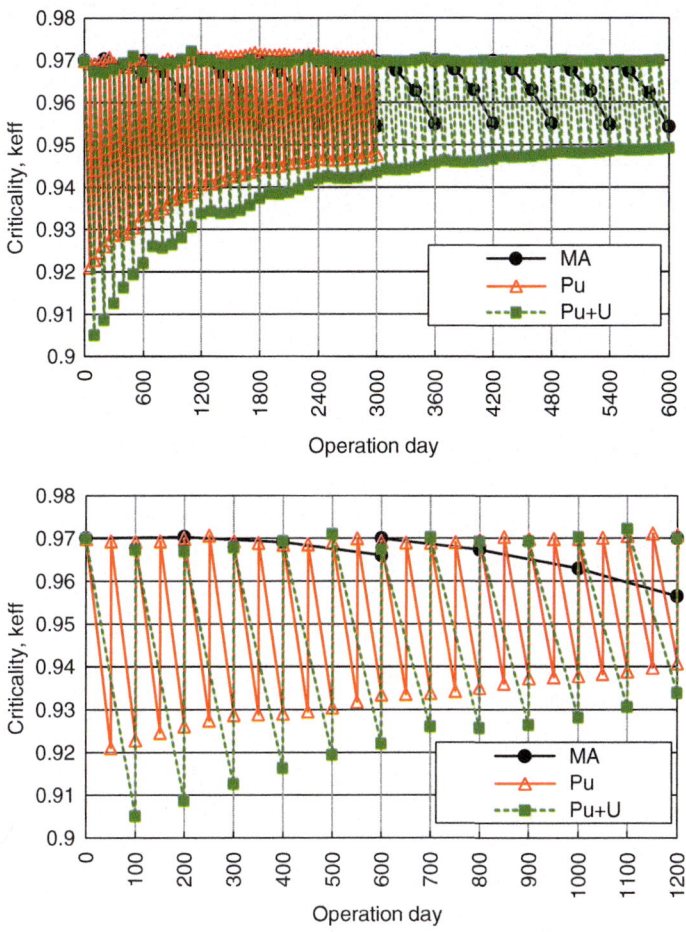

Fig. 19.4 Time evolution of criticality for six-batch design (*upper* = all operation, *lower* = expansion)

Table 19.6 ADS inventories for six-batch design (equilibrium core)

ADS case		MA (Ref., 1-batch)	Pu	Pu+U
Volume fraction of inert matrix (%)		69.8	85.4	62.5
Core inventory at BOC (t)	U	0.19	0.01	3.05
	Pu	1.83	1.84	2.04
	MA	2.37	0.19	0.20
Core inventory at EOC (t)	U	0.18	0.01	3.01
	Pu	1.79	1.80	2.00
	MA	1.92	0.19	0.20
Transmutation, BOC-EOC (t)	U	0.00	0.000	0.037
	Pu	0.04	0.041	0.043
	MA	0.45	0.000	0.000
Specific heat, h (MW/tHM)		182	393	151

Table 19.7 Assumption on maintenance schedule versus transmutation half-life for six-batch design (equilibrium core)

ADS case	MA	Pu			Pu+U		
Interval case	2 × 65 days	15 × 5 +60 × 1	30 × 5 +60 × 1	60 × 5 +60 × 1	15 × 4 +60 × 2	30 × 4 +60 × 2	60 × 4 +60 × 2
Batch	1	6	6	6	6	6	6
Operation (days)	600[a]	50	50	50	100	100	100
Short interval (days)[b]		15	30	60	15	30	60
Long interval (days)[c]	130[a]	60	60	60	60	60	60
ε_o (%)	82	69	59	45	77	71	63
In-core period (years)	2	1.19	1.40	1.81	2.01	2.22	2.63
Out-core period (years)	3	3	3	3	3	3	3
E_c (%)	40	28	32	38	40	43	47
$E_o*\varepsilon_c$ (%)	33	20	19	17	31	30	29
h (MW/tHM)	182	393	393	393	151	151	151
λtr (/years)	2.28E-02	2.93E-02	2.79E-02	2.55E-02	1.78E-02	1.75E-02	1.68E-02
Ttr (years)	30.5	23.7	24.8	27.2	39.0	39.7	41.3

[a]Two times of operation for 300 days and long interval for 65 days, in real
[b]Maintenance for fuel reloading of 1/6 core; short interval occurs five times for Pu-ADS and two times for Pu+U-ADS between long intervals
[c]Maintenance for accelerator and plant

Based on the foregoing assumptions, operation efficiency and cycle efficiency are determined as listed in Table 19.7, with specific heat and resulting transmutation half-life. Operation efficiency multiplied by cycle efficiency of the Pu-ADS is the poorest, but the transmutation half-life is the shortest because of the high specific heat. In the present study, a 30-day interval for fuel exchange is adopted as a nominal case. The transmutation half-life of the Pu-ADS is 24.8 years in the nominal case, which is applied to scenario analysis.

Another observation is that the impact of the out-core period on cycle efficiency is significant. The out-core period is presumed considering the half-life of ^{242}Cm of 126.8 days. If a shorter out-core period is accomplished by corresponding design of the reprocessing and fabrication plant, cycle efficiency and resulted transmutation half-life can be improved. Table 19.8 shows comparison of a 3-year and 1-year out-core period. An impact on the transmutation half-life of the Pu-ADS is a factor of around 2, and the transmutation half-life becomes as short as 13.5 years. Although 3 years of out-core period is applied as the nominal case, a shorter out-core period should be pursued in future study.

Table 19.8 Impact of out-core period on transmutation half-life

ADS case	MA	Pu	Pu+U
Interval case	2 × 65 days	30 × 5 + 60 × 1	30 × 4 + 60 × 2
In-core period (years)	2.0	1.4	30.0
Out-core period (years)	3.0	3.0	3.0
Ttr (years)	30.4	24.8	39.7
Out-core period (years)	1.0	1.0	1.0
Ttr (years)	18.3	13.5	24.5

19.4 Scenario Analysis

Based on the ADS design in Sect. 19.3 and FR design in Wakabayashi et al. [1], the impact of introducing transmutation is evaluated by mass-flow analysis. Table 19.9 lists analyzed scenarios in which two conventional scenarios and three transmutations are included. In the conventional once-through scenario identified as "LWR-OT," Pu and MA exist in spent fuel and are directly disposed of in an underground repository. The second conventional scenario identified as "LWR-PuT" is Pu utilization in a LWR. The spent uranium fuels from an LWR are reprocessed, and, separated Pu is fabricated as MOX and burned in the LWR. Spent MOX fuel is directly disposed.

In the FR scenario, both Pu and MA are mixed or co-extracted and transmuted in FR without any limit of MA content in the fuel. In the ADS scenario, Pu is transmuted in the present design (Pu-ADS) and MA are transmuted in the reference ADS (MA-ADS). In the FR+ADS scenario, MA content in FR is limited to less than 5 % and the remaining MA are transmuted in the ADS.

Characteristics of transmutation systems are listed in Table 19.10. FR has a twice larger thermal output than ADSs, although the specific heat is smaller. Therefore, initial inventory involving fuels in the core and in the fuel cycle of FR is much larger than Pu-ADS even if uranium is excluded; thus, the number of FR that can be introduced is limited. As a result, the transmutation half-life of FR is longer than ADSs by a factor of two.

19.4.1 Result of LWR-OT

Figure 19.5 illustrates a result of the LWR-OT scenario where time evolution of electricity generation, Pu inventory, and MA inventory are shown. The peak of 50 GWe appears in 2010 and decreases because of the Fukushima accident and closure after 40-year operations. All LWRs will be shut down in 2055. The Rokkasho reprocessing plant (RRP) will not be operated, but 7,100 tHM spent fuel has been reprocessed, mainly overseas. The year of reprocessing is not clear

Table 19.9 Scenarios

Scenario		Pu	MA
Conventional	LWR-OT (once-through)	Waste	Waste
	LWR-PuT	LWR	Waste
Transmutation	FR (Pu+MA15 %)	FR	FR
	ADS (TRU)	ADS	ADS
	FR (Pu+MA5 %)+ADS (TRU)	Mainly, FR	Mainly, ADS

Table 19.10 Characteristics of transmutation systems

		FR	Pu-ADS	MA-ADS
Power (thermal/electric)	GWe	1.6/0.6	0.8/0.264	
Pu ratio (in/out)	%	37.5/45	~100	~35
MA/HM ratio	%	<5	–	~65
Batch number		4	6	1
Operation period	Day	183	50	600
Operation efficiency	%	84 %	59 %	82 %
In-core period	Year	2.39	1.40	2.00
Out-core period	Year	3.00	3.00	3.00
Cycle efficiency	%	44 %	32 %	40 %
Burn-up	GWd/tHM	58.56	120	108
Specific heat	MW/tHM	80	400	180
λ_{tr}	/year	1.13E-02	2.84E-02	2.25E-02
T_{tr}	Years	61.3	24.4	30.9
Initial inventory[a]	t/unit	45.1	6.3	11.1

[a]Initial inventory involves fuel in core and in fuel cycle (cooling, reprocessing, and fabrication)

but assumed to be in the 1990s. A small amount of MOX fuel from this reprocessing will be utilized in LWRs.

Pu inventory mainly exists in UO_2-SF. "Pu" in the figure is not "separated" Pu, but Pu in MOX fresh fuel in this scenario. The total of plutonium is 350 t that is gradually disposed of to a repository from 2043 until 2105. The trend of MA inventory is almost the same, but it continues to increase after 2040 because ^{241}Pu becomes ^{241}Am with a half-life of 14.35 years.

19.4.2 Result of LWR-PuT

In MOX scenarios, the Rokkasho reprocessing plant (RRP) will be operated with annual capacity of 800 t and a MOX fabrication plant also (Fig. 19.6). The total amount of UO_2-SF reprocessed is 34,500 tHM, slightly larger than the planned amount of 32,000 tHM. Thus, the present analysis assumes an extension of the RRP by several years. The MOX loading to a usual LWR is limited to 30 %, although the

19 Transmutation Scenarios after Closing Nuclear Power Plants

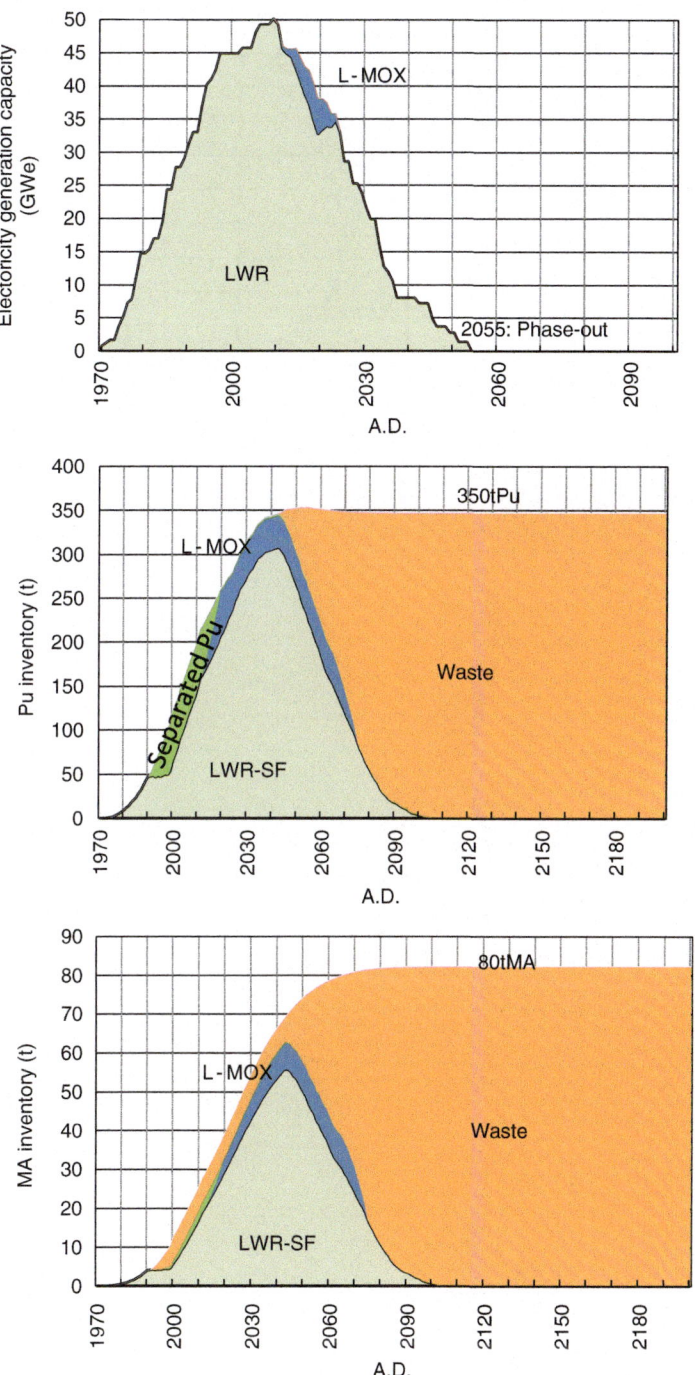

Fig. 19.5 Result of LWR-OT scenario

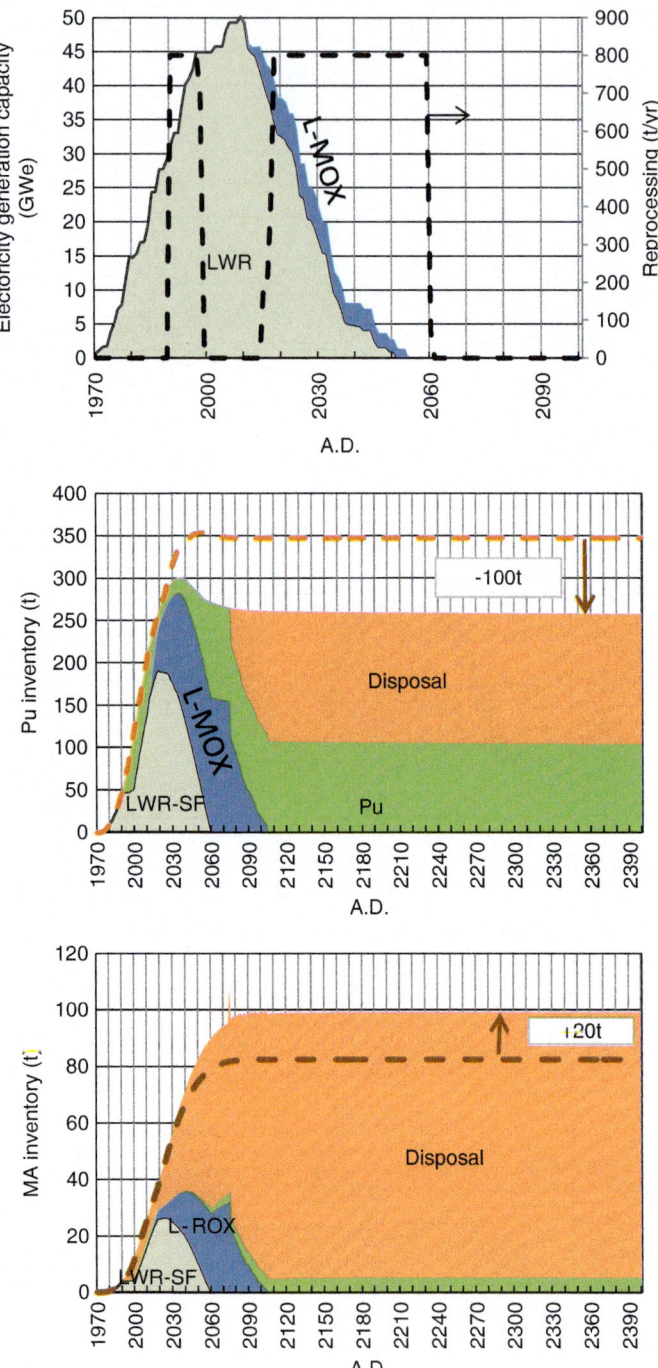

Fig. 19.6 Result of LWR-PuT scenario

Ohma full-MOX reactor starting in 2014 in this analysis can be operated only by MOX fuel. Because the RRP is operated after all LWRs are closed, part of the separated Pu cannot be burned. The total amount of Pu is reduced to 250 t, but that of MA is increased to 100 t.

19.4.3 Result of FR

In the FR scenario as well as other transmutation scenarios, Pu from the RRP is at first fabricated as LWR-MOX fuel and burned in LWR. Pu is co-extracted with same content of U in the current RRP, although MA is vitrified as waste. MA partitioning is assumed to be introduced in 2025 and stored until 2045. In 2045, before introduction of transmuters in 2050, reprocessing of LWR-MOX spent fuel will begin and provide Pu to the transmuters.

FRs are to be introduced in 2050 when 250 t plutonium and 100 t MA remains. MA of 20 t is vitrified by the RRP before 2025 and is not available for transmutation. Available TRU is 330 t. The required TRU to introduce an FR is approximately 25 t, if we assume 41 % of Pu content and 15 % of MA content and employ 45.1 t from Table 19.10. Theoretically, 14 ($=350/25$) FRs can be introduced in 2050, but only 8 can be deployed in practice because the plant life of an FR is assumed to be 60 years and sufficient TRU must be kept until 2110. Available TRU gradually decreases to 200 t in 2110 by transmutation. After 2110, FRs are replaced and reduced to 3 units corresponding to available TRU of 200 t that decreases to 130 t in 2170. Then, 2 FRs from 2170 to 2230 and 2 FRs from 2230 to 2290 will be deployed. After four generations of transmutation, the amounts of Pu and MA are reduced to 40 and 30 t, respectively.

MA content of FR is as high as 15 % (Fig. 19.7), which is above the design limit of 5 % in Wakabayashi et al. [1]. In the usual design of FBRs, MA accumulation is mitigated by a supply of fresh Pu from the blanket. Moreover, high Pu content of FR burner contributes to high MA content. High MA content generally causes deterioration of safety parameters (beta, Doppler coefficient, void reactivity) and difficulty in a reprocessing and fabrication plant.

19.4.4 Result of ADS

In the ADS scenario, transmuter is changed from FR to ADS. ADS can accept both Pu and MA; distribution is shown in Fig. 19.8. In 2050, 22 ADSs are to be introduced, corresponding to 140 t available TRU. Then, 7 and 3 ADSs are operated respectively from 2110 to 2170 and 2170 to 2230. After three generations, Pu and MA are reduced to 10 t and 3 t, respectively, excluding 16 t MA in vitrified waste.

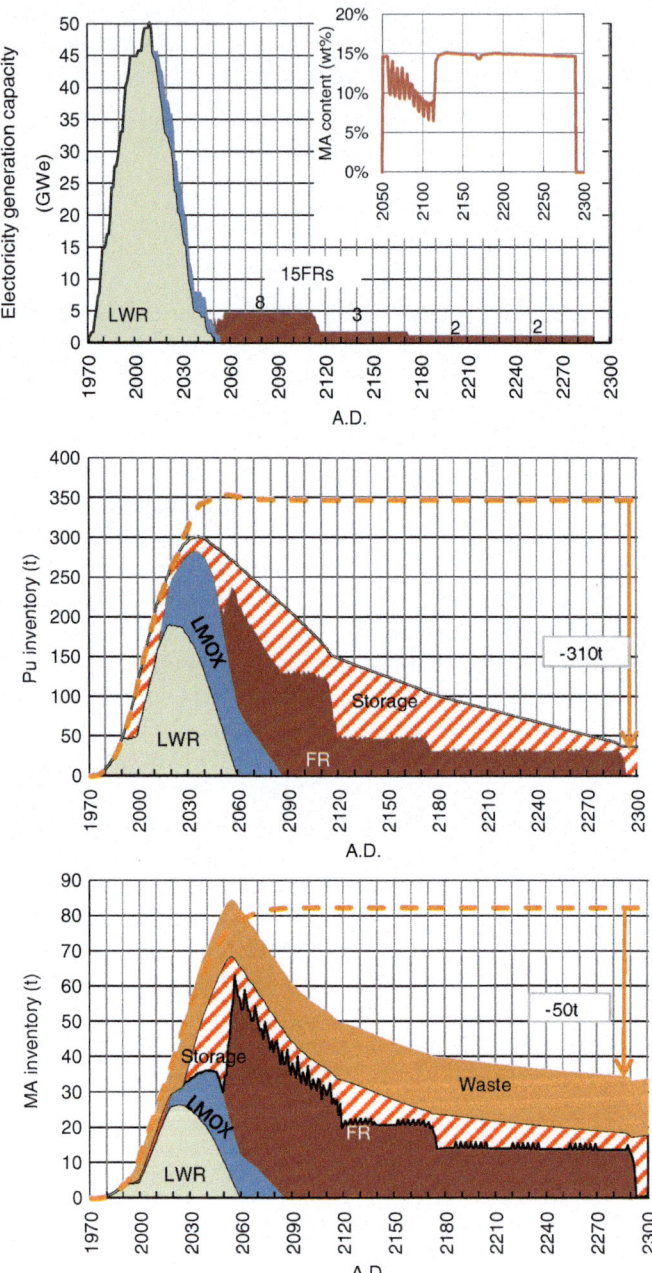

Fig. 19.7 Result of fast reactor (FR) scenario

19 Transmutation Scenarios after Closing Nuclear Power Plants

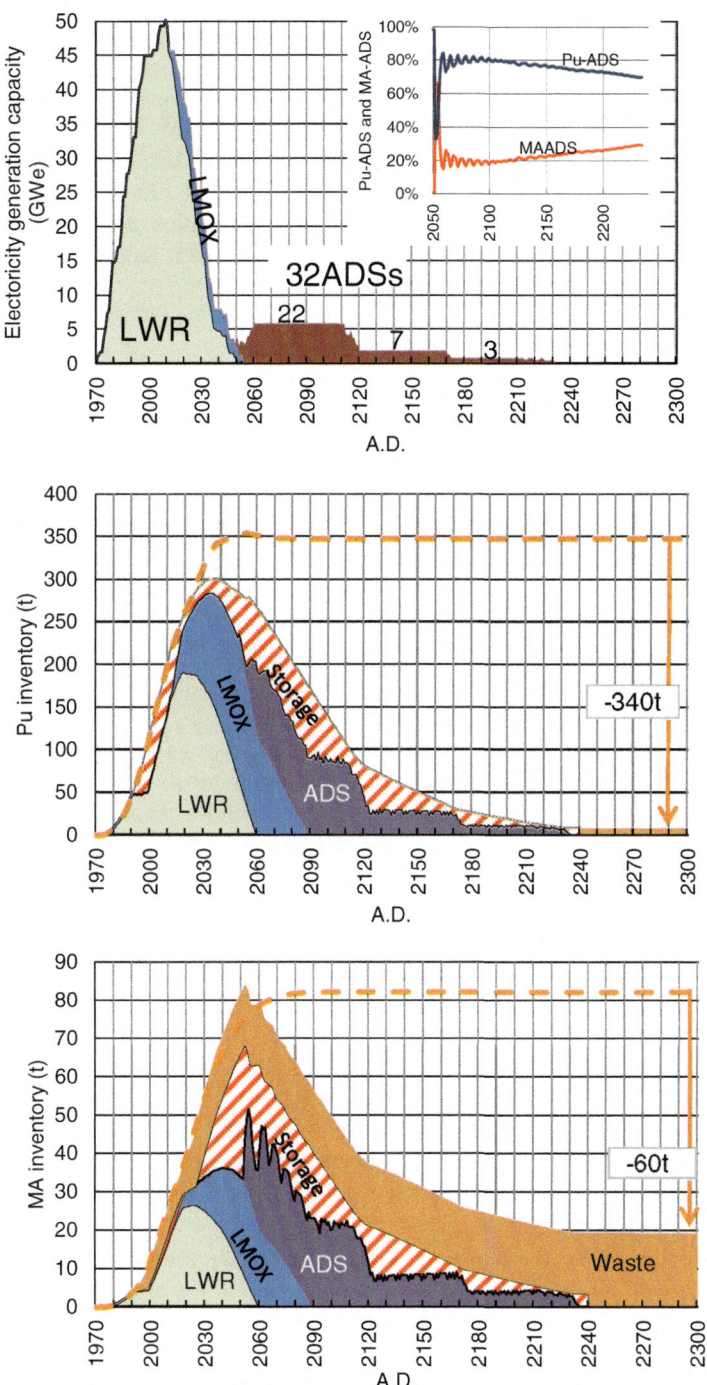

Fig. 19.8 Result of ADS scenario

19.4.5 Result of FR+ADS

In the FR+ADS scenario, MA content in FR is limited below 5 % with respect to design limit and the remaining MA is transmuted in the ADS. In the first generation of transmutation from 2050 to 2110, six FRs and three ADSs are deployed, then three FRs and two ADSs in the second generation, and two FRs and one ADS in the third generation are built (Fig. 19.9). In the fourth generation, only ADS is utilized as to reduce TRU rapidly. The total amount of Pu and MA is reduced to 20 and 10 t, respectively, excepting MA in vitrified waste.

19.4.6 Impact on the Repository

One of the impacts on the repository by transmutation is reduction of potential radiotoxicity, which is defined as total ingestion dose of the waste. Because waste is isolated from the public in the underground in reality, such direct ingestion never occurs and it is considered to be hypothetical, but it can represent the potential danger of waste. This toxicity of waste can be compared to that of uranium ore consumed for electricity generation causing radioactive wastes. Figure 19.10 illustrates those toxicities corresponding to whole operation of LWRs and transmuters. Consumed natural uranium is 370,000 t.

When wastes are generated, the toxicity becomes higher than corresponding uranium ore by three orders of magnitude. Fission products such as Sr and Cs are dominant in the early several hundreds of years, although actinides contribute to toxicity after that. Toxicity in the LWR-OT scenario decays to the level of uranium ore after 100,000 years. By reducing Pu in the LWR-PuT scenario, the decay time becomes shorter, to 70,000 years. In the transmutation scenarios, shortening of decay time depends on the remaining amount of TRU. The decay time is about 10,000 years in the ADS scenario in which the remaining TRU is approximately 30 t, including vitrified wastes. In comparison between the LWR-OT scenario and the ADS scenario, the amount of TRU is reduced by one order of magnitude, so toxicity is also reduced by same order. If MA in the vitrified wastes is retrievable, the amount of TRU will be reduced to around 10 t, which implies toxicity is reduced to 1/30 and the decay time is around 2,000 years. Thus, the impact on toxicity by transmutation is significantly affected by MA in the vitrified wastes. Early introduction of MA partitioning to the RRP and R&D for retrievability from the glass wastes is of importance in this aspect.

Another impact on the repository is reduction of repository size by partitioning and transmutation of heat-generating nuclides in the wastes. Repository size is represented by a repository footprint, which is defined as an area devoted for waste excluding aisles, ducts, utility area, surface facility, and other.

In the LWR-OT scenario, the footprint corresponding to 45,000 t spent fuel reaches almost 4 km^2, which is double the typical repository design for the glass

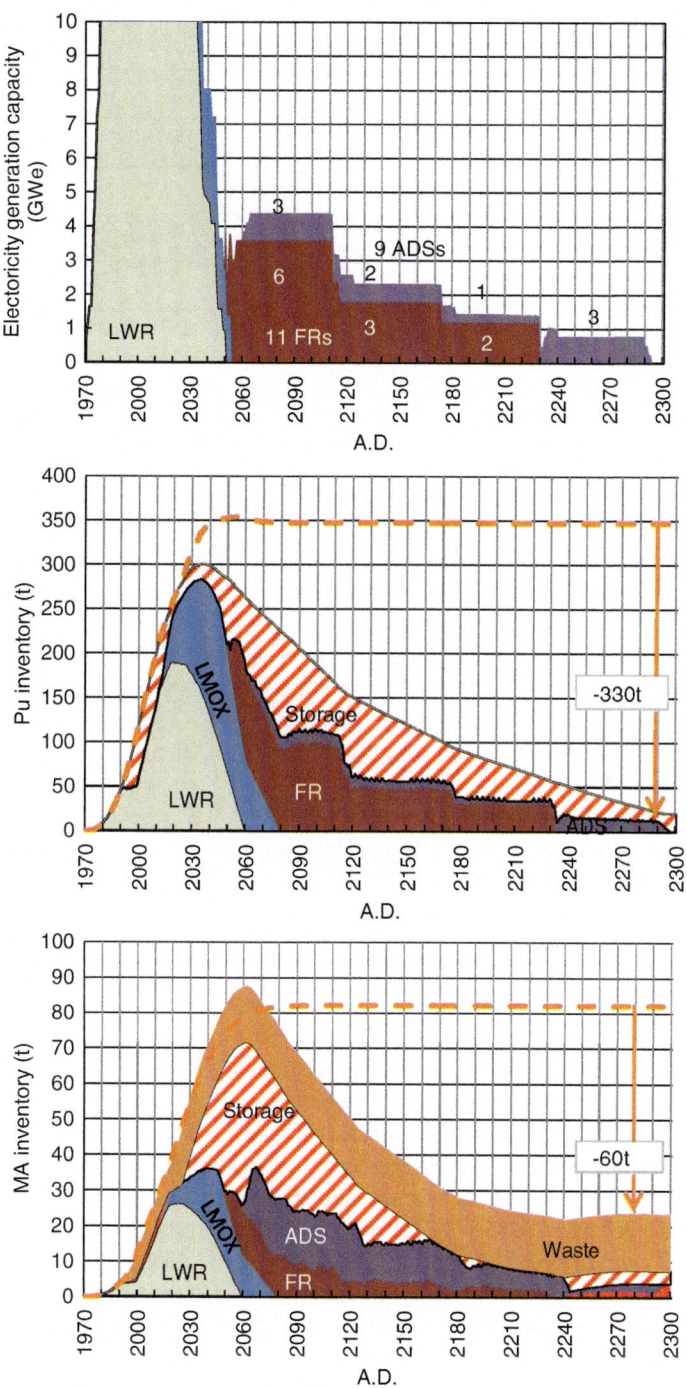

Fig. 19.9 Result of FR+ADS scenario

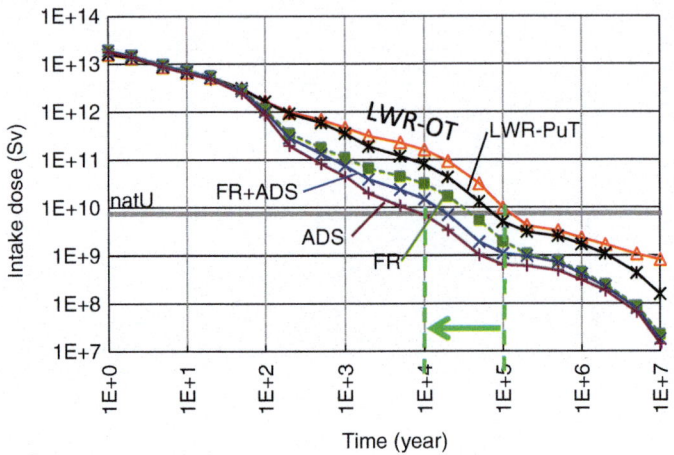

Fig. 19.10 Potential radiotoxicity of all wastes and uranium ore

waste corresponding to 40-year operation of the RRP, because the spent fuel assembly occupies more area and heat generation from the Pu in it also contributes.

In the LWR-PuT scenario, two kinds of waste form are produced: 37,000 glass waste forms containing FP and MA, and spent fuel assembly of MOX of 4,000 t. Each occupies 1.6 km^2, and the total is 3.3 km^2. Although an amount of MOX spent fuel is smaller than that of UO$_2$ spent fuel in the LWR-OT scenario by a factor of 11, it contains more heat-generating actinides such as Am and Pu, and its footprint is significant.

In the early several hundreds of years, ^{90}Sr and ^{137}Cs, whose half-life is around 30 years, are dominant for the footprint. They are separated in the RRP after 2025 as well as MA in the transmutation scenarios. They are absorbed by adsorbents such as zeolite and calcined to the waste form. Because half-life is rather short and the repository footprint is almost proportional to heat generation, long-term storage of the calcined waste is effective [9]. After 300 years of storage, an accumulated layout for the TRU wastes that is low heat generating and with long-term radioactive wastes becomes available. The footprint of this layout is smaller by two orders of magnitude than a typical layout for the vitrified waste. After separating ^{90}Sr and ^{137}Cs, ^{241}Am, whose half-life is 432.2 years, becomes dominant, but this nuclide is transmuted in the transmutation scenarios. Heat generation from other fission products that are vitrified quickly decays to the level of the TRU waste.

As result of the long-term storage and transmutation, the footprint becomes almost constant after 2025 (Fig. 19.11). The glass waste form that is produced before 2025 and contains MA occupies 0.5 km^2. In the ADS scenario, partitioning and long-term storage of Sr and Cs in the wastes produced from reprocessing of ADS spent fuel is not assumed because the impact is small. As a result, the footprint gradually increases to 0.8 km^2. Technologically, separation is possible in the reprocessing for ADS, and it will be applied if the increase becomes significant. Steps observed in 2230 and 2330 are caused by wastes of remaining TRU that will

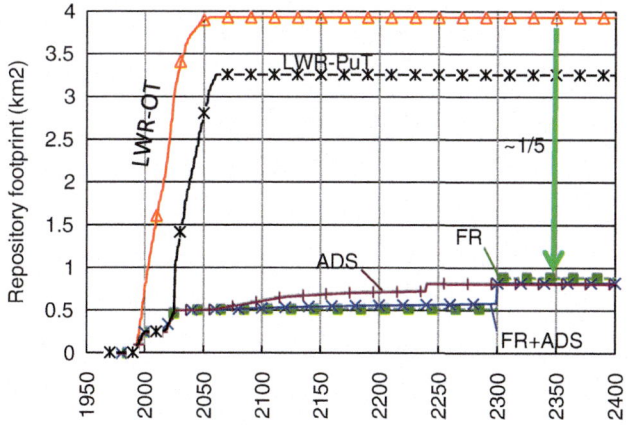

Fig. 19.11 Repository footprint when wastes are produced

be diluted to the glass waste, considering heat generation. The remaining TRU of the FR scenarios are more than that of the ADS scenario.

In the transmutation scenarios, the final footprint is around 0.8 km², which is a fifth of the LWR-OT scenario. As is the case of radiotoxicity, the time of introducing partitioning is significant because more than half of the repository is occupied by glass waste forms with MA.

19.4.7 Discussion

Table 19.11 summarizes the results of scenario analysis. In comparison between LWR-OT and LWR-PuT, reductions are observed in Pu amount, repository footprint, and decay time of toxicity, although they are not drastic. Important benefits of MOX utilization are Pu isotopic deterioration as a nuclear weapon and improved confinement of radionuclides by calcinations, as discussed by Nishihara et al. [7]. However, there remains 110 t of separated Pu that can raise concerns about proliferation.

In comparison between conventional and transmutation scenarios, significant reductions of TRU amount, repository area, and decay time of toxicity are observed. The remaining Pu undergoes several irradiations in the transmuter and is highly resistant to weapon utilization. Repository area is about one fifth and decay time is reduced by one tenth in the maximum case. To achieve such benefit, a total of 15–32 transmuters have to be introduced for 180–240 years with corresponding reprocessing and fabrication facilities. Cost and risk during operation of these facilities would be high compared to their reduction in the repository in the further future.

Comparing the transmutation scenarios, the number of units in the FR scenario is fewest owing to its high thermal output. However, transmutation performance is

Table 19.11 Summary of scenario analysis

Scenario		Transmuter		Period (years)[a]	Remaining Pu	Remaining MA	Repository	
		FR	ADS				Area (km^2)	Toxicity (years)[b]
Conventional	LWR-OT	–	–	–	350	80	3.9	100,000
	LWR-PuT	–	–	–	260	100	3.3	70,000
Transmutation	FR	15	–	240	40	17/16[c]	0.9	40,000
	ADS	–	32	180	8	3/16[c]	0.8	10,000
	FR+ADS	11	9	240	20	8/16[c]	0.8	20,000

[a] Necessary period for transmutation after closing LWRs
[b] Time to decay less than toxicity of corresponding uranium ore
[c] Separated MA/vitrified MA

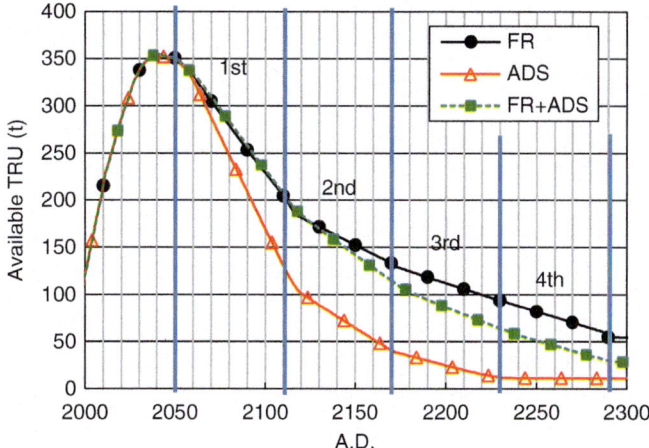

Fig. 19.12 Trans-uranium (TRU) amount available for transmutation in transmutation scenarios excluding minor actinides (MA) vitrified in waste

less than the ADS, resulting in longer transmutation era and larger remaining TRU. Figure 19.12 shows amounts of TRU excluding vitrified MA. An amount of the ADS scenario is reduced to 125 t after the first generation of the transmutation era from 2050 to 2110. In the FBR scenario, equal decrement is achieved after the second generation. Very high MA content up to 15 % in the FR fuel is also problematic. The decrement of the FR+ADS scenario is as same as that of the FR scenario until the second generation, but after that the decrement becomes faster because of ADS introduction.

Although transmutation performance of the FR is inferior to the ADS, cost including profit of electricity generation by transmuter would be much smaller than the ADS because the number of ADSs is doubled, accelerator cost is added, and thermal efficiency of the ADS is worse. Considering high MA content in the FR scenario, the FR+ADS scenario can be a modest solution, although the ADS scenario is preferable if rapid transmutation is required regardless of the cost.

19.5 Conclusion

With consideration of the phase-out option from NP utilization in Japan, an ADS for Pu transmutation was designed and scenario analysis introducing it was performed. The ADS was designed based on the existing design of the ADS for MA transmutation considering two options of Pu supply: pure Pu and a mixture of the same amount of Pu and U from the reprocessing plant for the LWR. After designing a one-batch core with large criticality drop, a six-batch core with a short operation day was analyzed. The criticality drop of the six-batch core was small enough in the equilibrium state. Several maintenance cases were assumed, and those effects on the transmutation half-life were surveyed. Finally, a core with pure Pu supply and 30-day fuel reloading was selected as the reference case for the scenario analysis. The transmutation half-life was estimated as 24.8 years, meaning that the amount of Pu is reduced to half after 24.8 years of operation, taking maintenance and cooling time of spent fuel into account.

In the scenario analysis, once-through scenario of LWR spent fuel was referred to as a conventional scenario. LWR-MOX utilization with reprocessing of LWR spent fuel was also considered. As the transmutation scenario, three cases of transmuters that are only-FR, only-ADS, and both-FR+ADS were analyzed. The numbers of necessary transmuters were obtained as 15 to 32 units, and the necessary period for transmutation as 180–240 years. Benefit to the repository by reduction of Pu and MA was reduction of repository area by a factor of five and of decay time of toxicity by one order of magnitude. It was shown that MA vitrified in the LWR reprocessing plant before introduction of the partitioning technology in 2025 considerably deteriorates both benefit. Therefore, early introduction of the partitioning process and retrievability of MA from vitrified waste should be investigated.

In comparison among transmutation scenarios, reduction of TRU in the ADS scenario is two times faster than that in the FR scenario. It was found that MA content in FR fuel in the FR scenario was 15 %, which is much higher than the design limit of 5 %. On the other hand, the cost of the FR scenario including profit of electricity generation by transmuter would be much smaller than that of the ADS scenario because the number of ADSs is double, the accelerator cost is added, and thermal efficiency of the ADS is worse. Considering high MA content in the FR scenario, the FR+ADS scenario can be a modest solution, although the ADS scenario is preferable if rapid transmutation is required regardless of cost.

The present scenario study revealed that the number of the transmuters and time necessary to transmute Pu and MA in the LWR legacy is considerably large. However, impact on the TRU amount in the repository related to the nonproliferation issue, repository size, and decay time of the potential radiotoxicity is also expected to be large. Assessments of increasing cost and risk to operate transmuters based on the present analysis are the next subject.

Open Access This chapter is distributed under the terms of the Creative Commons Attribution Noncommercial License, which permits any noncommercial use, distribution, and reproduction in any medium, provided the original author(s) and source are credited.

Nomenclature

w	Amount of heavy metal (t)
w_{tr}	Transmutation amount (t)
w_i	Initial amount (t)
P	Core thermal power (MW)
$h = \frac{P}{w_i}$	Specific heat (MW/t)

T_i	In-core period (year)
T_o	Out-core period (year)
ε_o	Operation efficiency
$\varepsilon_c = \frac{T_i}{T_i+T_o}$	Cycle efficiency
$E_{fiss} = 205 \text{ MeV} = 3.28 \cdot 10^{-11}$	Energy release per fission (J)
$N_A = 6.022 \cdot 10^{23}$	Avogadro number
$\overline{A}$	240 = averaged mass number
a	Constant (t/MW/year)
t	Time (year)
λ_{tr}	Effective transmutation rate (/year)
T_{tr}	Transmutation half-life (year)

References

1. Wakabayashi T et al (1997) Feasibility studies on plutonium and minor actinide burning in fast reactors. Nucl Technol 118:14–25
2. Sato T et al (2013) Particle and heavy ion transport code system: PHITS, version 2.52. J Nucl Sci Technol 50(9):913–923
3. Alcouffe et al (2005) PARTISN: a time-dependent, parallel neutral particle transport code system. LA-UR-05-3925, Los Alamos National Laboratory
4. Nakagawa M, Tsuchihashi K (1984) SLAROM: a code for cell homogenization calculation of fast reactor. JAERI 1294, Japan Atomic Energy Research Institute
5. Shibata K et al (2011) JENDL-4.0: a new library for innovative nuclear energy systems. J Nucl Sci Technol 48(1):1–30
6. Croff AG (1980) ORIGEN-2: a revised and updated version of Oak Ridge isotope generation and development code. ORNL-5621, Oak Ridge National Laboratory
7. Nishihara K et al (2013) Utilization of rock-like oxide fuel in the phase-out scenario. J Nucl Sci Technol 51(2):150–165
8. Nishihara K et al (2008) Neutronics design of accelerator-driven system for power flattening and beam current reduction. J Nucl Sci Technol 45:812–822
9. Nishihara K et al (2010) Impact of partitioning and transmutation on high-level waste disposal for the fast breeder reactor fuel cycle. J Nucl Sci Technol 47(12):1101–1117

Chapter 20
Sensitivity Analyses of Initial Compositions and Cross Sections for Activation Products of In-Core Structure Materials

Kento Yamamoto, Keisuke Okumura, Kensuke Kojima, and Tsutomu Okamoto

Abstract Sensitivity analyses of initial compositions and cross sections were conducted to quantitatively clarify the source elements and the nuclear reactions dominating the generation of activation products. In these analyses, the ORIGEN2.2 code was used with ORLIBJ40, a set of the cross-section libraries based on JENDL-4.0. Analyses were conducted for the activations of cladding tubes, end plugs, and spacers of fuel assemblies and channel boxes in BWR that are composed of zirconium alloy, stainless steel, and nickel-chromium-based alloy. From about 50 representative radioactive nuclides, several nuclides were selected as the targets of sensitivity analyses for the aspect of their large concentrations in the target materials.

The results of sensitivity coefficients clarified the source elements and the nuclear reactions dominating the generation of activation products even for the nuclides generated through complicated pathways. These results could be utilized to select the objectives of the impurity elements for measurements and of nuclear data for the improvement of accuracy. These results will contribute to improvements in the accuracy of numerical evaluations of activation product concentrations.

Keywords Activation products • Burn-up calculation • INCONEL alloy • ORIGEN2.2 • ORLIBJ40 • Sensitivity study • Stainless steel • Zircaloy

20.1 Introduction

In the research on the back-end of nuclear cycles, improvement of the accuracy of predicting concentrations of activation products is important for various evaluations. Providing the accurate initial compositions and nuclear data leading to the generation of activation products is necessary for accurate predictions of

---

K. Yamamoto (✉) • K. Okumura • K. Kojima • T. Okamoto
Japan Atomic Energy Agency, Ibaraki, Japan
e-mail: yamamoto.kento@jaea.go.jp

concentrations of the activation products. An effective first step to achieving this is to identify the dominant generation pathways of activation products. Sensitivity analyses of initial compositions and cross sections for activation products, which involve understanding the effects of initial compositions and cross sections on the concentrations of the target activation products, are powerful methods for quantitatively investigating the generation pathways. Thus, in the present study, sensitivity analyses focusing on the generation pathways for activation products were conducted.

The ORIGEN2.2 [1] code was used in the analyses; this code has been widely used for evaluating the concentrations of activation products. The one-group cross sections made with the appropriate neutron spectrum are required for the accuracy of the ORIGEN2.2 calculation. With respect to the activations of in-core structure materials, the existing ORIGEN2.2 cross-section libraries made with the in-core neutron spectrum are available. Thus, the target of the analyses in the present study is the activation of such in-core structure materials.

This chapter presents the method, calculation conditions, and results of the sensitivity analyses of initial compositions and cross sections for activation products in the materials of in-core structures, such as zirconium alloy, stainless steel, and nickel-chromium-based alloy. The results of the sensitivity analyses identify the elements and the nuclear reactions leading to the generation of activation products. These results will be effective in improving the accuracy of numerical evaluations of the concentrations of activation products.

20.2 Method of Calculating Sensitivity Coefficients

A sensitivity coefficient is defined as the ratio of the variation in concentration of the target activation product to the variation in the initial composition or cross section. The sensitivity coefficient of the initial composition and cross section is expressed by the following equations, respectively:

$$S = \frac{\Delta W/W_0}{\Delta X/X_0} \tag{20.1}$$

$$S = \frac{\Delta W/W_0}{\Delta \sigma/\sigma_0} \tag{20.2}$$

W_0: Concentration of the target activation product under normal condition
$\Delta W (= W' - W_0)$: Variation in concentration of the target activation product
X_0: Initial concentration of the element in the material under normal condition
$\Delta X (= X' - X_0)$: Variation in initial concentration of the element in the material
σ_0: Cross section under normal condition
$\Delta \sigma (= \sigma' - \sigma_0)$: Variation in cross section

For the calculation of concentration of activation products, ORIGEN2.2 was used with ORLIBJ40 [2], which is a set of the one-group cross-section libraries based on JENDL-4.0 [3]. The sensitivity coefficients are evaluated by executing two different burn-up calculations under normal condition and under composition-changed or cross-section-changed condition. In the former, the ORIGEN2.2 input files are changed; in the latter, the cross-section library files are changed. Utility programs to evaluate the sensitivity coefficients were prepared and used in these analyses.

20.3 Sensitivity Analyses

20.3.1 Analyses Conditions

As stated in Sect. 20.1, activations of in-core structure materials, such as cladding tubes, end plugs, and spacers of fuel assemblies and channel boxes, were investigated in this study. The materials of the in-core structures of PWR and BWR are shown in Table 20.1. The compositions of Zircaloy-2, Zircaloy-4, SUS304 stainless steel, and INCONEL alloy 718 are shown in Table 20.2. In Table 20.2, the average value of the upper and lower limits of the standard specification was applied to the calculation condition for additive elements and the upper limit was applied for impurity elements. The effect of impurity elements that are not specified in the standard are investigated in Sect. 20.3.4.

Typical conditions of BWR were assumed for the cross-section libraries and the irradiation condition, because the difference between the conditions of PWR and BWR is not so significant for the purpose of this study, which is clarifying the dominant generation pathways of activation products.

The cross-section libraries used in these analyses (Table 20.3) were chosen to correspond to the condition of the void ratio in the axial direction. A library made with an average void ratio (40 %) was applied to cladding tubes, spacers, and channel boxes for which the void ratio varies from 0 % to 70 %.

A BWR typical irradiation history consists of four cycles of irradiation of about 377 days with constant flux and 90 days of cooling time in the intervals of irradiation (Fig. 20.1). Considering the period for processing of radioactive wastes,

Table 20.1 Materials of in-core structure

	BWR	PWR
Cladding tube	Zircaloy-2	Zircaloy-4
Top end plug	SUS304	←
Bottom end plug	SUS304	←
Spacer	Plate: Zircaloy-2	Zircaloy-4 or
	Spring: INCONEL alloy 718	INCONEL alloy 718
Channel box	Zircaloy-4	–

Table 20.2 Compositions of materials

	Specification (wt%)			Value in analysis (wt%)
(a) Zircaloy-2 (JIS H 4751)				
H	0.0025	Max.		0.0025
B	0.00005	Max.		0.00005
C	0.027	Max.		0.027
N	0.008	Max.		0.008
Mg	0.002	Max.		0.002
Al	0.0075	Max.		0.0075
Si	0.012	Max.		0.012
Ca	0.003	Max.		0.003
Ti	0.005	Max.		0.005
Cr	0.05	–	0.15	0.10
Mn	0.005	Max.		0.005
Fe	0.07	–	0.20	0.135
Co	0.002	Max.		0.002
Ni	0.03	–	0.08	0.055
Cu	0.005	Max.		0.005
Zr		Balance		98.1456
Nb	0.01	Max.		0.01
Mo	0.005	Max.		0.005
Cd	0.00005	Max.		0.00005
Sn	1.20	–	1.70	1.45
Hf	0.01	Max.		0.01
W	0.01	Max.		0.01
U	0.00035	Max.		0.00035
(b) Zircaloy-4 (JIS H 4751)				
H	0.0025	Max.		0.0025
B	0.00005	Max.		0.00005
C	0.027	Max.		0.027
N	0.008	Max.		0.008
Mg	0.002	Max.		0.002
Al	0.0075	Max.		0.0075
Si	0.012	Max.		0.012
Ca	0.003	Max.		0.003
Ti	0.005	Max.		0.005
Cr	0.07	–	0.13	0.10
Mn	0.005	Max.		0.005
Fe	0.18	–	0.24	0.21
Co	0.002	Max.		0.002
Ni	0.007	Max.		0.007
Cu	0.005	Max.		0.005
Zr		Balance		98.1186
Nb	0.01	Max.		0.01

(continued)

Table 20.2 (continued)

	Specification (wt%)			Value in analysis (wt%)
Mo	0.005	Max.		0.005
Cd	0.00005	Max.		0.00005
Sn	1.20	–	1.70	1.45
Hf	0.01	Max.		0.01
W	0.01	Max.		0.01
U	0.00035	Max.		0.00035
(c) SUS304 stainless steel (JIS G 4303)				
C	0.08	Max.		0.08
Si	1.00	Max.		1.00
P	0.045	Max.		0.05
S	0.030	Max.		0.03
Cr	18.00	–	20.00	19.00
Mn	2.00	Max.		2.00
Fe		Balance		68.595
Ni	8.00	–	10.50	9.25
(d) INCONEL alloy 718 (UNS N07718)				
B	0.006	Max.		0.006
C	0.08	Max.		0.08
Al	0.20	–	0.80	0.50
Si	0.35	Max.		0.35
P	0.015	Max.		0.015
S	0.015	Max.		0.015
Ti	0.65	–	1.15	0.90
Cr	17.00	–	21.00	19.00
Mn	0.35	Max.		0.35
Fe		Balance		16.809
Co	1.00	Max.		1.00
Ni	50.00	–	55.00	52.50
Cu	0.3	Max.		0.30
Nb	4.75	–	5.50	5.125
Mo	2.80	–	3.30	3.05

Table 20.3 Cross-section libraries

	Specification in cross-section library
Cladding tubes, spacers, channel boxes	BWR STEP-III, void ratio 40 %
Top-end-plugs	BWR STEP-III, void ratio 70 %
Bottom-end-plugs	BWR STEP-III, void ratio 0 %

10 years of cooling time after irradiation was assumed in these analyses. The flux intensities at the center, top, and bottom in the axial direction are shown in Table 20.4. The flux intensity at the center corresponds to the average power in typical BWR fuel assemblies. The flux intensities at the top and bottom were

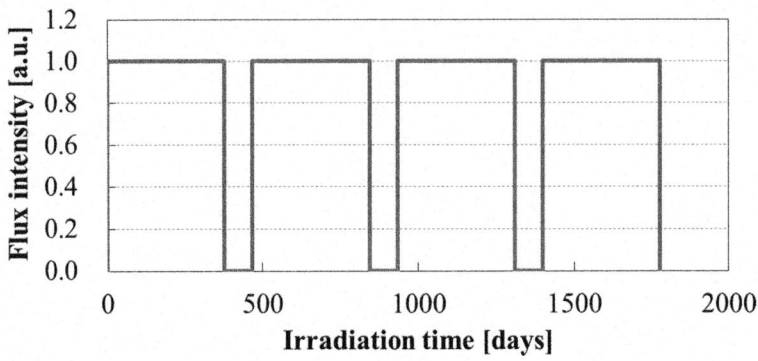

Fig. 20.1 Irradiation history

Table 20.4 Flux intensity at center, top, and bottom in axial direction

	Flux intensity (1/cm²s)
Center	1.994E + 14
Top and bottom	9.970E + 12

determined to be 5 % of that at the center, based on flux distribution evaluated by the one-dimensional neutron diffusion calculation.

20.3.2 Target Nuclides of Sensitivity Analyses

The representative radioactive nuclides in this study (Table 20.5) include not only the important nuclides for various evaluations of radioactive wastes but also the nuclides whose concentrations have been measured in the past, which will be useful for the validation of numerical evaluations.

Target nuclides of sensitivity analyses were selected on the basis of two criteria. The first was that the concentrations of activation products be larger than or comparable to the concentration of fission products generated from impurity uranium in the materials. The contents of impurity uranium in Zircaloy-2 and SUS304 stainless steel were 0.00035 wt% and 0.0001 wt% [5], respectively. This value in INCONEL alloy is unknown. The second criterion was that the concentrations of activation products be comparatively large. In these analyses, activation products with concentrations more than 1×10^{-9} g/t were chosen.

The concentrations of activation products larger than 1×10^{-9} g/t and fission products generated from impurity uranium are shown in Table 20.6. The fission products were calculated under the condition that the initial composition contains only the uranium impurity. Table 20.6 also shows the selected target nuclides that satisfy the foregoing criterion: 17 nuclides in Zircaloy-2 and Zircaloy-4, 8 nuclides in SUS304 stainless steel, and 16 nuclides in INCONEL alloy were selected as the target nuclides of sensitivity analyses.

Table 20.5 Representative radioactive nuclides[a,b]

No.	Nuclide	Half-life		No.	Nuclide	Half-life		No.	Nuclide	Half-life	
1	H-3	12.3	years	21	Se-79	377.0	E3 years	41	Cs-137	30.1	years
2	Be-10	1.5	E6 years	22	Rb-87	48.1	E9 years	42	Ba-133	10.5	years
3	C-14	5.7	E3 years	23	Sr-90	28.8	years	43	Sn-126	198.0	E3 years
4	Na-22	2.6	years	24	Nb-94	20.3	E3 years	44	Sb-124	60.2	days
5	Si-32	153.0	years	25	Nb-95	35.0	days	45	Sb-125	2.8	years
6	S-35	87.5	days	26	Mo-93	4.0	E3 years	46	Ce-141	32.5	days
7	Cl-36	301.0	E3 years	27	Mo-99	2.7	days	47	Ce-144	284.9	days
8	K-40	1.2	E9 years	28	Tc-99	211.1	E3 years	48	Eu-152	13.5	years
9	Sc-46	83.8	days	29	Tc-99m	6.0	hours	49	Eu-154	8.6	years
10	Ca-41	102.0	E3 years	30	Zr-93	1.5	E6 years	50	Eu-155	4.8	years
11	Cr-51	27.7	days	31	Zr-95	64.0	days	51	Gd-153	240.4	days
12	Mn-54	312.1	days	32	Ru-103	39.3	days	52	Ho-166m	1.2	E3 years
13	Fe-55	2.7	years	33	Ru-106	1.0	years	53	Hf-181	42.4	days
14	Fe-59	44.5	days	34	Ag-108m	438.0	years	54	Au-199	3.1	days
15	Co-57	271.7	days[c]	35	Ag-110m	249.8	days				
16	Co-58	70.9	days	36	I-129	15.7	E6 years				
17	Co-60	5.3	years	37	Cd-113m	14.1	years				
18	Ni-59	76.0	E3 years	38	I-131	8.0	days				
19	Ni-63	100.1	years	39	Cs-134	2.1	years				
20	Zn-65	244.1	days	40	Cs-135	2.3	E6 years				

[a]Half-life is referred to ORLIBJ40 decay library
[b]E3 years = 10^3 years, E6 years = 10^6 years
[c]Half-life of Co-57 is not contained in ORLIBJ40, so it is referred to [4]

Table 20.6 Concentration of activation products and fission products

Nuclide	Concentration of activation products (g/t)		Concentration of fission products (g/t)	Comparison (%)	Target nuclide
	①Zry-2	②Zry-4	③	③/(①+③)	
(a) Zircaloy-2 and Zircaloy-4					
Zr-93	2.0E+02	2.0E+02	1.2E-03	0	○
Ni-59	3.7E+00	4.7E-01	–	–	○
Ni-63	6.6E-01	8.9E-02	–	–	○
Co-60	5.1E-01	5.1E-01	–	–	○
C-14	4.0E-01	4.0E-01	–	–	○
Nb-94	3.0E-01	3.0E-01	3.2E-09	0	○
Sb-125	2.5E-01	2.5E-01	1.5E-06	0	○
Ca-41	3.0E-02	3.0E-02	–	–	○
K-40	2.2E-02	2.2E-02	–	–	○
Fe-55	2.1E-02	3.2E-02	–	–	○
Tc-99	9.1E-03	9.1E-03	1.7E-03	16	○
Mo-93	8.9E-03	8.9E-03	3.5E-14	0	○
Be-10	4.0E-05	4.0E-05	2.2E-08	0	○
Sr-90	2.3E-05	2.3E-05	6.0E-04	96	–
Mn-54	3.9E-06	6.1E-06	–	–	○
Ag-108m	3.3E-07	3.3E-07	5.6E-12	0	○
Rb-87	1.1E-07	1.1E-07	3.7E-04	100	–
H-3	3.0E-08	3.0E-08	5.8E-08	66	○
I-129	6.4E-09	6.4E-09	3.9E-04	100	–
Zn-65	2.8E-09	2.8E-09	–	–	○
(b) SUS304 stainless steel					
Nuclide	Concentration of activation products (g/t)		Concentration of fission products (g/t)	Comparison (%)	Target nuclide
	①Bottom	②Top	③	③/(①+③)	
Ni-59	4.8E+01	2.6E+01	–	–	○
Ni-63	7.7E+00	4.1E+00	–	–	○
Fe-55	7.0E-01	3.9E-01	–	–	○
Co-60	5.3E-03	5.0E-03	–	–	○
Mn-54	9.9E-05	9.9E-05	–	–	○
Be-10	5.7E-06	5.7E-06	2.0E-10	0	○
C-14	3.2E-06	2.3E-06	–	–	○
Cl-36	1.6E-06	4.4E-07	–	–	○

(continued)

Table 20.6 (continued)

Nuclide	Concentration of activation products (g/t)		Concentration of fission products (g/t)	Comparison (%)	Target nuclide
	①Zry-2	②Zry-4	③	③/(①+③)	

(c) INCONEL alloy 718

Nuclide	Concentration of activation products (g/t)	Target nuclide
Ni-59	3.5E+03	○
Ni-63	6.2E+02	○
Co-60	2.6E+02	○
Nb-94	1.6E+02	○
Mo-93	5.4E+00	○
Tc-99	5.4E+00	○
Fe-55	3.4E+00	○
Zr-93	1.3E-01	○
Mn-54	4.8E-04	○
Be-10	2.8E-04	○
Cl-36	1.7E-04	○
C-14	5.4E-05	○
Zn-65	1.7E-07	○
Sr-90	1.3E-08	○
Si-32	7.9E-09	○
H-3	1.8E-09	○

20.3.3 Results of Sensitivity Analyses

Sensitivity analyses were conducted for several selected nuclides in Zircaloy-2, SUS304 stainless steel, and INCONEL alloy. Analyses in Zircaloy-4 were skipped because the sensitivity coefficients were thought to be almost the same as that in Zircaloy-2 because calculation conditions were similar. For SUS304 stainless steel, activations using the cross-section library of void ratio 0 % were evaluated because the concentrations in the case of void ratio 0 % were larger than that of void ratio 70 %.

The sensitivity coefficients of initial compositions are shown in Table 20.7. As defined in Eq. (20.1), the value shows the relative amount of variation in concentration of the target nuclide when the initial composition of element varies by a unit amount. Therefore, the source elements leading to the generation of target nuclides was clarified from the results. For example, Table 20.7a shows that Fe-55 is generated from both iron and nickel and that the contribution from iron is dominant. The results can also be useful in the evaluation of the error propagated from the measurement uncertainty of initial composition.

As defined in Eq. (20.2), a sensitivity coefficient of a cross section shows the relative amount of variation in the concentration of the target nuclide when the

Table 20.7 Sensitivity coefficients of initial composition

(a) Zircaloy-2

	Target nuclide																	
	Zr-93	Ni-59	Ni-63	Co-60	C-14	Nb-94	Sb-125	Ca-41	K-40	Fe-55	Tc-99	Mo-93	Be-10	Mn-54	Ag-108 m	H-3	Zn-65
H	0	0	0	0	0	0	0	0	0	0	0	0	0	0	0	0.9995	0
B	0	0	0	0	0	0	0	0	0	0	0	0	0.0334	0	0	0	0
C	0	0	0	0	5E-05	0	0	0	0	0	0	0	0.9666	0	0	0	0
N	0	0	0	0	1.0000	0	0	0	0	0	0	0	0	0	0	0.0003	0
Mg	0	0	0	0	0	0	0	0	0	0	0	0	0	0	0	0	0
Al	0	0	0	0	0	0	0	0	0	0	0	0	0	0	0	0	0
Si	0	0	0	0	0	0	0	0	0	0	0	0	0	0	0	0	0
Ca	0	0	0	0	0	0	0	1.0000	1.0000	0	0	0	0	0	0	0	0
Ti	0	0	0	0	0	0	0	0	0	0	0	0	0	0	0	0	0
Cr	0	0	0	0	0	0	0	0	0	0	0	0	0	0	0	0	0
Mn	0	0	0	0	0	0	0	0	0	0	0	0	0	0.0032	0	0	0
Fe	0	0	0	0.0004	0	0	0	0	0	0.9564	0	0	0	0.9967	0	0	0
Co	0	0	0	0.9980	0	0	0	0	0	0	0	0	0	0	0	0	0
Ni	0	1.0000	0.9904	0.0015	0	0	0	0	0	0.0436	0	0	0	0	0	0	0.0001
Cu	0	0	0.0096	0.0001	0	0	0	0	0	0	0	0	0	0	0	0	0.9999
Zr	1.0000	0	0	0	0	0.0001	0	0	0	0	0.0289	0	0	0	0	0	0
Mo	0	0	0	0	0	0.0001	0	0	0	0	0.9711	1.0000	0	0	0	0	0
Nb	0	0	0	0	0	0.9998	0	0	0	0	0	0	0	0	0	0	0
Cd	0	0	0	0	0	0	0	0	0	0	0	0	0	0	1.0000	0	0
Sn	0	0	0	0	0	0	1.0000	0	0	0	0	0	0	0	0	0	0
Hf	0	0	0	0	0	0	0	0	0	0	0	0	0	0	0	0	0
W	0	0	0	0	0	0	0	0	0	0	0	0	0	0	0	0	0

(b) SUS304 stainless steel

	Target nuclide							
	Ni-59	Ni-63	Fe-55	Co-60	Mn-54	Be-10	C-14	Cl-36
C	0	0	0	0	0	1.0000	1.0000	0
Si	0	0	0	0	0	0	0	0
P	0	0	0	0	0	0	0	0
S	0	0	0	0	0	0	0	1.0000
Cr	0	0	0	0	0	0	0	0
Mn	0	0	0	0	0.0027	0	0	0
Fe	0	0	0.9890	0.0819	0.9974	0	0	0
Ni	1.0000	1.0000	0.0110	0.9181	0	0	0	0

(c) INCONEL alloy 718

	Target nuclide															
	Ni-59	Ni-63	Co-60	Nb-94	Mo-93	Tc-99	Fe-55	Zr-93	Mn-54	Be-10	Cl-36	C-14	Zn-65	Sr-90	Si-32	H-3
B	0	0	0	0	0	0	0	0	0	0.5830	0	0	0	0	0	0
C	0	0	0	0	0	0	0	0	0	0.4170	0	1.0000	0	0	0	0
Al	0	0	0	0	0	0	0	0	0	0	0	0	0	0	0	0.0012
Si	0	0	0	0	0	0	0	0	0	0	0	0	0	0	0.7059	0.0011
P	0	0	0	0	0	0	0	0	0	0	0	0	0	0	0.2926	0.0002
S	0	0	0	0	0	0	0	0	0	0	1.0000	0	0	0	0.0015	0.0005
Ti	0	0	0	0	0	0	0	0	0	0	0	0	0	0	0	0.0007
Cr	0	0	0	0	0	0	0	0	0	0	0	0	0	0	0	0.0121
Mn	0	0	0	0	0	0	0	0	0.0019	0	0	0	0	0	0	0.0001
Fe	0	0	0.0001	0	0	0	0.7411	0	0.9982	0	0	0	0	0	0	0.0258
Co	0	0	0.9970	0	0	0	0	0	0	0	0	0	0	0	0	0.0004
Ni	1.0000	0.9994	0.0028	0	0	0	0.2589	0	0	0	0	0	0.0014	0	0	0.9557
Cu	0	0.0006	0	0	0	0	0	0	0	0	0	0	0.9985	0	0	0.0012
Nb	0	0	0	0.9999	0	0	0	0.9839	0	0	0	0	0	0.9841	0	0.0003
Mo	0	0	0	0.0001	1.0000	1.0000	0	0.0161	0	0	0	0	0	0.0159	0	0.0008

Table 20.8 Sensitivity coefficients of cross sections

Target nuclide	Sensitivity coefficient of cross section								
	First largest			Second largest			Others		
(a) Zircaloy-2									
Zr-93	Zr-92	(n, γ)	0.98	Zr-94	$(n, 2n)$	0.02			
Ni-59	Ni-58	(n, γ)	0.99						
Ni-63	Ni-62	(n, γ)	0.97						
Co-60	Co-59	$(n, \gamma)_m$	0.46	Co-59	(n, γ)	0.42			
C-14	N-14	(n, p)	1.00						
Nb-94	Nb-93	(n, γ)	1.00						
Sb-125	Sn-124	(n, γ)	0.56	Sn-124	$(n, \gamma)_m$	0.48			
Ca-41	Ca-40	(n, γ)	1.00						
K-40	Ca-40	(n, p)	1.00						
Fe-55	Fe-54	(n, γ)	0.95	Ni-58	(n, α)	0.04			
Tc-99	Mo-98	(n, γ)	1.00	Mo-97	(n, γ)	0.03	Zr-96	(n, γ)	0.03
Mo-93	Mo-92	(n, γ)	0.99						
Be-10	C-13	(n, α)	0.97	B-10	(n, p)	0.03			
Mn-54	Fe-54	(n, p)	1.00						
Ag-108 m	Cd-106	(n, γ)	1.00	Ag-107	$(n, \gamma)_m$	0.97			
H-3	H-2	(n, γ)	1.00	H-1	(n, γ)	0.78	He-3	(n, p)	0.01
Zn-65	Zn-64	(n, γ)	1.00	Cu-63	(n, γ)	1.00			
(b) SUS304 stainless steel									
Ni-59	Ni-58	(n, γ)	1.00						
Ni-63	Ni-62	(n, γ)	1.00						
Fe-55	Fe-54	(n, γ)	0.99	Ni-58	(n, α)	0.01			
Co-60	Ni-60	(n, p)	0.92	Fe-58	(n, γ)	0.08	Co-59	(n, γ)	0.04
							Co-59	$(n, \gamma)_m$	0.04
Mn-54	Fe-54	(n, p)	1.00						
Be-10	C-13	(n, α)	1.00						
C-14	C-13	(n, γ)	1.00						
Cl-36	S-34	(n, γ)	1.00	Cl-35	(n, γ)	1.00			
(c) INCONEL alloy 718									
Ni-59	Ni-58	(n, γ)	0.99						
Ni-63	Ni-62	(n, γ)	0.98						
Co-60	Co-59	$(n, \gamma)_m$	0.46	Co-59	(n, γ)	0.42			
Nb-94	Nb-93	(n, γ)	1.00						
Mo-93	Mo-92	(n, γ)	0.99						
Tc-99	Mo-98	(n, γ)	1.00						
Fe-55	Fe-54	(n, γ)	0.74	Ni-58	(n, α)	0.26			
Zr-93	Nb-93	(n, p)	0.98	Mo-96	(n, α)	0.02			
Mn-54	Fe-54	(n, p)	1.00						
Be-10	B-10	(n, p)	0.58	C-13	(n, α)	0.42			
Cl-36	S-34	(n, γ)	1.00	Cl-35	(n, γ)	0.97			
C-14	C-13	(n, γ)	1.00						

(continued)

Table 20.8 (continued)

Target nuclide	Sensitivity coefficient of cross section							
	First largest			Second largest			Others	
Zn-65	Zn-64	(n, γ)	0.99	Cu-63	(n, γ)	0.99		
Sr-90	Zr-93	(n, α)	1.00	Nb-93	(n, p)	0.98		
Si-32	Si-31	(n, γ)	1.00	Si-30	(n, γ)	0.71	P-31	(n, p)	0.29
H-3	H-2	(n, γ)	1.00	H-1	(n, γ)	1.00	Ni-58	(n, p)	0.95
							He-3	(n, p)	0.01

$(n, \gamma)_m$ means the (n, γ) reaction yielding to meta-stable state

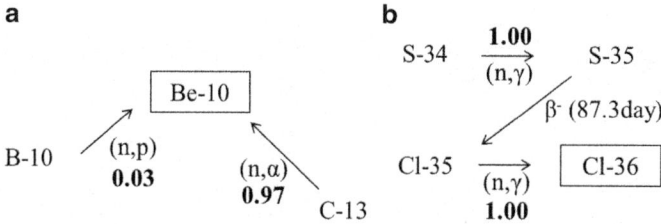

Fig. 20.2 Examples of complicated generation pathways of activation products. The values in *bold font* in the figures show the sensitivity coefficients of the cross sections. **a** Be-10 generation in Zircaloy-2. **b** Cl-36 generation in SUS304 stainless steel

cross section varies by a unit amount. Therefore, a positive value of this coefficient indicates that the target activation product is generated through the nuclear reaction. Thus, if a sensitivity coefficient is positive and large, the cross section of the nuclear reaction is significant for the generation of the target activation products. In the analyses, the objectives of reaction were six reactions treated in ORLIBJ40 library; the reaction of (n, γ), $(n, 2n)$, (n, α), and (n, p) yielding to nuclides of ground state and the reaction of (n, γ) and $(n, 2n)$ yielding to nuclides of meta-stable state. The summary of the results of sensitivity analyses of cross sections are shown in Table 20.8, where the sensitivity coefficients that are positive and more than 0.01 are extracted from all the results and listed in descending order. The results clarified the nuclear reaction dominating the generation of target nuclides. For example, it is thought that Fe-55 in Zircaloy-2 can be generated from the (n, γ) reaction of Fe-54, the (n, α) reaction of Ni-58, and the $(n, 2n)$ reaction of Fe-56. Table 20.8a clearly shows the (n, γ) reaction of Fe-54 is dominant in the generation of Fe-55.

It was remarkable that the dominant generation pathways were clarified even for the target nuclides generated through complicated pathways. Some of the examples are shown in Fig. 20.2.

Figure 20.2a shows an example of nuclides generated with some contributing pathways. Be-10 is generated in Zircaloy-2 mainly through two pathways, the (n, p) reaction of Be-10 and the (n, α) reaction of C-13. It is not predictable which pathway is dominant from the initial composition of the material. The sensitivity

Table 20.9 Initial composition of SUS304 stainless steel

Element	Value based on measurement data	Value based on the standard specification
C	–	0.08
N	0.05	–
Si	–	1.00
P	–	0.045
S	0.004	0.030
Cl	0.001	–
K	4.0E-05	–
Cr	–	19.00
Mn	–	2.00
Fe	72	68.60
Co	0.1	–
Ni	9.25	9.25
Cu	0.16	–
Zr	0.00032	–
Nb	0.02	–
Mo	0.13	–
Th	2.0E-07	–
U	2.0E-07	0.0001

coefficients clearly showed that the (n, α) reaction of C-13 is the dominant pathway for Be-10 generation in Zircaloy-2.

Figure 20.2b shows an example of nuclides generated through long and complicated generation chains. The source nuclide of Cl-36 generated in SUS304 stainless steel is ambiguous because the initial composition in this analysis does not contain chlorine, which could be the dominant source element of Cl-36. The sensitivity coefficients quantitatively clarified that S-34 is the source nuclide of Cl-36 even for the long and complicated chain.

20.3.4 Sensitivity Analysis Using the Initial Composition Based on Measured Data

The sensitivity coefficients shown in Sect. 20.3.3 are valid within the assumed analysis conditions in Sect. 20.3.1. However, the impurity elements that are not specified in the standard specification can be possibly present in the material. To know the effect of the difference in the initial composition on sensitivity coefficients, additional analyses were conducted using the initial composition based on measured data. The evaluation of activation products in SUS304 stainless steel is described here.

Table 20.10 Concentration of activation products in SUS304 stainless steel

Nuclide	Concentration of activation products (g/t)
Ni-59	4.8E+01
Ni-63	7.8E+00
Co-60	1.7E+00
Fe-55	7.4E-01
C-14	1.7E-01
Cl-36	5.4E-02
Nb-94	3.4E-02
Tc-99	1.2E-02
Mo-93	1.1E-02
K-40	1.8E-04
Mn-54	1.0E-04
Zr-93	6.6E-05

Except for the initial composition, the analysis conditions described in Sect. 20.3.1 were assumed. The composition data reported by the Atomic Energy Society of Japan [6] were applied in this analysis. In this reference, the concentration distributions of some elements with their mean values and standard deviations have been determined based on several measured data. The initial composition based on measured data is shown in Table 20.9 together with that based on the standard specification.

The concentration of activation products using the initial composition based on measured data is shown in Table 20.10. As a matter of course, the concentrations were changed from those in Table 20.6b because the different initial compositions were assumed. It was found that Nb-94, Tc-99, Mo-93, K-40, and Zr-93 appeared in Table 20.10 because of the presence of niobium, molybdenum, and potassium in the initial composition. For the comparison with the sensitivity coefficients shown in Sect. 20.3.3, sensitivity analyses of cross sections were conducted for the several nuclides Ni-59, Ni-63, Fe-55, Co-60, Mn-54, C-14, and Cl-36, which were listed in both Tables 20.6b and 20.10.

The sensitivity coefficients of cross sections using the initial composition based on measured data are shown in Table 20.11. It was found that the results of Co-60, C-14, and Cl-36 are much different from those in Table 20.8b, which indicates that the dominant generation pathways of these nuclides were changed. Figure 20.3 shows the comparison of dominant generation pathways of Co-60, C-14, and Cl-36 between different analysis conditions. The source nuclides of Co-60, C-14, and Cl-36 were Co-59, N-14, and Cl-35, respectively, under the conditions based on measurement data, whereas those were Ni-60, C-13, and S-34, respectively, under the conditions based on the standard specification.

As shown in the foregoing example, the dominant generation pathway can be changed corresponding to the initial composition. The reliable measured data of initial impurity elements should be used if they are available. For any condition, sensitivity analyses on the basis of the methodology stated in this study can systematically identify the dominant generation pathways of activation products.

Table 20.11 Sensitivity coefficient of cross section in SUS304 stainless steel

Target nuclide	Sensitivity coefficient of cross section					
	First largest			Second largest		
Ni-59	Ni-58	(n, γ)	1.00			
Ni-63	Ni-62	(n, γ)	1.00			
Fe-55	Fe-54	(n, γ)	0.99	Ni-58	(n, α)	0.01
Co-60	Co-59	(n, γ)	0.46	Co-59	$(n, \gamma)_m$	0.46
Mn-54	Fe-54	(n, p)	1.00			
C-14	N-14	(n, p)	1.00			
Cl-36	Cl-35	(n, γ)	1.00			

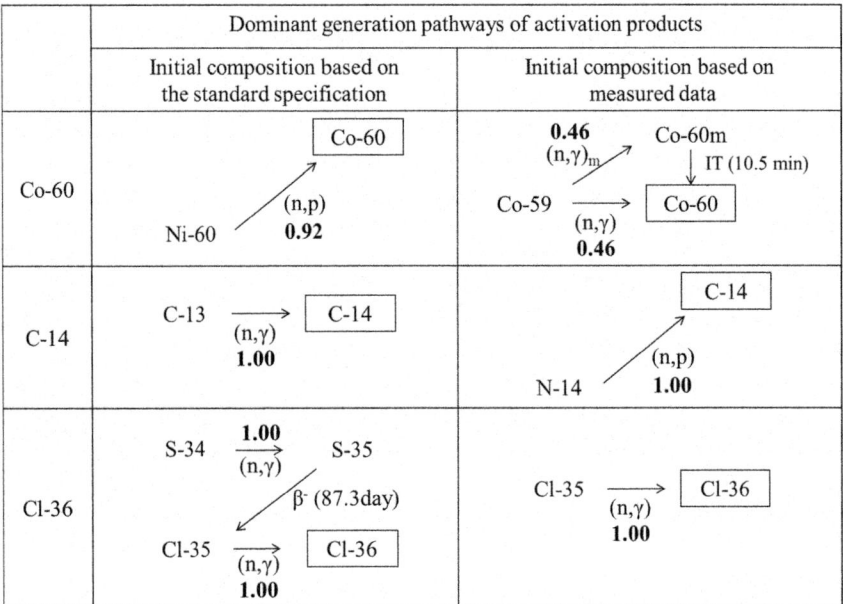

Fig. 20.3 Dominant generation pathways of activation products in SUS304 stainless steel

20.4 Conclusion

This study shows the sensitivity analyses of initial compositions and cross sections for activation products of in-core structure materials. The results clarified the source elements and nuclear reactions dominating the generation pathways of the activation products even for the nuclides generated through complicated pathways. The sensitivity coefficients of initial compositions are beneficial for the evaluation of the error propagated from the uncertainty of the initial composition of target materials. The sensitivity coefficients of cross sections are effective in selecting the objectives of nuclear reactions for the improvement of nuclear data. These results will contribute to improvement of the accuracy of numerical evaluations for the concentration of activation products.

The methodology of sensitivity analyses stated in this study is efficient for acquiring information about important impurity elements and nuclear reactions to evaluate the activation product concentrations. This methodology can be applied to the activations of ex-core structure materials if the appropriate one-group cross sections are prepared with a corresponding neutron spectrum.

Open Access This chapter is distributed under the terms of the Creative Commons Attribution Noncommercial License, which permits any noncommercial use, distribution, and reproduction in any medium, provided the original author(s) and source are credited.

References

1. Ludwig SB, Croff AG (1998) Revision to ORIGEN2: version 2.2. Transmittal memo of CCC-0371/17. Oak Ridge National Laboratory, Oak Ridge
2. Okumura K, Sugino K, Kojima K, Jin T, Okamoto T, Katakura J (2012) A set of ORIGEN2 cross section libraries based on JENDL-4.0: ORLIBJ40. JAEA Data/Code 2012-032. Japan Atomic Energy Agency [in Japanese], Ibaraki
3. Shibata K, Iwamoto O, Nakagawa T, Iwamoto N, Ichihara A, Kunieda S, Chiba S, Furukawa K, Otuka N, Ohsawa T, Murata T, Matsunobu H, Zukaran A, Kameda S, Katakura J (2011) JENDL-4.0: a new library for nuclear science and engineering. J Nucl Sci Technol 48(1):1–30
4. Namekawa M, Fukahori T (2012) Tables of nuclear data (JENDL/TND-2012). JAEA Data/Code 2012-014. Japan Atomic Energy Agency, Ibaraki
5. Evans JC, Lepel EL, Sanders RW, Wilkerson CL, Silker W, Thomas CW, Abel KH, Robertson DR (1984) Long-lived activation products in reactor materials. NUREG/CR-3474. U.S. Nuclear Regulatory Commission, Washington
6. Atomic Energy Society of Japan (2010) AESJ-SC-F015: basic procedure to determine the activity concentration of sub-surface disposal waste [in Japanese]

Chapter 21
Options of Principles of Fuel Debris Criticality Control in Fukushima Daiichi Reactors

Kotaro Tonoike, Hiroki Sono, Miki Umeda, Yuichi Yamane, Teruhiko Kugo, and Kenya Suyama

Abstract In the Three Mile Island Unit 2 reactor accident, a large amount of fuel debris was formed whose criticality condition is unknown, except the possible highest ^{235}U/U enrichment. The fuel debris had to be cooled and shielded by water in which the minimum critical mass is much smaller than the total mass of fuel debris. To overcome this uncertain situation, the coolant water was borated with sufficient concentration to secure the subcritical condition. The situation is more severe in the damaged reactors of Fukushima Daiichi Nuclear Power Station, where the coolant water flow is practically "once through." Boron must be endlessly added to the water to secure the subcritical condition of the fuel debris, which is not feasible. The water is not borated relying on the circumstantial evidence that the xenon gas monitoring in the containment vessels does not show a sign of criticality. The criticality condition of fuel debris may worsen with the gradual drop of its temperature, or the change of its geometry by aftershocks or the retrieval work, that may lead to criticality. To avoid criticality and its severe consequences, a certain principle of criticality control must be established. There may be options, such as prevention of criticality by coolant water boration or neutronic monitoring, prevention of the severe consequences by intervention measures against criticality, etc. Every option has merits and demerits that must be adequately evaluated toward selection of the best principle.

Keywords Criticality control • Fuel debris • Fukushima Daiichi

21.1 Introduction

In normal nuclear facilities, the goal of criticality control is to secure subcritical conditions of fissile materials, which is achieved by regulating the composition, geometry, or mass of the fissile materials [1]. In the accident of Three Mile Island

K. Tonoike (✉) • H. Sono • M. Umeda • Y. Yamane • T. Kugo • K. Suyama
Japan Atomic Energy Agency, 2-4 Shirakata Shirane, Tokai, Ibaraki, 319-1195, Japan
e-mail: tonoike.kotaro@jaea.go.jp

Unit 2 reactor (TMI-2), heavily damaged and melted fuel assemblies formed a large amount of fuel debris whose composition was unknown except the possible highest ^{235}U/U enrichment, 3 wt%, whose geometry is uncertain, and whose mass is larger than the minimum critical mass derived from the enrichment. Moreover, the fuel debris had to be cooled and shielded by water. To overcome this uncertain situation, the coolant water was borated with a concentration, >4,350 ppm, sufficient to secure the subcritical condition [2].

The situation of the damaged reactors in Fukushima Daiichi Nuclear Power Station (1FNPS) is more severe than that of TMI-2 because of the water issue. The most major difference is that the coolant water flow is practically "once through." Boron should be ceaselessly added in the water to maintain its lowest concentration necessary to secure the subcritical condition, which is not feasible. The water is not borated relying on the circumstantial evidence that the xenon gas monitoring in the containment vessels (CVs) does not show a sign of criticality. Although the fuel debris will not be touched for a while, its condition may change because of a gradual drop of its temperature or change of its geometry by aftershocks. The condition will be intentionally changed when the fuel debris is retrieved. Every such change may lead to the criticality of fuel debris [3].

To avoid criticality and its severe consequences, a certain principle of criticality control must be established. There may be options, such as prevention of criticality by coolant water boration or by neutronic monitoring, prevention of the severe consequences of criticality, etc. Each has merits and demerits.

It is necessary to understand the actual condition of the fuel debris regarding the selection of an appropriate principle from those options and the realization of certain criticality control following the selected principle. Adequate observation, sample taking, and analysis of the fuel debris must be conducted.

21.2 Present Condition of 1FNPS Fuel Debris

Fuel assemblies with the design called "BWR STEP 3" had been loaded in the reactors. Each new fuel assembly contains six kinds of uranium dioxide (UO_2) fuel (Fig. 21.1, Table 21.1). The most popular initial ^{235}U/U enrichment in the fuels is 4.4 wt%, whose inventory per assembly is 76.8 kgU. The fuel of 9.6 kgU per assembly has the highest initial enrichment of 4.9 wt%. The initial uranium inventory in total is 170.9 kgU per assembly, including fuels of other enrichments and of the UO_2-gadolinium oxide (Gd_2O_3) composite [4].

The Unit 1 reactor in 1FNPS had 400 assemblies, which consisted of six batches of burn-up. Each of the Unit 2 and 3 reactors had 548 assemblies of five batches. Among these assemblies, 64 in the Unit 1 reactor, and 116 in the Unit 2 reactor, had a low burn-up of only 3–5 GWD/t (Table 21.2). Other assemblies of the same number are older but still have a burn-up as low as 15–16 GWD/t. The oldest assemblies have a burn-up of about 40 GWD/t [5].

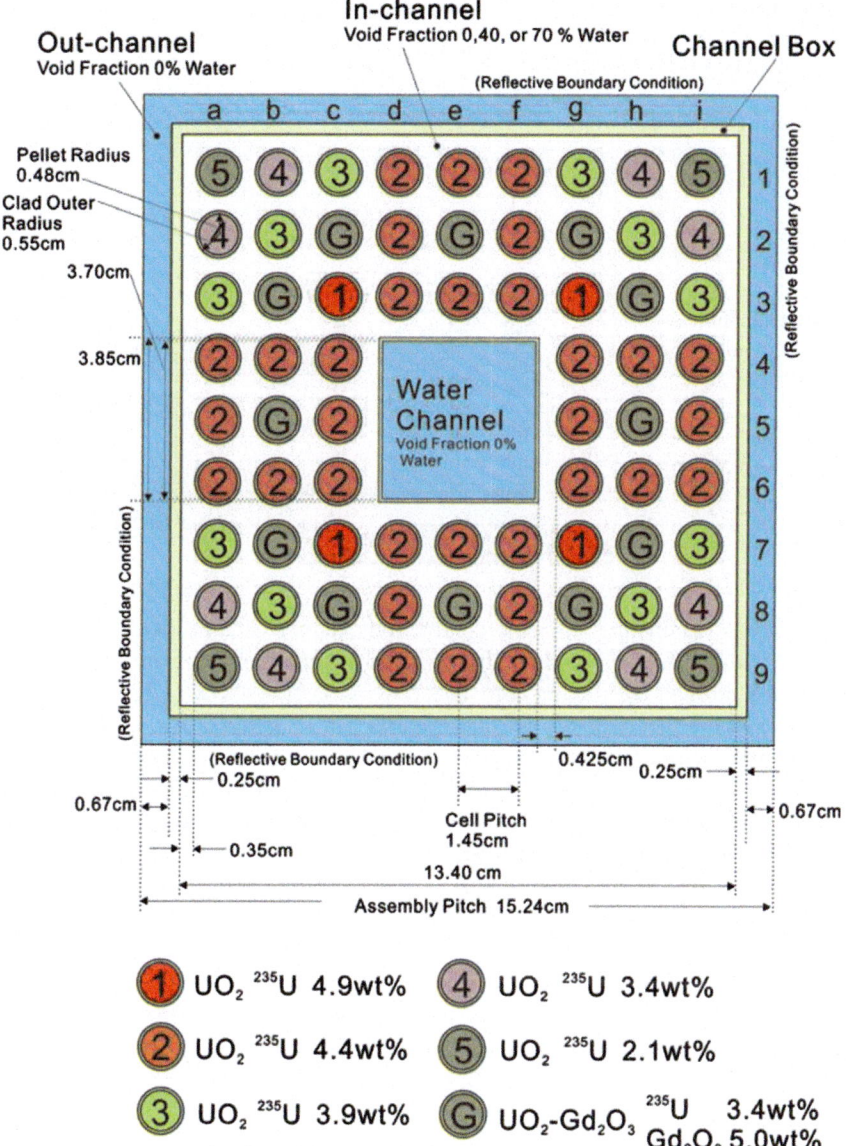

Fig. 21.1 Benchmark model of the BWR STEP3 fuel assembly

The condition of the fuel debris has not yet been identified in any reactor except estimations by severe accident analysis codes. Study of the TMI-2 fuel debris [6], however, suggests that various kinds of fuel debris may also be produced in the 1FNPS reactors, such as hard and loose debris. Especially, loose debris may show a wide variety of composition including structural materials such as Zircaloy and

Table 21.1 Initial uranium inventory in a boiling water reactor (BWR) STEP 3 fuel assembly

^{235}U/U enrichment	Mass (kgU)
4.9 wt%	9.6
4.4 wt%	76.8
3.9 wt%	28.8
3.4 wt%	19.2
2.1 wt%	9.6
3.4 wt% (with Gd$_2$O$_3$)	26.9
Total	170.9

Table 21.2 Burn-ups of fuel assemblies in the 1FNPS reactors

Unit 1	Unit 2	Unit 3
5.2:64	3.3:116	4.7:148[a]
15.2:64	15.8:116	15.5:112
24.2:80	26.0:120	28.5:140
33.3:68	35.2:120	36.2:112
37.5:64	40.6:76	40.5:36
40.2:60	(GWD/t, number of assemblies)	

[a]16 MOX assemblies included

steel. Boron originating from the control rods cannot be expected necessarily to coexist with the fuel debris. It is also possible that the fuel debris in CVs has been generated through the molten core–concrete interaction (MCCI). It must be considered that the fuel debris is not uniform and will be found at various locations.

The fuel debris is being cooled with nonborated water although it is highly preferable to add neutron poison and to maintain enough concentration in the water to secure the subcritical condition such as was performed after the TMI-2 accident. Boration is not realistic at present because of the coolant water leakage from CVs and underground water inflow to the coolant water circulation. Boron will be injected only in the event of re-criticality [7].

21.3 Criticality Characteristics of Fuel Debris

The criticality safety handbook shows the minimum critical masses of homogeneous uranium-water mixtures, 36 and 53 kg, respectively for the ^{235}U/U enrichments of 5 and 4 wt%. Mass control limits that can avoid criticality are also given for heterogeneous UO$_2$-water composites, that is, 28 kg for the 5 wt% enrichment. Even for the 3 wt% enrichment, its mass limit is still 67 kg [8]. These numbers are small compared to the possible uranium inventory in each fuel assembly with low burn-up.

Fuel debris may exist as composites of UO$_2$ and structural materials such as Zircaloy and steel in the pressure vessels (PVs). Zircaloy does not greatly affect the criticality characteristics of fuel debris because of its small neutron absorption cross section, but the iron in steel may increase the critical mass of fuel debris because it has strong neutron absorption.

The MCCI product would be a composite of UO_2 and concrete. The major content of concrete is silicon dioxide, which has also a small neutron absorption cross section and neutron moderation capability. The critical mass of the UO_2–concrete composite has been evaluated as 400 kg for the fresh UO_2 of 5 wt% $^{235}U/U$ enrichment. For the fuel burned up to 12 GWD/t, the critical mass can be as small as 800 or 2,000 kg, depending on how the effect of fission products is considered. Only the water bonded in concrete is considered in the evaluation; therefore, the critical masses can be smaller when the MCCI product is submerged in the coolant water [9]. The mass of 2,000 kg is equivalent to 12 fuel assemblies. It is also known that a certain cluster of 16 assemblies in the Unit 2 reactor has an average burn-up of about 14 GWD/t. Thus, this evaluation is not far from reality.

Before knowing the actual condition of fuel debris, it is possible to compute critical conditions. Such work has been already conducted for many years to produce a handbook or a database for criticality safety. It is easy to extend these standards to wider conditions such as UO_2–steel composite or UO_2–concrete composite. The computation will supply a new set of "criticality maps of fuel debris." These maps will indicate (Fig. 21.2) subcritical and critical conditions, and supercritical conditions that would likely bring severe consequences. In Fig. 21.2, the horizontal line represents variation of composition, and the vertical line represents variation of geometry. Composition on the right has higher reactivity and smaller critical volume. On the left, the composition is certainly subcritical, which can be excluded from the criticality control.

The actual criticality situation will be assessed by placing onto the map the fuel debris condition revealed by observations or sample analyses. It is also necessary to study how the condition can move on this map from expected changes such as temperature drop in the fuel debris or geometry changes caused by retrieval work of fuel debris, etc.

21.4 Options of Criticality Control Principles

21.4.1 Prevention of Criticality by Poison or Dry Process

The boration of coolant water was practiced in TMI-2 and is most preferable. Borated water bounds the criticality characteristics of all debris into a small region, indicated as "Boration" in Fig. 21.3, and keeps the region far from critical condition no matter how much temperature or geometry changes. By securing the lowest boron concentration in water, the subcritical condition can be guaranteed as well. The water issue, however, must be fixed to implement this option. Moreover, a structure made of carbon steel or aluminum will act as the water boundary when a

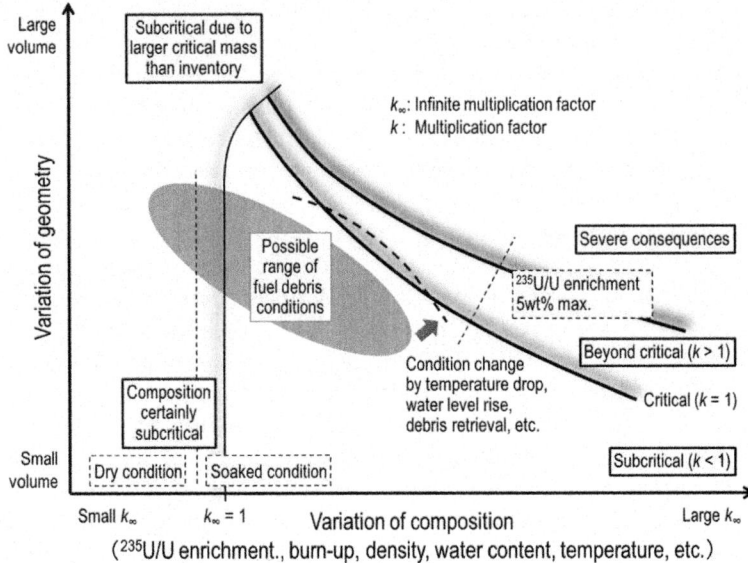

Fig. 21.2 Criticality map of fuel debris

CV is filled with water. Then, corrosion of such material by boron must be studied to prevent recurrence of the water issue.

The dry process without using coolant water will be also a certain criticality control method (Fig. 21.3). There will be, however, other engineering challenges. CVs must be sealed to avoid unexpected intrusion of water. It will be necessary as well to shield radiation and to suppress airborne migration of radioactive materials without water during fuel debris retrieval work.

21.4.2 Prevention of Criticality by Monitoring

Utilization of borated water may not be feasible if the water issue cannot be remedied. An alternative may be subcriticality monitoring. It is necessary to detect the signs of approach to the critical condition across the defense line set in the subcritical region in Fig. 21.4, and an intervention measure must be deployed quickly before the critical condition is reached. Detection may be possible by setting neutron counters near the fuel debris.

There are key natures of the intervention measure to be understood. The injection of neutron poison is the only way, and it will be realistic only if the actual condition of fuel debris is far from critical condition. It will be, however, difficult to make the defense line effective if the buffer zone is small. To retain the effect of intervention even after the event, the neutron poison concentration must be maintained in the coolant water.

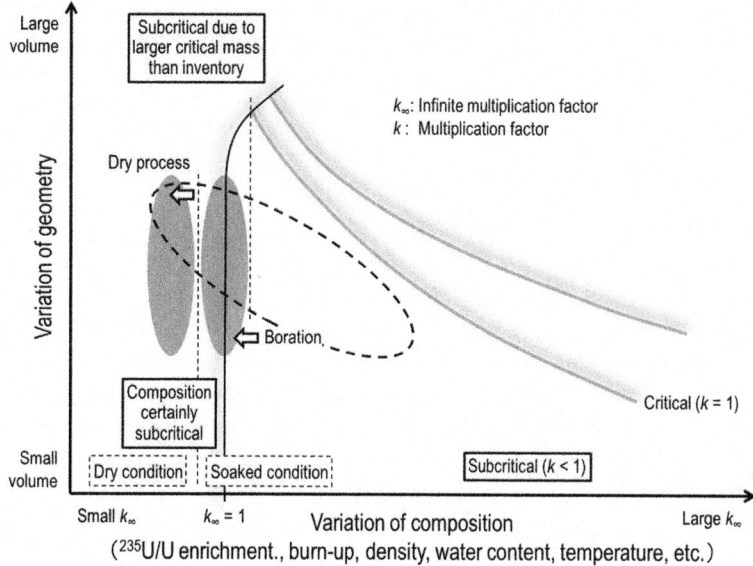

Fig. 21.3 Prevention of criticality by boration or dry process

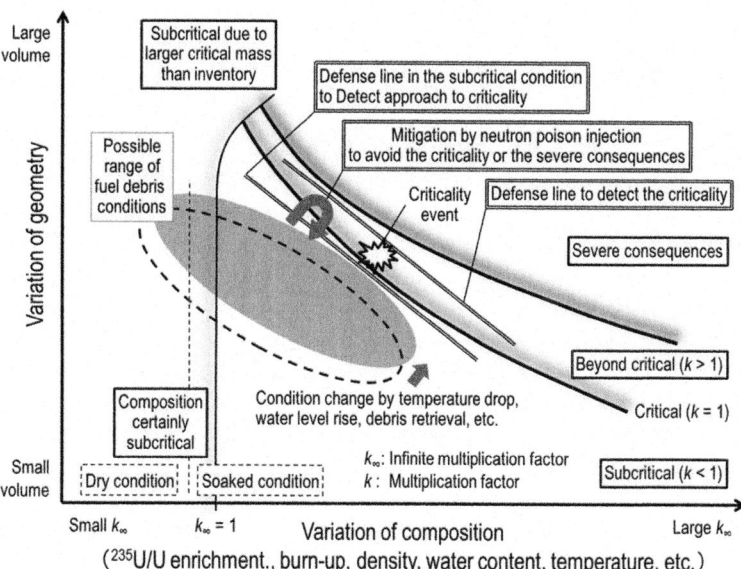

Fig. 21.4 Prevention of criticality and the severe consequences by monitoring

Thus, this option does not differ, essentially, from the first option, which is prevention of criticality by poison. Monitoring still makes sense if we integrate it with the first option and use it as an implementation of the "double contingency principle."

21.4.3 Prevention of Severe Consequence

The last option is, in fact, being currently applied. The defense line consists of xenon gas monitoring and the injection of borated water. The monitoring sensitivity is not sufficient to measure subcriticality but can detect the event beyond the occurrence of critical condition before severe consequences result. The borated water on standby will be injected when the monitoring detects the criticality. A study is under way to improve the monitoring sensitivity to make the detection and intervention quicker and to reduce the risk of this option.

A much bolder idea is also being brought up, which is to consider such quick detection and intervention as a regular reactivity control. A small-scale, controlled chain reaction is permissible in the concept, and the resumption of fuel debris retrieval is allowed after suppressing the criticality. To realize this kind of criticality control, its risk must be fully understood.

21.4.4 Risk Assessment

The risk study is necessary regardless of which option is chosen because the subcritical condition is not secured at present. Even though the fuel debris will not be touched for a while, the temperature of the fuel debris may drop gradually in time, which slowly increases reactivity. The risk of "low probability and high consequence events" must be also evaluated. An aftershock of large magnitude may change the fuel debris geometry greatly. The extreme event would be the fall of fuel debris in the PV onto the other in CV.

The fuel debris retrieval must be assessed carefully, of course, if it is conducted under nonborated water. The first step of the risk analysis is to understand the actual conditions of fuel debris. Exhaustive observation of the fuel debris should be conducted as early as possible, which enables us to complete the maps described in the previous sections.

According to each option, engineering work should be performed in parallel to establish design requirements. For the prevention of criticality by borated water, its required lowest concentration must be established. For the prevention of criticality by monitoring, requirements of sensitivity and time response of the monitoring and time response of an intervention measure must be clarified. For the prevention of severe consequences, an allowable limit of fission number must first be set. Then, the time response of detection and intervention must be defined to regulate fission numbers of supposed criticality events within the limit.

21.5 Conclusions

In 1FNPS, fuel debris conditions in the three damaged reactors are still unknown and uncertain. The water issue also affects criticality control, as the coolant water is not borated. Although fortunately no sign of criticality has yet been seen, the subcritical condition is not secured. There are options of principles to pursue a certain critical control of the fuel debris: prevention of criticality by poison, by dry process, or by monitoring, and prevention of the severe consequences resulting from criticality. Engineering research and development is to be conducted regarding any of these options.

Open Access This chapter is distributed under the terms of the Creative Commons Attribution Noncommercial License, which permits any noncommercial use, distribution, and reproduction in any medium, provided the original author(s) and source are credited.

References

1. Japan Atomic Energy Research Institute (2005) Working Group on Nuclear Criticality Safety Data. Nuclear criticality safety handbook, version 2. JAERI-1340. Japan Atomic Energy Research Institute (in Japanese)
2. Stratton WR (1987) Review of the state of criticality of the Three Mile Island Unit 2 core and reactor vessel. DOE/NCT-01, Lawrence Livermore National Laboratory, California, United States of America
3. Tonoike K et al (2013) Major safety and operational concerns for fuel debris criticality control. In: Proceedings of the GLOBAL 2013: International nuclear fuel cycle conference, Salt Lake City, UT, September 29–October 3, 2013
4. Suyama K et al (2012) OECD/NEA burnup credit criticality benchmark phase IIIC. Nuclide composition and neutron multiplication factor of BWR spent fuel assembly for burnup credit and criticality control of damaged nuclear fuel. OECD/NEA/WPNCS/EGBUC, Paris, France
5. Nishihara K et al (2012) Estimation of fuel composition in Fukushima-Daiichi nuclear power plant. JAEA-Data/Code 2012–018, Japan Atomic energy Agency (in Japanese), Ibaraki, Japan
6. Akers DW et al (1992) TMI-2 examination results from the OECD-CSNI Program. NEA/CSNI/R(91)9. Committee on the safety of nuclear installations, Organization for Economic Cooperation and Development, Paris, France
7. Tokyo Electric Power Company, Inc. (2011) Status of Fukushima Daiichi nuclear power station. http://www.tepco.co.jp/en/nu/fukushima-np/index-e.html
8. Okuno H et al (2009) Second version of data collection part of nuclear criticality safety handbook (Contract Research). JAEA-Data/Code 2009–010, Japan Atomic Energy Agency, Ibaraki, Japan
9. Izawa K et al (2012) Infinite multiplication factor of low-enriched UO_2-concrete system. J Nucl Sci Technol 49(11):1043

Chapter 22
Modification of the STACY Critical Facility for Experimental Study on Fuel Debris Criticality Control

Hiroki Sono, Kotaro Tonoike, Kazuhiko Izawa, Takashi Kida,
Fuyumi Kobayashi, Masato Sumiya, Hiroyuki Fukaya, Miki Umeda,
Kazuhiko Ogawa, and Yoshinori Miyoshi

Abstract For the decommissioning of the Fukushima Daiichi Nuclear Power Stations, fuel debris involving molten structural materials should be retrieved from each reactor unit. The fuel debris, which is of uncertain chemical composition and physical state, needs to be treated with great care from the standpoint of criticality safety. For developing criticality control for the fuel debris, the Japan Atomic Energy Agency (JAEA) has been planning to modify the Static Experiment Critical Facility (STACY) and to pursue critical experiments on fuel debris. STACY, a facility using solution fuel, is to be converted into a thermal critical assembly using fuel rods and a light water moderator. A series of critical experiments will be conducted at the modified STACY using simulated fuel debris samples. The simulated fuel debris samples are to be manufactured by mixing uranium oxide and reactor structural materials with various chemical compositions. This report summarizes a facility development project for an experimental study on criticality control for fuel debris using the modified STACY and simulated fuel debris samples.

Keywords Critical facility • Criticality control • Criticality safety • Fuel debris • Fukushima Daiichi • Simulated fuel debris sample • STACY

22.1 Introduction

In the severe accident at the Fukushima Daiichi Nuclear Power Stations (NPS), most of the fuel loaded in the cores of Units 1, 2, and 3 was seriously damaged and melted, resulting in a considerable amount of fuel debris [1]. It is believed that some parts of the fuel debris involve molten structural materials such as zircaloy, stainless steel, and concrete. This fuel debris, which contains much burned fuel,

---

H. Sono (✉) • K. Tonoike • K. Izawa • T. Kida • F. Kobayashi • M. Sumiya
H. Fukaya • M. Umeda • K. Ogawa • Y. Miyoshi
Japan Atomic Energy Agency, Tokai-mura, Naka-gun, Ibaraki, 319-1195, Japan
e-mail: sono.hiroki@jaea.go.jp

© The Author(s) 2015
K. Nakajima (ed.), *Nuclear Back-end and Transmutation Technology for Waste Disposal*, DOI 10.1007/978-4-431-55111-9_22

still continues to emit radiation and heat. For decommissioning of the Fukushima Daiichi NPS, all the fuel debris should be retrieved from the pressure and containment vessels of each reactor unit.

In preparation for retrieval of the fuel debris from the Fukushima Daiichi NPS, however, there remain the following serious problems: (1) leakage of cooling water from containment vessels, (2) inflow of groundwater into reactor buildings, (3) maintenance of subcritical state of the fuel debris, and (4) shielding of radiation from the fuel debris [2]. The cooling water of the fuel debris concerns these four problems. In a similar accident that occurred at Three Mile Island NPS Unit 2 (TMI-2), where its pressure vessel was not seriously damaged, all these problems were settled or did not arise because the pressure vessel could be filled with cooling water containing highly concentrated boron as a neutron absorber and radiation shield [3]. In contrast, all four problems make it extremely difficult to retrieve the fuel debris from each reactor unit of the Fukushima Daiichi NPS.

The fuel debris, which has uncertain chemical composition and physical state, needs to be treated with great care from the aspect of criticality safety. In particular, large blocks of fuel debris can cause a change in physical state, such as size and water content, when they are broken into fragments to be retrieved in cooling water. Furthermore, a recent study on fuel debris resulting from the molten core–concrete interaction has revealed its potential for criticality [4]. There will probably be no risk of a criticality accident if it is possible to keep a high concentration of boron in the cooling water and take the criticality control measures that were used in the TMI-2 accident. However, these measures will be difficult unless both (1) the leakage of cooling water and (2) the inflow of groundwater are completely stopped. If not, retrieval of the fuel debris will require alternative approaches to criticality control in cooling water or dry retrieval with radiation shielding.

The authors focus on the former approach: new criticality control measures for fuel debris, such as criticality safety standards and criticality monitoring methodology [5]. This report summarizes a facility development project for an experimental study on criticality control for fuel debris.

22.2 Experimental Study on Criticality Control for Fuel Debris

22.2.1 Modification of STACY

To implement the new criticality control measures for fuel debris, the Japan Atomic Energy Agency (JAEA) has been carrying a project to modify the Static Experiment Critical Facility (STACY) to pursue critical experiments on fuel debris [6]. STACY, a facility using solution fuel (low-enriched uranyl nitrate), is to be converted into a thermal critical assembly using fuel rods and a light water moderator.

In the modified STACY, the core configuration consists of fuel rods loaded in the core tank (up to 900 rods) and light water fed as moderator. Because the maximum thermal power is only 200 W, fuel burn-up is negligibly small and cooling water is unnecessary. The reactivity of the core is controlled not with control rods but by water level, and with safety plates (cadmium) in the case of emergency shutdown, similar to the present STACY. The fuel rods contain 5 wt.%-enriched UO_2 pellets and have zircaloy cladding. A soluble neutron poison (boron) can be added to the light water moderator. Major core specifications and a schematic diagram of the modified STACY are shown in Table 22.1 and Fig. 22.1, respectively.

Table 22.1 Major core specifications of the modified Static Experiment Critical Facility (STACY)

Item			Present STACY	Modified STACY
Core tank			Closed tank	Open tank
			Replaceable (cylinder, slab, heterogeneous, interaction)	Cylinder (1.8 m in diameter, 1.9 m in height)
Core size			Same as each core tank	Maximum 60 cm × 90 cm
			Critical height 40–140 cm	Critical height 40–140 cm
Maximum thermal power			200 W	200 W
Maximum integrated power			100 W·h/operation, 300 W·h/week, 3 kW·h/year	100 W·h/operation, 300 W·h/week, 3 kW·h/year
Fuel	Fuel solution		6-, 10 wt.%-enriched uranyl nitrate solution	Not used
	Maximum concentration		500 gU/l	
	Fuel rods		5 wt.%-enriched UO_2 pellets	5 wt.%-enriched UO_2 pellets (<10 wt.% available)
	Cladding		Zircaloy cladding (9.5 mm in diameter, 150 cm in length)	Zircaloy cladding (9.5 mm in diameter, 150 cm in length)
	Maximum loading		400 rods	900 rods
Volume ratio of moderator to fuel (lattice pitch of fuel rods)			1.9–15 (13.0–29.0 mm)	0.9–11 (10.9–25.5 mm)
Moderator			Solution fuel	Light water
Temperature			Room temperature ~40 °C	Room temperature ~70 °C
Reactivity control			Solution level	Water level
Maximum excess reactivity			0.2 dollar in normal operation	0.3 dollar in normal operation
			0.8 dollar in abnormal transient	0.8 dollar in abnormal transient
Maximum reactivity addition rate			3 cent/s	3 cent/s
Emergency shutdown			Insertion of safety plates	Insertion of safety plates
			Drain of fuel solution	Drain of light water moderator
Shutdown margin			<0.985 in k_{eff}	<0.985 in k_{eff}
One-rod stuck margin			<0.995 in k_{eff}	<0.995 in k_{eff}

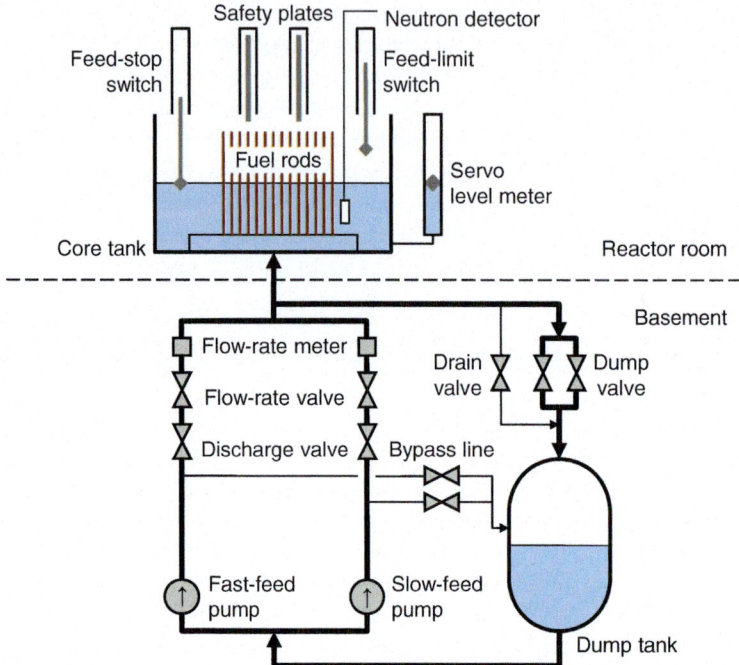

Fig. 22.1 Schematic diagram of the modified Static Experiment Critical Facility (STACY)

22.2.2 Critical Experiments on Criticality Safety for Fuel Debris

The JAEA research program includes computation of criticality characteristics covering a wide range of fuel debris conditions and validation of the computation by critical experiments. In the former activity, several data sets will be systematically obtained by calculation to establish new criticality safety standards for fuel debris. The new standards will be provided as "criticality maps" that indicate subcritical and critical conditions. The maps also show supercritical conditions that would likely lead to a significant threat of human injury [7]. In the latter activity, the new standards (including computation models) will be validated regarding reactivity worth, coefficients of reactivity, and critical mass by critical experiments with simulated fuel debris samples. A criticality monitoring methodology will also be studied to improve the criticality control measures for fuel debris.

To pursue the aforementioned critical experiments, the core of the modified STACY has a widely distributed neutron energy spectrum between thermal reactor spectra and intermediate reactor spectra. The neutron energy spectrum of the core can be varied by the lattice pitch of the fuel rods, which range from 10.9 to 25.5 mm, corresponding to a moderator-to-fuel volume ratio ranging from 0.9 to 11. Typical neutron energy spectra of the modified STACY are shown in Fig. 22.2 [8].

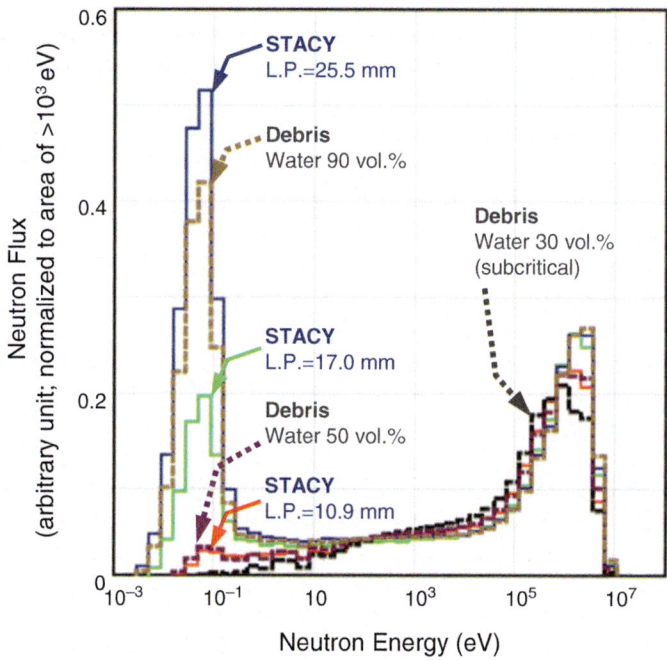

Fig. 22.2 Neutron energy spectrum of the modified STACY core

This figure also shows typical spectra of hypothetical fuel debris of a BWR fuel pellet (3.7 wt.% ^{235}U, 27.5 GWd/t, 5-year-cooled), for comparison. Both spectra were calculated using a burn-up code, ORIGEN2 [9], and a Monte Carlo code, MVP2 [10], with a nuclear data library, JENDL-3.3 [11]. It can be seen in Fig. 22.2 that the core spectrum with a lattice pitch of 10.9 mm is equivalent to the debris spectrum in 50 vol.% water. The core spectrum of the modified STACY can cover relatively hard spectra of the fuel debris likely to become critical.

For the measurement of the neutronic characteristics of fuel debris, two sets of experimental equipment should be prepared: one includes reactor material structures simulating fuel debris (zircaloy, stainless steel, concrete, etc.), which are pin-, plate-, or box type and are loaded between fuel rods. The other is a sample-loading device to measure its reactivity and which is installed at a test region in the core tank. The experimental equipment is shown in Fig. 22.3.

22.2.3 *Manufacturing and Analytical Equipment for Simulated Fuel Debris Samples* [12]

The simulated fuel debris samples (sintered pellets) are to be manufactured by mixing UO_2 and reactor structural materials (Zr, Fe, Si, Gd, B, etc.) with various chemical compositions. These debris materials will be mixed in the form of oxide

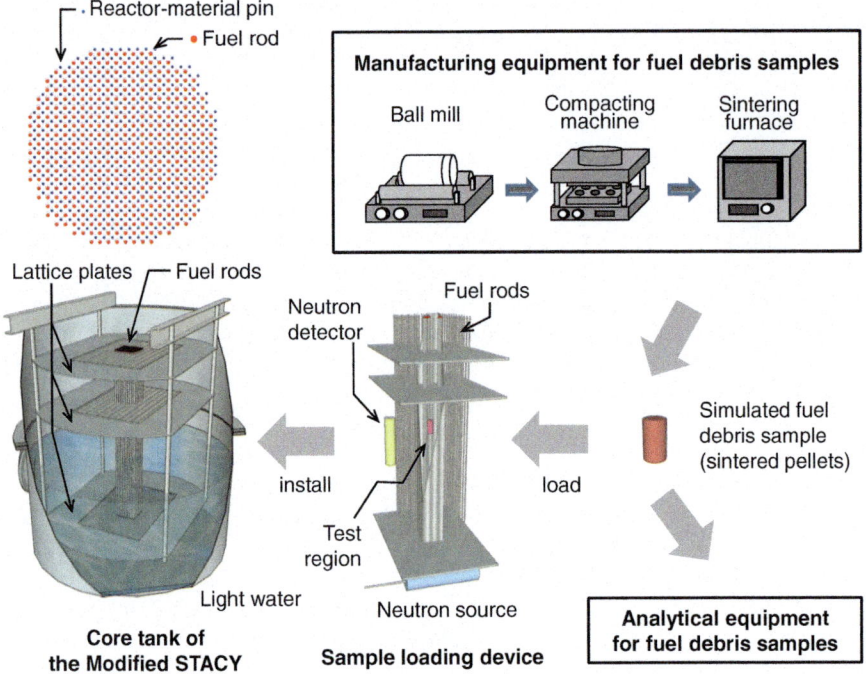

Fig. 22.3 Experimental equipment for simulated fuel debris samples

powders. The manufacturing equipment for the debris samples is composed of a ball mill, compacting machine, and sintering furnace. The debris samples will be analyzed destructively or nondestructively to determine nuclide composition, O/U ratio, density, and impurities. The manufacturing ability is to be 300 pellets a month. The analytical precision is still a matter under consideration. The manufacturing and analytical equipment are to be installed in glove boxes in the experimental building adjoining the modified STACY.

22.3 License Application and Schedule of the STACY Modification

The license application for the STACY modification was sent in February 2011 and has been under safety review by the Nuclear Regulation Authority (NRA) of Japan to comply with new safety standards for research reactors enforced in December 2013 [13]. In particular, the NRA will strictly demand prevention measures against natural disasters such as a tsunami from all reactors located at a low altitude. The modified STACY, the reactivity of which is controlled by water level, has a risk of criticality accidents for the duration of tsunami attacks. The prevention measures

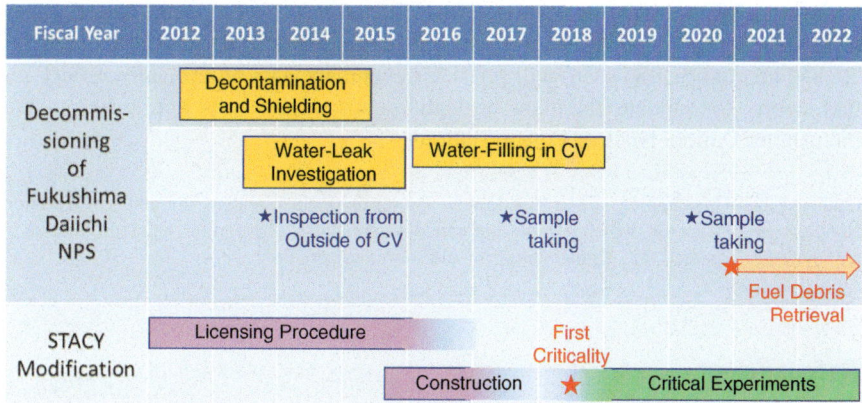

Fig. 22.4 Schedule of the STACY modification. *CV* containment vessel

against criticality accidents are important requirements for the modified STACY: for example, limitation of the core configuration together with the safety plates inserted so as to keep a subcritical state during submersion.

A schedule of the STACY modification is shown in Fig. 22.4. The first criticality experiment in the modified STACY is scheduled for 2018. The modified STACY will provide benchmark data on criticality safety for fuel debris to validate the criticality control measures applicable to the Fukushima Daiichi NPS. The new criticality control measures need to be established by the time the fuel debris begins to be retrieved from each reactor unit of the Fukushima Daiichi NPS. According to the governmental council, retrieval of the fuel debris is scheduled to start as early as 2020, depending on the progress in the decommissioning of each reactor unit [2].

22.4 Concluding Summary

For the decommissioning of the Fukushima Daiichi Nuclear Power Station Units 1, 2, and 3, research and development activities have been pursued to retrieve fuel debris from the pressure and containment vessels of each reactor unit. In preparation for the retrieval, however, there remain serious problems concerning the cooling water of fuel debris from the aspect of criticality safety.

To study the new criticality control measures for the fuel debris, the Japan Atomic Energy Agency has carried forward a project to modify the Static Experiment Critical Facility (STACY) and to pursue critical experiments regarding the fuel debris. STACY, a facility using solution fuel, is to be converted into a thermal critical assembly using fuel rods and a light water moderator. A series of critical experiments will be conducted in the modified STACY using simulated fuel debris samples. These samples are to be manufactured by mixing UO_2 and reactor structural materials with various chemical compositions.

The license application for the STACY modification has been under safety review. The first criticality experiment in the modified STACY is scheduled for 2018. The modified STACY will provide benchmark data on criticality safety for fuel debris to validate the new criticality control measures applicable to the Fukushima Daiichi Nuclear Power Stations.

Open Access This chapter is distributed under the terms of the Creative Commons Attribution Noncommercial License, which permits any noncommercial use, distribution, and reproduction in any medium, provided the original author(s) and source are credited.

References

1. Tokyo Electric Power Company, Inc. (2012) Fukushima nuclear accident analysis report
2. Nuclear Emergency Response Headquarters, Council for the Decommissioning of TEPCO's Fukushima Daiichi Nuclear Power Station (2013) Mid- and-long-term roadmap towards the decommissioning of TEPCO's Fukushima Daiichi nuclear power station unit 1–4
3. Stratton WR (1987) Review of the state of criticality of the Three Mile Island unit 2 core and reactor vessel. DOE/NCT-01. Lawrence Livermore National Laboratory
4. Izawa K et al (2012) Infinite multiplication factor of low-enriched UO_2-concrete system. J Nucl Sci Technol 49(11):1043
5. Tonoike K et al (2013) Major safety and operational concerns for fuel debris criticality control. In: Proceedings of the GLOBAL 2013: international nuclear fuel cycle conference, Salt Lake City, September 29–October 3, 2013
6. Miyoshi Y (2013) STACY modification program and critical experiments using pseudo fuel debris. Working party on nuclear criticality safety, October 11, 2013, OECD/NEA, Paris
7. Tonoike K et al (2013) Principle options of fuel debris criticality control in Fukushima Daiichi Reactors. In: International symposium on nuclear back-end issues and the role of nuclear transmutation technology after the accident of TEPCO's Fukushima Daiichi Nuclear Power Stations, Kyoto, November 28, 2013
8. Izawa K et al (2012) Feasibility study on critical experiment with fuel debris in modified STACY. 3. Neutronic characteristics analysis of modified STACY core. In: Proceedings of the annual meeting of AESJ, Fukui, March 19–21, 2012, E37 (in Japanese)
9. Croff AG (1980) ORIGEN2: a revised and updated version of the Oak ridge isotope generation and depletion code. ORNL-5621. Oak Ridge National Laboratory
10. Nagaya Y et al (2005) MVP/GMVP II: general purpose Monte Carlo codes for neutron and photon transport calculations based on continuous energy and multigroup methods. JAERI-1348. Japan Atomic Energy Research Institute
11. Shibata K et al (2002) Japanese evaluated nuclear data library, version 3, revision 3: JENDL-3.3. J Nucl Sci Technol 39(11):1125–1136
12. Umeda M et al (2012) Feasibility study on critical experiment with fuel debris in modified STACY. 2. Fabrication and analyses of pseudo fuel debris at nuclear fuel usage facility in NUCEF. In: Proceedings of the annual meeting of AESJ, Fukui, March 19–21, 2012, E36 (in Japanese)
13. Kida T et al (2013) Evaluation of nuclear characteristics and safety design examination of modified STACY for critical experiments on fuel debris. 1. Outline of critical experiments on fuel debris and safety design principle. In: Proceedings of the annual meeting of AESJ, Higashi-Osaka, March 26–28, 2013, H30 (in Japanese)

Part VII
Nuclear Fuel Cycle Policy and Technologies: National Policy, Current Status, Future Prospects and Public Acceptance of the Nuclear Fuel Cycle Including Geological Disposal

Chapter 23
Expectation for Nuclear Transmutation

Akito Arima

Abstract It is my great honor and pleasure to speak to you this morning on the occasion of the International Symposium on Nuclear Back-end Issues and the Role of Nuclear Transmutation Technology after the accident of TEPCO's Fukushima Daiichi Nuclear Power Stations. I would like to thank the organizers, especially Professor Hirotake Moriyama and Professor Hajimu Yamana, for inviting me to this Symposium.

I believe that this Symposium is very important and well timed to solve urgent problems concerning nuclear back-end issues and to develop nuclear transmutation technology. I myself am a nuclear theoretical physicist and am ignorant of nuclear technology. However, I believe that nuclear energy is indispensable for the future of human beings and that nuclear engineering must be further developed.

My talk consists of the following four subjects:

1. Demand for primary energy and electricity is increasing year by year.
2. Global warming is becoming a more serious problem.
3. Development of renewable energy must be promoted. However, it will require sufficient resources of time and budget.
4. Human beings cannot avoid depending on nuclear energy as well as other energy resources that do not emit CO_2.
5. Nuclear technology must be developed.

    (a) The safety technology of nuclear energy has to be developed for the future.
    (b) The technology for the back-end of the nuclear fuel cycle has to be enhanced. The site for final disposal of nuclear wastes has to be determined as soon as possible in Japan, which is a responsibility of the Central Government.
    (c) The research and development of innovative technologies, such as accelerator-driven systems, must be promoted to encourage the progress of final disposal.

A. Arima (✉)
Japan Radioisotope Association, 2-28-45, Honkomagome, Bunkyo-ku,
Tokyo 113-8941, Japan
e-mail: arima@musashi.jp

(d) Research and development of nuclear technologies for reactor decommissioning, safety technology, back-end, etc., must be promoted intensively through international cooperation.

Keywords Accelerator-driven system • Decommissioning • Final disposal • Nuclear back-end • Nuclear energy • Nuclear transmutation

23.1 Demand for Primary Energy and Electricity Is Increasing Year by Year

Figure 23.1 shows a prediction of the world population together with its past history. This figure shows that the world population had already reached 7 billion in 2011 and will be 9.2 billion in 2050. Another prediction indicates that the population of the world would be 11 billion by the end of this century.

It is not easy to predict the future demand for primary energy. Let me estimate it taking an extremely naive way. Figure 23.2 shows how much primary energy per capital is consumed annually in each country in terms of tons of oil equivalent. In 2009, the average of consumption of primary energy was 1.8 t/year and the world population was 7 billion. It is a reasonable assumption that everybody in the world hopes to enjoy the American life using 7 t/year, or at least the average of OECD countries by using 4 t/year. Let us assume that in the near future the average will become 4 t/year and the world population will be 10 billion in 2100. Then, simple arithmetic tells us that the total demand for primary energy will be 3.2 times $[=(10 \times 4)/(7 \times 1.8)]$ more than the present consumption.

More realistically, the International Energy Agency (IEA) predicted the future demand for primary energy. The demand in countries other than OECD in 2035 will be 1.8 times more than in 2010. The demand for primary energy in the world in 2035 will be 1.35 times more than in 2010. We should be careful because this increase of 35 % will occur only 25 years from now. If this increase continues linearly for the next 100 years, we find a 140 % increase, that is, altogether 2.4 times more than the present consumption. According to IEA, the demand for electricity in the world in 2035 will be 1.73 times more than in 2011, which is an increase 2 times as fast as that for primary energy.

23.2 Global Warming Is Becoming a More Serious Problem

To prepare for further increase of the demand for primary energy and to stop global warming by reducing CO_2 emissions into the air, we need to develop renewable energy as well as nuclear energy.

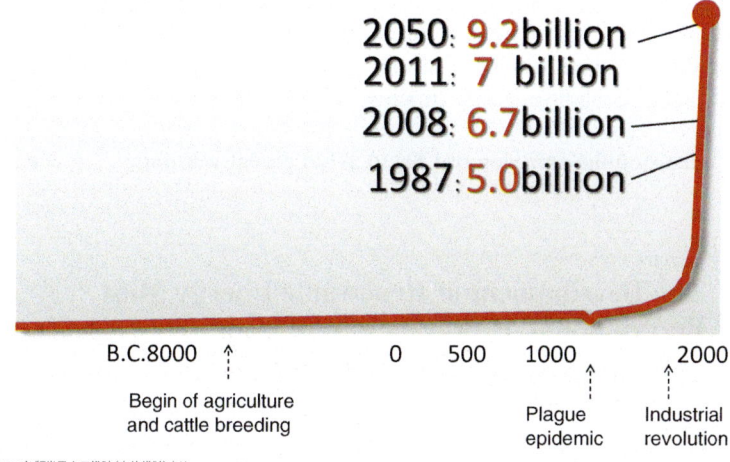

Fig. 23.1 Explosion of world population

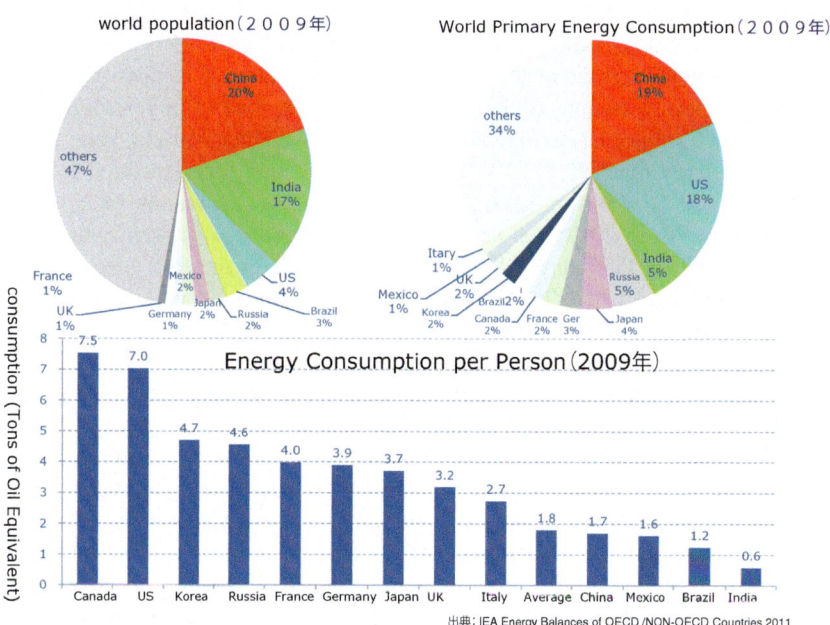

Fig. 23.2 Regional population and energy consumption per person

Warming of the climate system is unequivocal, and since the 1950s, many of the observed changes are unprecedented over decades to millennia. The atmosphere and ocean have warmed, the amounts of snow and ice have diminished, sea level has risen, and the concentrations of greenhouse gases have increased.

Professor Akimasa Sumi and his collaborators have carried out computer simulations using climate models for many years.

According to their results, it seems very very clear that the anthropological emission of greenhouse gases (mainly CO_2) is a main contributor to global warming.

We must stop the emission of CO_2 to avoid global warming.

23.3 The Development of Renewable Energy Must Be Promoted. However, It Will Require Sufficient Resources of Time and Budget

According to the world energy outlook of IEA, the total electric power generation will be increased as shown in Fig. 23.3. The electric power generated by renewable energy is predicted as shown in Fig. 23.4. The electric power generated by renewable energy other than water power will increase very slowly, from about 4 % in 2010 to only 15 % in 2035, whereas the electric power generated by nuclear energy will be kept almost constant from 13 % in 2010 to 12 % in 2035.

We must try to increase renewable energy more as quickly as possible.

In this respect, I commend Germany, which has strived to increase the development of renewable energy (Fig. 23.5) after 2000. In 2010, electricity generated by renewable energy reached 103.5 billion kWh. Deducting that generated by water power, we have 82.9 billion kWh. Total electric power generation in Japan was 976.2 billion kWh in 2010; that is, electric power generated by renewable energy other than water in Germany in 2010 was only 8.5 % of the total electric power generation in Japan in the same year. The electric power generated by nuclear energy in Japan was 300.4 billion kWh in 2010. Therefore, electric power generated in Germany by renewable energy sources other than water in 2010 is only 28 % of this amount. Germany has striven so much in these 10 years from 2000 to 2010, and the average price of electricity per house has doubled; that is, Germany has invested a large budget. It takes many years to increase renewable energy, and the result is still not satisfactory. Even if Japan tries as much as Germany, it will takes at least 30 years to replace nuclear energy by renewable energy. Meanwhile, Japan must depend on fossil fuel, which increases CO_2 emissions into the atmosphere. To import fossil fuel, the deficit in foreign trade of Japan, which is now already more than 4 trillion yen (about $40 billion), will continue as the result of the decrease in nuclear energy.

When we stop all nuclear power stations in Japan, renewable energy must be increased, not only to replace nuclear energy but also the energy produced by fossil fuel. Is this really possible in the near future? It is time for us to deliberate upon the future of energy in Japan to guarantee energy security, to avoid global warming, and to stabilize the economy of Japan.

23 Expectation for Nuclear Transmutation

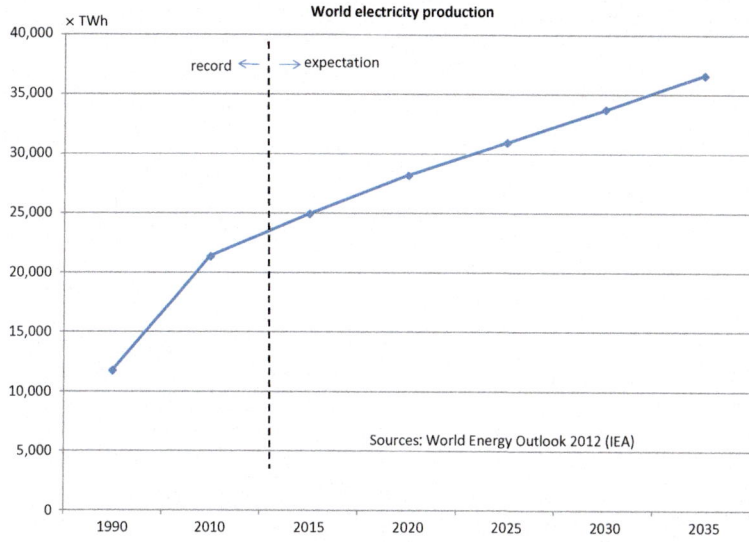

Fig. 23.3 World energy outlook of the International Energy Agency (IEA)

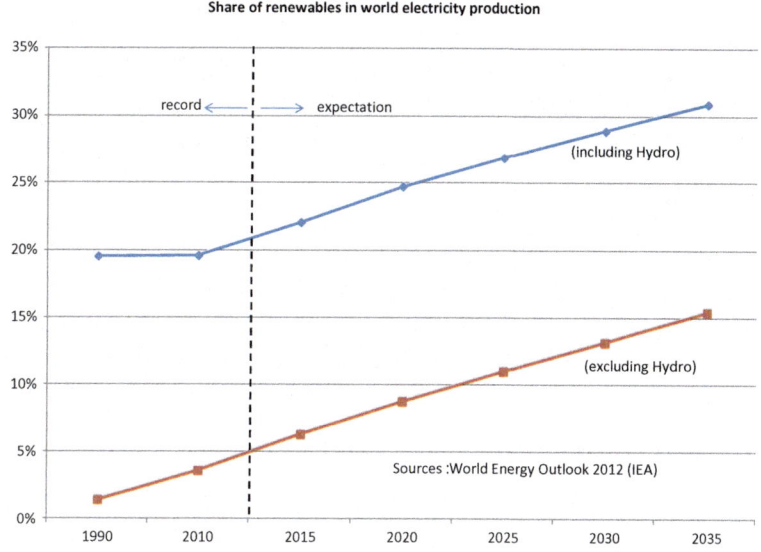

Fig. 23.4 Electric power generated by renewable energy (prediction by IEA)

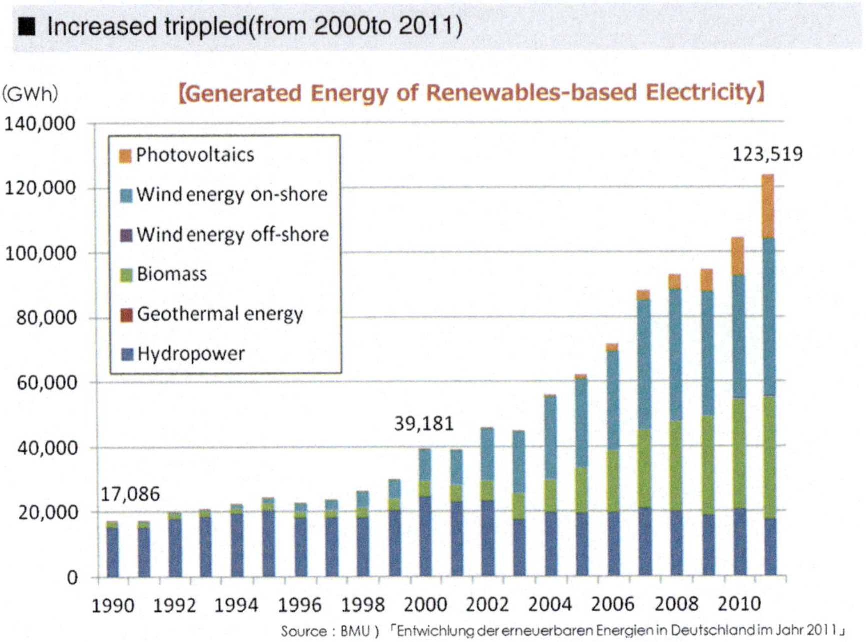

Fig. 23.5 Development of renewables-based electricity generation in Germany

23.4 Human Beings Cannot Avoid Depending on Nuclear Energy as Well as Other Energy Resources, Including Renewable Energy, Which Do Not Emit CO_2 into the Air

It is now very clear that it is almost impossible for renewable energy to replace fossil fuel in the near future. Both nuclear energy and renewable energy are necessary, not only in Japan but also in the world. At the same time we must develop a new technology to compensate for CO_2 emissions from fossil fuels.

23.5 Nuclear Technology Must be Developed

23.5.1 Safety Technology of Nuclear Energy Must Be Developed for the Future

Concerning nuclear energy, we must not stop researching and developing new advanced reactors in which greater safety is guaranteed against natural calamities as well as manmade disaster. Small-scale nuclear reactors also should be developed to decentralize electric power stations. If economical problems are overcome, smaller-scale reactors might be easier to guarantee safety.

23.5.2 Technology for the Back-end of the Nuclear Fuel Cycle Must Be Enhanced. The Site for Final Disposal of Nuclear Wastes Must be Determined as Soon as Possible in Japan, Which is a Responsibility of the Central Government

I have learned that technology for the back-end of the nuclear fuel cycle has already been well developed, but it still does not seem to be working well.

I hope that the solutions will be realized as soon as possible. Especially, the location for final disposal of nuclear wastes must be determined as soon as possible, and this is really a responsibility of the National Government to determine the location for the final disposal.

Not only Japan, but almost all countries including Germany, USA, Britain, and Russia, have not yet decided the location for final disposal, excepting Finland and Sweden. This decision must be made irrespective of whether nuclear power stations are to be continued.

23.5.3 Research and Development of Innovative Technologies, Such as Accelerator-Driven Systems, Must Be Promoted to Encourage the Progress of Final Disposal

It is extremely important to shorten the lifetimes of many radioactive nuclei in nuclear wastes. The role of nuclear transmutation technology is one of the main themes of this Symposium. The accelerator-driven system is one of the most promising methods to transmute radioactive nuclei to those of shorter lifetimes.

In Japan, the Omega project, which includes an accelerator-driven system, has been discussed for more than 10 years. I have helped to establish the J-PARC because one of its purposes is to develop the transmutation technology.

I expect that Dr. Hiroyuki Oigawa will tell us about the accelerator-driven system.

I would like to learn about the present situation of the transmutation technology in Japan and in the world.

23.5.4 The Research and Development of Nuclear Technologies for Reactor Decommissioning, Safety Technology, Back-end, etc., Must Be Promoted Intensively Through International Cooperation

Nuclear technologies for reactor decommissioning, safety technology, back-end, etc., must be urgently developed. They are very important, especially in Japan after the Fukushima Daiichi Accident.

These technologies, however, are also desired in all countries that already have nuclear power stations, and also in countries which are planning nuclear power stations. These technologies therefore should be researched and developed through international cooperation. Fukushima would be a very good candidate for us to construct an international center for researching and developing technologies for reactor decommissioning.

23.6 Conclusion

For the future of human beings, nuclear technology is indispensable to guarantee the safety of energy and to reduce CO_2 in the atmosphere, which causes global warming.

For promoting nuclear technology, we must encourage young researchers to be interested in nuclear science and engineering. Education is very important for this purpose.

You who are experts in nuclear science and technology should be very proud of your specialty. It is the most important time for you to solve very difficult problems after the accident of the Fukushima Daiichi Nuclear Power Station. I sincerely hope that you will overcome this crisis caused by the Fukushima accident.

Let us change the misfortune into good luck for the future of human beings.

I hope that this Symposium will succeed in producing good fruits.

Open Access This chapter is distributed under the terms of the Creative Commons Attribution Noncommercial License, which permits any noncommercial use, distribution, and reproduction in any medium, provided the original author(s) and source are credited.

Chapter 24
Issues of HLW Disposal in Japan

Kenji Yamaji

Abstract Concerning the disposal of high-level radioactive waste (HLW) in Japan, the Nuclear Waste Management Organization of Japan (NUMO) has been making efforts toward beginning a literature survey, a first step of HLW disposal according to fundamental policies and final disposal plan based on the "Designated Radioactive Waste Final Disposal Act." However, a difficult situation continues in which responses from municipalities, which are necessary for beginning a literature survey, are not being made.

In September 2010 the Science Council of Japan (SCJ) received a deliberation request from the Chairman of the Japan Atomic Energy Commission, and SCJ formed a Review Committee for Disposal of High-Level Radioactive Waste. The Review Committee made a Reply on Disposal of High-Level Radioactive Waste in September 2012, in which six proposals are made including safe temporal storage and management of the total amount of HLW. In this chapter, an outline of the current HLW disposal policy in Japan and the contents of the Reply are introduced.

Keywords Geological disposal • High-level radioactive waste (HLW) • Risk • Temporal safe storage

24.1 Concerns on HLW

HLW stands for high-level radioactive waste. Concern about the safety of HLW disposal is another important element for the public in deciding their choice of nuclear power along with the safety issues related to nuclear power plant operation. Former Prime Minister Koizumi changed his political stance clearly after the Fukushima nuclear accident in March 2011, from pro-nuclear to anti-nuclear, mainly on the basis of his concern about the safety of HLW disposal.

K. Yamaji (✉)
Research Institute of Innovative Technology for the Earth (RITE), 9-2 Kizugawadai, Kizugawa-shi, Kyoto 619-0292, Japan
e-mail: yamaji@rite.or.jp

24.2 Current Status of HLW

HLW contains very toxic fission products. Fission products in the spent nuclear fuels are highly radioactive. Some countries such as Finland, Sweden, and USA directly dispose spent nuclear fuels as HLW after cooling at spent fuel storage. According to the conventional nuclear fuel cycle policy, spent nuclear fuels in Japan are reprocessed for separating fission products from uranium and plutonium, and the separated fission products are vitrified and then contained in canisters made of stainless steel. The option of direct disposal of spent nuclear fuels was seriously discussed in the first time in Japan at the process for formulating the 2005 Framework for Nuclear Energy Policy, and after the Fukushima accident, direct disposal of the spent fuel is becoming a more realistic option.

Right now, 1,984 HLW canisters (vitrified wastes) are stored in Japan. Among the 1,984, 1,442 HLW canisters were sent back from France and UK according to the contracts for the reprocessing commissioned to these countries; the rest are the HLW canisters produced by domestic reprocessing (295 from the test operation of the Rokkasho reprocessing plant and 247 from the Tokai pilot reprocessing plant). An additional 770 HLW canisters will be sent back from the UK, and high-level liquid waste, which is equivalent to 630 HLW canisters, is stored at the Tokai pilot plant.

In addition to the HLW canisters produced by reprocessing, about 17,000 t of spent nuclear fuels is stored at nuclear power plants (about 14,000 t in total) and the Rokkasho reprocessing plant (around 3,000 t). If all these spent fuels are reprocessed at the Rokkasho reprocessing plant, about 21,250 HLW canisters would be added. Thus, even if Japan decided to no longer operate nuclear reactors, we still must dispose HLW equivalent to 24,634 HLW canisters. We cannot run away from HLW issues.

24.3 HLW Disposal Program in Japan

Japan's research and development program for HLW disposal started in 1976 (Fig. 24.1). The first progress report was released in 1992 by PNC (Power Reactor and Nuclear Fuel Development Corporation). PNC was reorganized as JNC (Japan Nuclear Fuel Cycle Development Institute) in 1998, then merged with JAERI (Japan Atomic Energy Research Institute) to be JAEA (Japan Atomic Energy Agency) in 2005).

In 1999, JNC released the second progress report, and more importantly, in 2000 the Specified Radioactive Waste Final Disposal Act (Final Disposal Act, hereinafter) was legislated.

The process for the legislation of the Final Disposal Act is shown in Fig. 24.2. As shown here, the Special Panel on Disposal of High-Level Radioactive Waste formed under the Japan Atomic Energy Commission (AEC) played an important

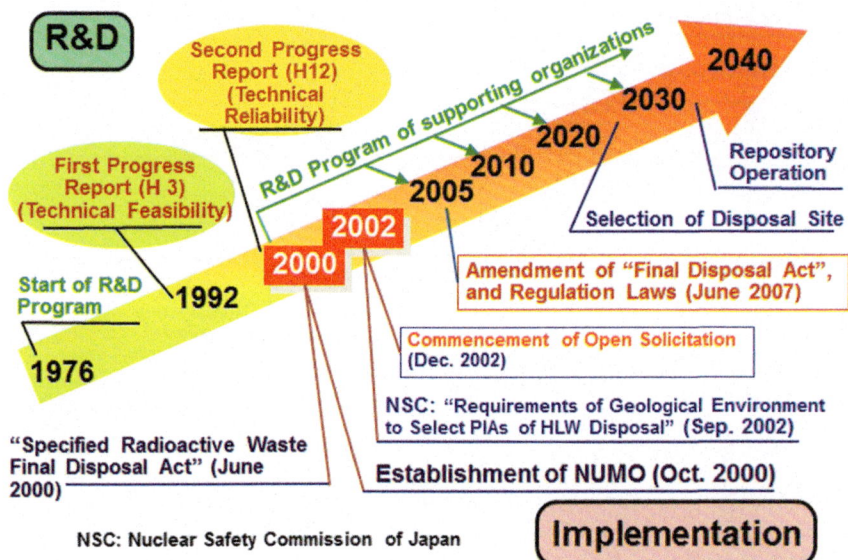

Fig. 24.1 Evolution of high-level radioactive waste (HLW) disposal in Japan (Modified from ANRE/METI and JAEA [1])

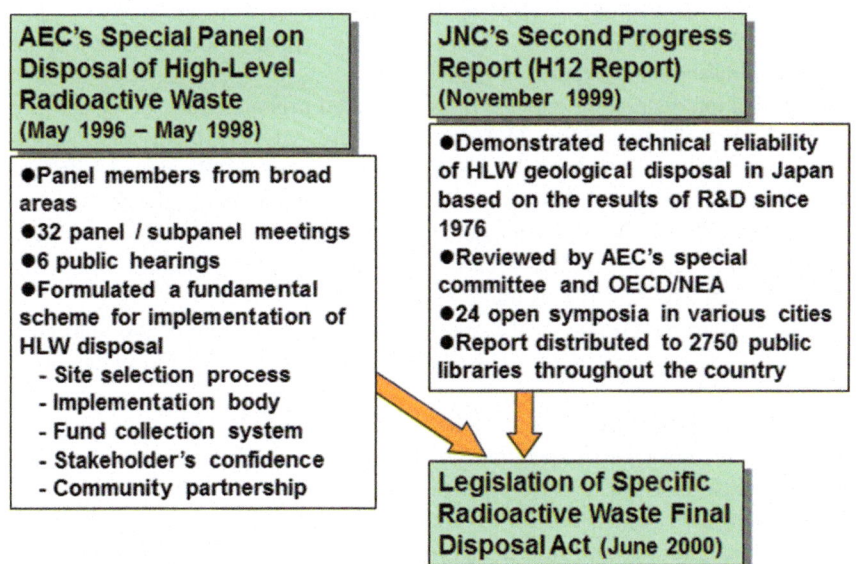

Fig. 24.2 Legislation of specific radioactive waste final disposal act (June 2000) (Private communication from NUMO on November 13, 2013)

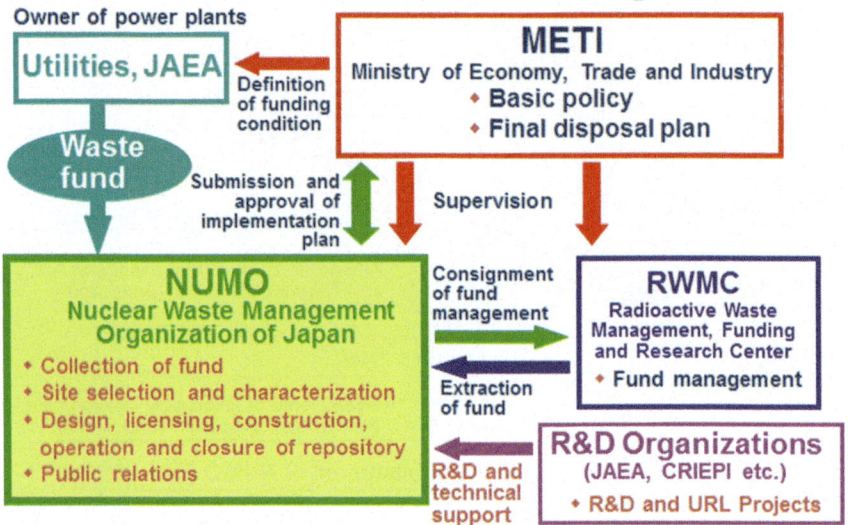

Fig. 24.3 Organizations and roles in the HLW disposal program in Japan (*CRIEPI* Central Research Institute of Electric Power Industry, *URL* Underground Research Laboratory) (From NUMO [2])

role along with the second progress report of JNC to set the contents of the Final Disposal Act.

Under the act, geological disposal is chosen for HLW disposal, and NUMO (Nuclear Waste Management Organization of Japan) was established for implementing the final disposal of HLW.

Organizational structure and the roles of related organizations set by the Final Disposal Act are shown in Fig. 24.3. As shown here, METI (Ministry of Economy, Trade and Industry) decides a basic policy and supervises all related activities. Owners of nuclear power plants provide a waste fund, which is collected from the electricity tariff, and the fund management is done by RWMC (Radioactive Waste Management, Funding and Research Center), while implementation of HLW disposal including site selection is borne by NUMO.

According to the current final disposal plan (Fig. 24.1), site of the final HLW disposal is to be selected in the 2020s and the final disposal will start in the middle of 2030s.

The Final Disposal Act was amended in 2007 to include TRU (trans-uranium) waste as a second type of specified waste (first type is HLW canisters, vitrified waste) because TRU waste is also to be disposed by geological disposal technology.

Although open solicitation for volunteer municipalities was employed for site selection, there has been no case except for a failed attempt by Toyo Town in Kochi Prefecture in 2007. Taking into account the failed attempt, METI added another scheme by the government to invite municipalities. The difficult situation, however, has continued, and after the Fukushima accident, the difficulties are increasing greatly.

24.4 Concept of Geological Disposal and Risk

Geological disposal is a globally common technology of HLW disposal for either vitrified HLW canisters or the spent nuclear fuel itself. Figure 24.4 shows the HLW disposal scheme employed in Japan, which incorporates the multi-barrier concept in the scheme. The first barrier is the vitrified HLW canister itself; the solubility of vitrified waste is very low and it is contained in a canister made of stainless steel. The second barrier is a thick package made of carbon steel, the third is a buffer made of bentonite and sand, and last, the multiply packaged waste is placed in stable host rock located deep underground.

Difficulty in securing the safety of HLW disposal comes from the requirement that risks associated with HLW disposal must be maintained below an acceptable level for a very long period, beyond 10,000 years. Whatever technical measures are taken, risks would remain. This is basically the same problem as the case of safety measures for severe accidents of nuclear power plants. The safety issue of HLW disposal, however, is more difficult because of the very long time period in which human intervention for maintaining safety cannot be expected.

24.5 Difficulty in Site Selection

According to the current basic policy for HLW disposal in Japan, the siting process is to be carried out with three stages (Fig. 24.5). The first stage is "literature survey," the second is "preliminary investigation," and the third is "detailed investigation." Then, construction of the repository will start. At each stage,

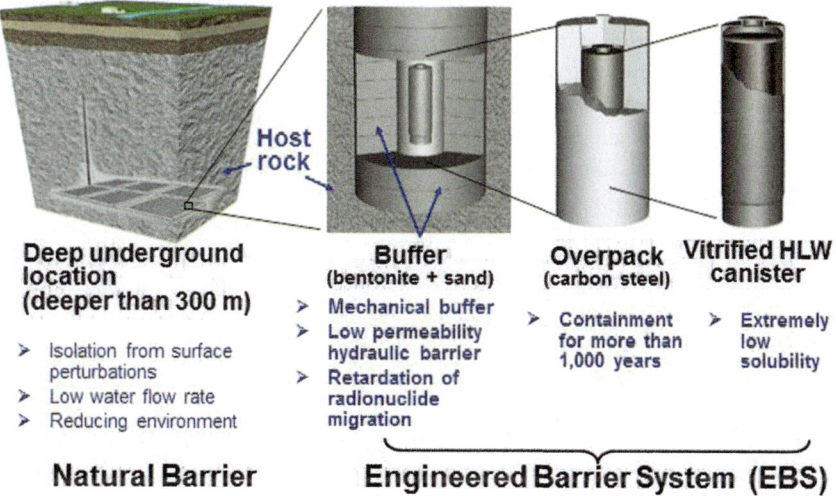

Fig. 24.4 HLW disposal scheme in Japan (multi-barrier concept) (Modified from NUMO [2])

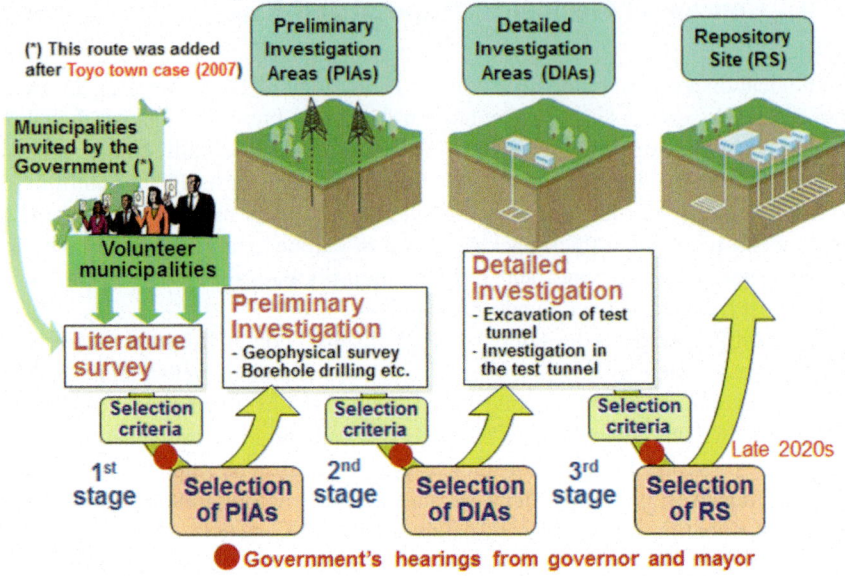

Fig. 24.5 Three stages of site selection process for HLW disposal in Japan (Modified from NUMO [2])

decisions will be made by selection criteria, taking into account the opinions of the local mayor (municipality) and local governor (prefecture).

In reality, there has been no occurrence of the first literature survey, although more than 10 years have passed since the siting process started. As mentioned before, a scheme of open solicitation was adopted for volunteers to apply for the literature survey, but after the failed attempt of Toyo Town in 2007, another scheme was added in which the government invited municipalities for the literature survey. However, the situation did not improve; rather, after the Fukushima accident the situation is becoming worse.

Facing these difficult situations, the government of Japan decided to take a more positive role in site selection. It is expected that a promising area could be more narrowly defined by screening sites on the basis of existing geological and geographical information.

24.6 Six Proposals by the Science Council of Japan

In September 2010 the Science Council of Japan (SCJ) received a deliberation request from the Chairman of the Japan Atomic Energy Commission, and SCJ formed a Review Committee for Disposal of High-Level Radioactive Waste. The author participated in the Review Committee as a member of SCJ. The Review

Committee made a Reply on Disposal of High-Level Radioactive Waste [3] in September 2012.

The Review Committee of SCJ pointed out the following six proposals to search for a path toward consensus formation: (1) fundamental reconsideration of policies related to disposal of HLW with extended definition, which includes spent nuclear fuels as well as vitrified HLW canisters; (2) awareness of the limits of scientific and technical abilities and securing scientific autonomy; (3) rebuilding a policy framework centered on temporal safe storage and management of the total amount of HLW; (4) necessity of persuasive policy decision procedures for fairness of burdens; (5) necessity of multiple-stage consensus formation by establishing opportunities for debate; and (6) awareness that long-term persistent undertakings are necessary for problem resolution.

Considering the SCJ report, the Japan Atomic Energy Commission, however, expressed its intention to maintain a policy of implementing the geological disposal on December 2012 with extension of the scope to include the direct disposal of spent nuclear fuel [4].

24.7 Setting a Moratorium Period by "Temporal Safe Storage"

Proposals of the SCJ report, particularly, the concepts of temporal safe storage and management of total amount, triggered many discussions widely concerning the issue of HLW disposal.

The temporal safe storage is characterized by securing a moratorium period of several dozen or several hundred years to establish appropriate handling measures for the problem. It provides the advantages of using this period to refine technological developments and scientific knowledge, guaranteeing the possibility of creating handling measures that target a longer period; for example, improvement of the durability of containers, development of nuclear transmutation technology to reduce volume and toxicity of HLW, and research related to the stability of geological layers.

In addition, the temporal safe storage makes it possible to keep various options for future generations to choose for final disposal of HLW.

The concept of safe storage, however, still has a wide range of uncertainties in technical specifications; for example, duration of storage, location characteristics such as on ground or underground, and number of storage facilities. The concept ranges from currently available interim storage of spent fuel to retrievable geological disposal. In fact, the response of Japan Atomic Energy Commission mentioned retrievable geological disposal in the context of temporal safe storage.

SCJ had set up a Follow-up Committee as an extension of the Review Committee in August 2013 to clarify the concept of the temporal safe storage.

24.8 "Management of the Total Amount" of HLW

As clearly stated in the SCJ report, "management of the total amount" has two connotations: "setting an upper limit for the total amount" and "controlling increases of the total amount." "Setting an upper limit for the total amount" corresponds to the withdrawal from nuclear power, and the level of upper limit depends on the tempo of that withdrawal. On the other hand, "controlling increases of the total amount" corresponds to keeping nuclear power in the future with strictly controlling increases of the total amount, and the amount of disposed waste per unit of generated power must be controlled to the smallest amount possible. There are many technical options to control the increase of the total amount of HLW, for example, increasing burn-up of fuels, transmutation of radioactive nuclides, and longer temporal storage of HLW, which secure time for radioactivity to decay.

However, in fact, many readers of the SCJ report mistakenly recognized that management of the total amount means setting an upper limit for the total amount, and thus believed that SCJ proposed withdrawal from nuclear power: this is a complete misunderstanding. At the background of the proposal of management of the total amount, there is recognition that we should respond to the concerns on the limitless increase of HLW.

24.9 Awareness of the Limits of Scientific and Technical Abilities

The Review Committee of SCJ consists of various experts from wide-ranging academic fields from physical science, engineering, life science, social science, and humanities. The proposal concerning awareness of the limits of scientific and technical abilities was formed through interdisciplinary discussions among the experts. Some readers of the SCJ report seem to have felt uneasiness with this proposal because this proposal apparently cast a scientific doubt on the feasibility of the geological disposal of HLW. To the author's understanding, this proposal is a rather general statement that there is no perfect scientific evidence to support the safety of HLW disposal for more than 10,000 years.

Having heard the discussions related to this proposal, the author recognized there are many different academic approaches depending on the field of science. For example, natural scientists seek truths in natural phenomena, whereas engineers try to make things and/or systems that are valuable and acceptable for human society. HLW issues are related not only to various fields of science, but also to value systems shared by society. Here again, the author was convinced that we need to reflect more deeply on the relationship between science and society.

Open Access This chapter is distributed under the terms of the Creative Commons Attribution Noncommercial License, which permits any noncommercial use, distribution, and reproduction in any medium, provided the original author(s) and source are credited.

References

1. Agency of Natural Resources and Energy (ANRE), METI, JAEA (2009) Total plan of research and development on HLW geological disposal (in Japanese). Agency of Natural Resources and Energy (ANRE), METI, JAEA, Tokyo, Japan
2. NUMO (2008) Geological disposal of radioactive waste in Japan. NUMO, Tokyo, Japan
3. Science Council of Japan (2012) Reply on disposal of high-level radioactive waste. Science Council of Japan (in Japanese), Tokyo, Japan
4. Japan Atomic Energy Commission (2012) Renewing approaches to geological disposal of HLW (statement). Japan Atomic Energy Commission, Tokyo, Japan, December 18

Chapter 25
Considering the Geological Disposal Program of High-Level Radioactive Waste Through Classroom Debate

Akemi Yoshida

Abstract Although nuclear power has become recognized as a social issue—one that concerns us all—there is still, in Japan, insufficient public debate on the problems posed by this form of energy. In particular, interest among the younger generation on this and many other issues is limited, a situation reflected in the low turnout of young people at elections. The disposal of high-level radioactive waste is an issue that cannot be simply solved by shutting down nuclear reactors. Yet, in spite of the need to urgently find a solution to the problem of nuclear waste, many young people appear to be apathetic. Part of the reason for this lack of interest is that students majoring in the so-called humanities do not feel confident approaching the issue. As a way to raise such students' interest in the issue of nuclear waste disposal, debating courses were held in the social science departments of two universities located in Aichi Prefecture, Japan. This chapter reports on these courses, discusses the value and effectiveness of debate in raising awareness of social issues, and assesses potential problems with implementing debating in educational contexts.

Keywords Active learning • Classroom debate • Communication training • Fundamental literacy for members of society • Nuclear power • Radioactive waste

25.1 Introduction

25.1.1 The Situation Now

Although the problems surrounding nuclear power have certainly become a social issue, it cannot be said that the present discussion on this issue is always calm and rooted in science. Elsewhere in the world, including in the USA and various European countries, the importance of public debate has been emphasized and reports published on the efficacy of specific examples. In Japan, however, particularly since the Fukushima Daiichi nuclear accident, sensational media reporting

---

A. Yoshida (✉)
Sugiyama Jogakuen University, 3-2005 Takenoyama, Nissin-City, Aichi, Japan
e-mail: ayoshida@sugiyama-u.ac.jp

© The Author(s) 2015
K. Nakajima (ed.), *Nuclear Back-end and Transmutation Technology for Waste Disposal*, DOI 10.1007/978-4-431-55111-9_25

has not served to encourage sound debate on the issue of nuclear power. Moreover, this is not limited to the issue of nuclear power, nor is it a phenomenon that dates from the Fukushima disaster: issues around food safety and gender equality have been similarly characterized by emotive media reporting. Debate around these issues has not been informed by current scientific knowledge; indeed, what has prevailed is argument based on emotion engaged in without even an understanding of the relevant laws.

Being a democratic society entails that the path society follows is set according to the wishes of its citizens. These citizens exchange their divergent views and do their utmost to reach a consensus. If an agreement cannot be reached, society acts in accordance with the opinion of the majority. Necessary for such a process is that people think about an issue and express their views. In modern society, however, because many issues are complicated and difficult for people to adequately understand, often such issues are left to the "experts."

Although the citizens' right of self-determination should go hand in hand with responsibility, entrusting decision making to the experts has resulted in responsibility for these decisions being thrust upon them. Entrusting all responsibility to the experts could be expected to expedite decision making, but this has not been the case; instead, the emotional response of the public has contributed further to the deferring of decision making.

25.1.2 Why has Such a Situation Occurred?

One reason is that expressing one's opinion is not necessarily viewed in a favorable light at school and in the home. The way of thinking in Japan, encapsulated in the saying *wa wo motte totoshi* (harmony is of utmost importance), leads to virtue being placed on conforming to the views of others rather than asserting one's own opinions. It is partly because of this feeling that Japanese people in general have had relatively little practice in expressing their ideas and opinions to others. In recent years, debate-based lessons have been introduced by some elementary schools, but the spread of such lessons through elementary school education as a whole has not been sufficient. The second reason is rooted in the pivotal role given at the elementary level to *sakubun*, or essay writing. Pupils practice exploring their emotions and putting them down on paper, but not how to think in a logical way and support their opinion with evidence. A third reason given for the situation described here is the complexity of social problems nowadays, making understanding difficult for "ordinary people," who just give up even thinking about the issue.

The more complex issues become, however, the greater the necessity for citizens to engage in discussion, express their opinions, and then make decisions. It is, therefore, incumbent on each citizen to grapple with and discuss such issues. Judging from the present situation, it does not seem that staging public debates, having experts offer explanations, or other simple methods can serve as a substitute for real engagement by the citizens. Although workshops aimed at citizens have been held, it is doubtful that participation in debates that occur in such forums is built on a sufficient understanding of the issue at hand. There also appear to be cases when debate is based only on information that fits the administration's agenda. The information is simplified, and any exchange of views is debate in name only, without substance. It is necessary to look for effective ways to discuss issues to make such workshops and public debates productive and to change the explanatory meetings from superficial gesturing into something more substantive.

Moreover, in spite of the fact that many of the social problems we face are not just a concern for now, but also for the future, many of those interested in social issues and who vote in elections are elderly. Listening to the voices of the elderly is, of course, important, but, for a healthy democracy, it is necessary for young people to engage in debate and be involved in making decisions that affect their society. But how can such engagement by young people be promoted? The way that the author would like to suggest in this chapter is through debating lessons in school.

In a narrow sense, debate can be defined as follows: "discussion on a specific issue involving two groups of speakers, with one group taking a position of supporting the topic and the other arguing against it. Each of the groups seeks to persuade a third party." [1] (e.g., Yomiuri Shimbun 2013:2). Many of the topics for discussion are chosen from policy issues. To encourage participants to approach an issue from new and multiple perspectives, they are allocated (usually by the teacher) to either one of the groups, that is, they do not choose for themselves which side of the issue to support. This chapter describes a case study of a debating course, taught by the author, and examines its effects and any issues that emerged. Aimed at undergraduate students in a university, the topic of debate in the course was "the problem of high-level atomic waste disposal."

25.1.3 Deciding the Topic

The founding of the All Japan Educational Debate Association, of which the author is a committee member, was the catalyst for a national debating contest, which was started in the *Tokai*, or central region, of Japan. This contest, having been sponsored for some years by Chubu Electric Power Company, debates energy-related issues. High school students have faced each other over topics such as "Japan should abandon nuclear power: for or against?"; "Television broadcasting time in Japan should be limited to save energy: for or against?" Although the topic of energy, as well as many other policy issues, needs to be thought about by the next generation, it is just such issues that the younger generation does not appear eager to tackle face on. This is where debate, with its game-like, competitive element can serve an important role. In the context of debate, young people have been shown to engage seriously with such issues.

With this in mind, the author organized a debate for her class of university students. They debated the following motion, suggested originally by Chiba University's Assistant Professor, Daisuke Fujikawa: Japan should scrap the plan to store high-level radioactive waste underground: Do you agree or disagree? [1] This paper describes the procedure of the debate in the classroom, assesses its effectiveness, and discusses certain problems that emerged from this activity.

25.2 Research Method

In 2013, the author compared and examined debate-focused courses held in two universities located in Aichi, central Japan; details of the courses are as follows.

1. "*Enshu* 1" (Seminar 1), a first-semester course, was taught to third- and fourth-year students at Sugiyama Jogakuen University's Department of Human Sciences.
2. "Introduction to debate," a first-semester course, was taught to second- and third-year students at Aichi Shukutoku University's Faculty of Global Culture and Communication.
3. "Introduction to debate," this time a second-semester course, was taught to second- and third-year students at Aichi Shukutoku University's Faculty of Global Culture and Communication.

By studying these classes, the author hoped, first, to elucidate how students understanding of the debate issue—the disposal of high-level radioactive waste—would be affected by the lessons; and, second, to assess the effectiveness of debating classes on issues related to natural science facing modern society. In addition, students completed a questionnaire, "Fundamental Literacy for Members of Society" [2]. Through analysis of the results of this survey, the author sought to gain new insights into methodology to promote a deeper understanding among students of issues facing modern society.

The course as listed here had four main features. First, the theme of the debate was announced at the beginning of the course. Second, rather than having students choose the subject for debate, the topic was assigned to the students. The fact that the topic was a science-related one was the third feature of the course. Because the students were from a humanities/social sciences background, their basic knowledge of science was, on the whole, rather limited. Because there was some concern that students would not be able to cope with debate, efforts were made to deepen students' understanding of the issues involved before the actual debating contest. For example, Hajimu Yamana, professor at the Kyoto University Research Reactor Institute (KURRI), and Tomohisa Kakefu of the Japan Science Foundation were invited as guest speakers, and students also visited the Mizunami Underground Research Laboratory and the visitor facilities at Hamaoka Nuclear Power Station.

The fourth feature of the course was, therefore, that students were not left to research the topic by themselves, but were supported by, for example, being given the opportunity to listen and talk to experts. In addition, there was an element of experiential learning incorporated into the course in the form of, for example, the visit to Hamaoka Nuclear Power Station just mentioned.

25.2.1 Outline of the Courses

(a) *"Enshu I"* (Seminar 1) comprised the following lessons:

Lesson 1: Orientation

Lessons 2–5: Discussion of selected readings from a 2009 book (*"Buraika Suru Onatachi"* [The Making of the Rowdy Women]) that was unrelated to the theme of the debate.

Lessons ~6–8: Entitled "Introduction to debate," these lessons covered the theory and practice of debating, including debating techniques.

Lesson 9: Visit to Mizunami Underground Research Laboratory.

Lesson 10: Lecture on "High-level radioactive waste" by Hajimu Yamana (see above).

Lesson 11: Lecture on the geological disposal of radioactive waste by a member of staff of NUMO (the Nuclear Waste Management Organization of Japan).

Lesson 12–13: Preparation for debate.

Lesson 14–15: Staging of a debating contest.

In addition, during the course (on June 29, 2013) students toured visitor facilities at Hamaoka Nuclear Power Station in Shizuoka Prefecture.

Yoshida [2] (Yoshida, A. 2014:123)

(b) "Introduction to debate" comprised the lessons shown below.

Lesson 1: Orientation

Lessons 2–5: Entitled "Introduction to debate," these lessons covered theory and practice of debating, including debating techniques.

Lesson 6: Lecture on the geological disposal of radioactive waste by a member of staff of NUMO (the Nuclear Waste Management Organization of Japan).

Lesson 7: Lecture on "High level radioactive waste" by a guest speaker, Hajimu Yamana, of KURRI (see above).

Lesson 8: Focus on geological disposal of radioactive waste

Lesson 9–10: Students engaged in "communication training."

Lesson 11–12: Preparation for debate.

Lesson 13–15: Staging of a debating contest.

During the course students were taken to Mizunami Underground Research Laboratory (June 22) and to the visitor facilities at Hamaoka Nuclear Power Station (June, 29).

Yoshida [2] (Yoshida, A. 2014:123)

(c) The second semester "Introduction to debate" course comprised the following 11 lessons:

Lesson 1: Orientation

Lessons 2–3: Entitled "Introduction to debate," these lessons covered the theory and practice of debating, including debating techniques.
Lesson 4: Visit to Hamaoka Nuclear Power Station in Shizuoka Prefecture.
Lessons 5–6: Entitled "Introduction to debate 2," in these classes students participated in a model debate and watched a DVD of the debating competition held in the first semester.
Lesson 7: Lecture on the geological disposal of radioactive waste by a member of staff of NUMO (the Nuclear Waste Management Organization of Japan).
Lesson 7–8: Students engaged in "communication training."
Lesson 9: Tomohisa Kakefu gave a talk on radiation, which incorporated demonstrations.
Lesson 10: Lecture on the geological disposal of radioactive waste by a member of staff of NUMO (the Nuclear Waste Management Organization of Japan).
Lesson 11: Visit to Mizunami Underground Research Laboratory.

25.3 Reflections on the Courses

In each of the courses, the students reached a level that enabled them to maintain a debate during the competition held during at the end of the course. Owing in large part to Fujikawa's well-constructed topic, the debate proved to be a balanced one, with the result of the debate not overly biased either for or against the proposition.

Perhaps because of their humanities/social sciences background, the students had little background knowledge of radiation. Regarding the storage pool for spent nuclear fuel, for example, a number of students mistakenly thought that the spent fuel was dissolved in the water of the pool (the spent fuel is stored in the form of solid rods). However, because preparing for a debate necessitates clarifying which areas of one's understanding are lacking, students' grasp of the subject area gradually improved as the course progressed. During the period of the course various news stories appeared in the press related to nuclear power and NUMO. Students not only responded to this news in class, but also actively gathered information reported in the media, and were able to use such up-to-date information in the debate competition.

Despite the fact that nuclear power was an issue directly related to the provision of energy for *their* lives, some considered that nuclear energy concerned no more than Fukushima; in other words, their awareness of nuclear power as an issue pertinent to them was low. However, through doing the course, students' understanding deepened and they also began to appreciate that the issue was one that directly affected them.

The input from experts was also important for helping students to understand the issues. At the beginning of the course, there were students who, not understanding fundamental facts—for example, the difference between radial rays and radioactivity—practically "gave up thinking" about the issues. However, with the tuition of the guest speakers and their demonstrations of a cloud chamber (a simple device that allows the decay of radioactive materials to be observed) and other experiments, students gradually learned more about the science involved, leading them to become more proactive in thinking about issues for themselves.

25.4 Results of the Questionnaire Survey

The results of the "Fundamental literacy for members of society" questionnaire are summarized in this section. Responses were obtained from a total of 42 students (Enshu I): responses from 10 of the 12 students registered for the course; Introduction to Debate (first semester): 26 of the 27 students; Introduction to Debate (second semester): 6 of the 8 students. The composition of the classes was as follows: Enshu 1: third- and fourth-year female students; Introduction to Debate (first semester) and Introduction to Debate (second semester): second- and third-year female and male students. Because the sample was small, no comparison of data based on gender and year of study was made; instead, the results of the group as a whole is shown (see Fig. 25.1 and Table 25.1).

To gain a firm understanding of the disposal of high-level radioactive waste, and to be able to consider the issues, requires specialist knowledge as well as critical and logical thinking skills. In this respect, the overall results of the questionnaire were positive, but the scores were particularly high for items

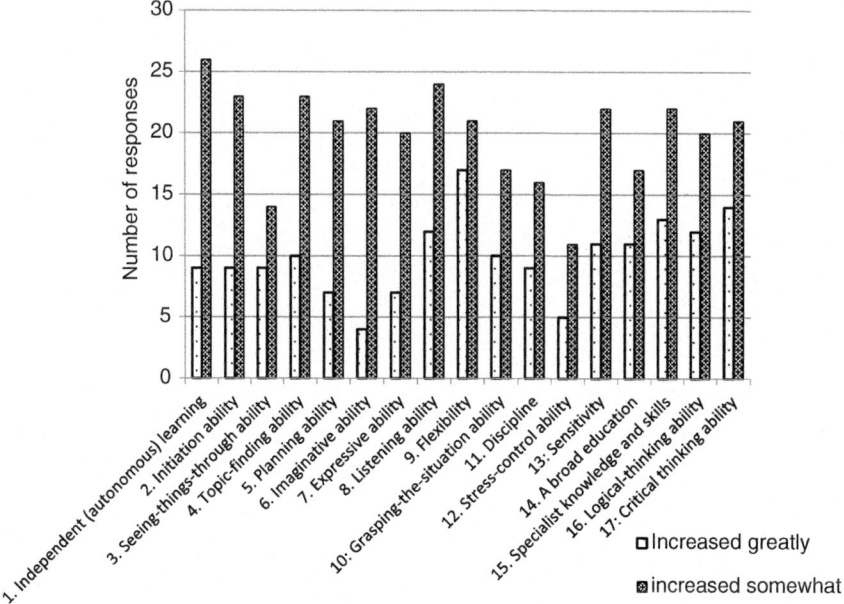

Fig. 25.1 Responses that indicated an increase in fundamental social-literacy skills

Table 25.1 The fundamental social-literacy skills gained through this course

	Increased greatly	Increased somewhat	No change	Slight decrease	Decreased greatly	No response
1. Independent (autonomous) learning	9	26	7	0	0	0
2. Initiation ability	9	23	10	0	0	0
3. Seeing-things-through ability	9	14	19	0	0	0
4. Topic-finding ability	10	23	9	0	0	0
5. Planning ability	7	21	13	0	0	1
6. Imaginative ability	4	22	16	0	0	0
7. Expressive ability	7	20	15	0	0	0
8. Listening ability	12	24	6	0	0	0
9. Flexibility	17	21	4	0	0	0
10. Grasping-the-situation ability	10	17	15	0	0	0
11. Discipline	9	16	17	0	0	0
12. Stress-control ability	5	11	25	1	0	0
13. Sensitivity	11	22	9	0	0	0
14. A broad education	11	17	14	0	0	0
15. Specialist knowledge and skills	13	22	7	0	0	0
16. Logical thinking ability	12	20	10	0	0	0
17. Critical thinking ability	14	21	7	0	0	0

15, 16, and 17 (that is, Specialist knowledge and Skills, Logical thinking Ability, and Critical thinking ability). The results overall indicated that debate can be an effective activity for shedding light into, and for examining, social issues that require specific background knowledge and judgment.

25.5 Issues for the Future

The results of this study suggest that when dealing in class with social issues such as the disposal of high-level radioactive waste that require a certain level of basic scientific knowledge, debate and other kinds of active learning may be more effective than lecture-style, noninteractive pedagogical methods. When implementing active learning, however, time and cost and many other factors need to be considered. For the courses described in this chapter, the author was able to gain the cooperation of NUMO, which made it possible for her to integrate into the course lectures by experts and visits by students to relevant facilities. But when such outside help cannot be obtained, alternatives must be considered.

To take debate on social issues beyond the walls of the university classroom, a DVD of the 2013 students' debate competition may be a valuable resource that could be shown to the general public. In addition, there is a need for a more conducive learning environment for the debating, which could perhaps be achieved by creating scenarios for model debates and organizing workshops.

This course also pointed to the need for consumer education. It is important to not only think about the short-term consequences of the things we use and their disposal, but also to consider from a global perspective how our consumption will affect future generations. In debate for education, it is important to tackle a range of social issues from multiple points of view. In debating the issue of radioactive waste, in particular, it must be conveyed to students that the issue is not one that is solved just by shutting down nuclear reactors; whether or not we use nuclear energy in the future, the waste has already been produced, a by-product of our consumption.

This chapter has focused on courses aimed at undergraduate students. If possible, in the future, the author would like to run similar debating courses with middle- and high-school students and with adults, and then assess, as was done in this study, the impact of debate as an educational activity.

25.6 Notes

1. Professor Daisuke Fujikawa, with the support of NUMO (as was had in this present study), conceived this theme for a debate held in 2012 for undergraduate students on a teacher-training course at Chiba University [3] (Fujikawa, D.2013:5).
2. Sugiyama Jogakuen University, the university to which the author is affiliated, is a participant in a project supported financially by the Ministry of Education, Culture, Sports, Science and Technology. Called the "Project of Educational Reform and Structural Improvement to Respond to the Needs of Industry," it is composed of 23 universities (including junior colleges) in the central region of Japan. As part of the project, each university engages in career education, and Sugiyama Jogakuen University for its part has focused on active learning to raise its educational performance. The "Fundamental Literacy for Members of Society Questionnaire" was implemented as part of this effort to effectively utilize active learning. The questionnaire consisted of the following 17 items, together with a short definition.

    1. Independent (autonomous) learning: The capacity to engage in an activity on one's own volition.
    2. Initiation ability: Appealing and encouraging others to become involved.
    3. Seeing-things-through ability: The capacity to set a target and act to achieve it.
    4. Topic-finding ability: The capacity to analyze the situation and clarify aims and issues.

5. Planning ability: The capacity to clarify the process that can lead to a solution; the ability to plan.
6. Imaginative ability: The capacity to create new value.
7. Expressive ability: The capacity to convey your views clearly in an easily understandable way.
8. Listening ability: The capacity to listen attentively to what the other person is saying.
9. Flexibility: The capacity to understand other people's opinions and positions.
10. Grasping-the-situation ability: The capacity to understand the relationship between you and the people and the situation around you.
11. Discipline: The capacity to follow society's rules and keep promises made with others.
12. Stress-control ability: the capacity to respond appropriately to sources of stress.
13. Sensitivity: The capacity to respond to stimulus from the external environment.
14. A broad education: The possession not just of knowledge, but also an ability to understand and process knowledge creatively.
15. Specialist knowledge and skills: The possession of in-depth knowledge and skills in a particular academic or other field.
16. Logical thinking ability: the capacity to think coherently and logically.
17. Critical thinking ability: the capacity to analyze and judge the suitability and validity of issues and arguments.

Sugiyama Jogakuen University [4] (Sugiyama Jogakuen University 2013) Sugiyama Jogakuen University et al. [5] (Sugiyama Jogakuen University, Special Committee on Career Education, Whole-Faculty Faculty Development Committee 2013)

Open Access This chapter is distributed under the terms of the Creative Commons Attribution Noncommercial License, which permits any noncommercial use, distribution, and reproduction in any medium, provided the original author(s) and source are credited.

References

1. Shimbun Y, The All Japan Educational Debate Association (2013) Dibeeto koshien staartobukku (The Startbook for the National Debating Competition for Junior and Senior High Schools). http://nade.jp/files/uploads/startbook2013.pdf/. Accessed 12 December 2013
2. Yoshida A (2014) Possibility of debate class as active learning in University. Ningenkankeigakukenkyu. (Research Journal of the Department of Human Relations), No. 12, Sugiyama Jogakuen University
3. Fujikawa D (2013) Kyoinkeidaigaku ni okeru dibeito jugyou ni oite gendaiteki torikumi wo okonau kokoromi: koureberu houshyasei haikibutsu no shori mondai wo toriagete (An attempt to deal with a contemporary issue in a debate education class of faculty on teacher training.

On the issue of high-level radioactive waste management). Project Report of the Chiba University School of Humanities and Social Sciences
4. Sugiyama Jogakuen University (2013) Shakaijin kisoryoku ni kansuru ankeeto no onegai. (Request to complete "Fundamental literacy for members of society" questionnaire)
5. Sugiyama Jogakuen University, Special Committee on Career Education, Whole-Faculty Faculty Development Committee (2013) Shakaijin kisoryoku to akutibu raaningu ni kansuru ankeeto chosa kaitou youshi. (Response sheet for the questionnaire survey on fundamental literacy for members of society and active learning)

Part VIII
Environmental Radioactivity: Development of Radioactivity Measurement Methods and Activity of Radionuclides in the Environment Monitored After the Accidents at TEPCO's Nuclear Power Stations

Chapter 26
Environmental Transfer of Carbon-14 in Japanese Paddy Fields

Nobuyoshi Ishii, Shinichi Ogiyama, Shinji Sakurai, Keiko Tagami, and Shigeo Uchida

Abstract It has been recognized that carbon-14 (^{14}C) is one of the dominant radionuclides affecting dose from transuranic (TRU) wastes. This radionuclide has a decay half-life of 5,730 years, and ^{14}C organic materials have very low sorption properties to clay and rock in the environment, which raises some concerns about the releases of ^{14}C to the biosphere from radioactive waste repositories. For the safety assessment of TRU waste disposal, we studied the behavior of ^{14}C in rice paddy field soils. We also determined key parameters such as soil–soil solution distribution coefficients (K_ds) and soil-to-rice plant transfer factors (TFs) of ^{14}C in the field soils. The TFs were obtained in laboratory and field experiments. In our laboratory experiments, we used [1,2-^{14}C] sodium acetate as a source of ^{14}C because it has been suggested that low molecular weight organic-^{14}C compounds are released from metallic TRU wastes. The results showed that ^{14}C-bearing sodium acetate in irrigated paddy soils was rapidly decomposed by indigenous bacteria. Although some of the ^{14}C was assimilated into the bacterial cells, most of the ^{14}C was released into the air as gaseous compounds. The main chemical species of ^{14}C gases was $^{14}CO_2$, and a part of the released $^{14}CO_2$ gas was used by rice plants during photosynthesis. Only a negligible amount of ^{14}C was absorbed through the roots. Therefore, the contamination of rice plants is mainly caused by gasification of ^{14}C, and microorganisms are responsible for driving this process. The activity of microorganisms is a key issue in the behavior of ^{14}C in paddy fields.

Keywords Bacteria • Behavior • Degradation • Radiocarbon • Rice paddy fields • Safety assessment • TRU wastes

26.1 Introduction

Transuranic (TRU) wastes contain a variety of radionuclides, for example, Np, Pu, and long-lived radionuclides such as ^{14}C and ^{129}I. In Japan these wastes are categorized into four groups in accordance with their physical properties and the

N. Ishii (✉) • S. Ogiyama • S. Sakurai • K. Tagami • S. Uchida
Research Center for Radiation Protection, National Institute of Radiological Sciences, 4-9-1 Anagawa, Inage-ku, Chiba 263-8555, Japan
e-mail: nobu@nirs.go.jp

concentration of radioactive materials. Group two waste includes hull and end piece wastes with relatively high amounts of ^{14}C, and leaching of low molecular weight ^{14}C organic materials from simulated hull wastes has been reported [1]. The ^{14}C organic materials have very few sorption properties to clay and rock, and ^{14}C has a relatively long half-life of 5,730 years. These properties raise concerns about releases of ^{14}C to the biosphere from radioactive waste repositories.

Rice is a major agricultural crop throughout Asia, and thus human exposure to ^{14}C through rice intake must be considered. To reduce the risk of the internal radiation dose from ^{14}C, it is important to clarify the behavior of ^{14}C in rice paddy fields. In this study, we determined transfer pathways of ^{14}C through the rice paddy fields to rice grains. Environmental parameters such as soil–soil solution distribution coefficients (K_ds) and soil-to-rice plant transfer factors (TFs) of ^{14}C were also determined, because these parameters are often used in transfer models to predict the behavior of radionuclides in the environment. From a series of our experimental results, we describe the behavior of ^{14}C in rice paddy field soils and the importance of microbial activity.

26.2 Partitioning of ^{14}C into Solid, Liquid, and Gas Phases

We carried out batch sorption experiments using 63 Japanese rice paddy soil samples to clarify the transfer pathways of ^{14}C in rice paddy fields. The soil samples were collected throughout Japan and taken to our laboratory where they were air dried and sieved (<2 mm). These sieved soils were mixed with a [1,2-^{14}C] sodium acetate solution at the ratio of soil : solution = 0.5 g : 5 ml, and the flooded soil samples were incubated at 25 °C for 7 days [2]. During the incubation period, the ^{14}C atoms of the sodium acetate were partitioned into solid, liquid, and gas phases. Each partitioning ratio is shown in Fig. 26.1. Approximately 63 % of the total ^{14}C on average was released into the air as gaseous compounds. Partitioning ratios into solid and liquid phases were 34 % and 3 %, respectively. These results suggest that gasification is an important pathway in the environmental transfer of ^{14}C in Japanese rice paddy fields.

When ^{14}C is released into the air, ^{14}C-bearing gases must pass through the soil solution. Because soil solution pH affects chemical reactions such as hydrolysis and degassing of CO_2, chemical forms of ^{14}C-bearing gases may change in the soil solution. We, therefore, investigated relationships between pH and partitioning ratios of ^{14}C into the liquid phase at day 7 of incubation (Fig. 26.2). The partitioning ratio increased with increasing in pH, and a significant correlation ($r = 0.7$) was found. These data fit well with the solubility curve of total carbonic acid in water, which refers to the sum of dissolved carbon dioxide and the carbonic acid. This observation suggested that the dominant chemical species of ^{14}C in gas forms was carbon dioxide. To confirm the effect of pH on the partitioning of ^{14}C into the liquid phase, a soil sample was suspended in MES [2-(N-morpholino)ethanesulfonic acid] buffers with the initial pH value adjusted to 5.5, 6.5, and 7.5 (Fig. 26.3). A control

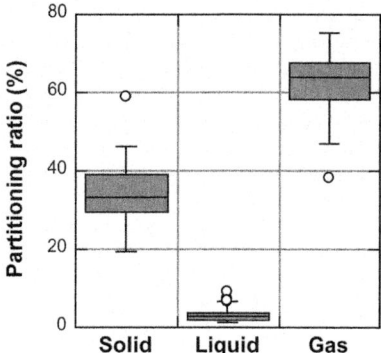

Fig. 26.1 Box plots for each partitioning ratio of ^{14}C into solid, liquid, and gas phases

sample was prepared consisting of the soil and deionized water (pH unadjusted). The partitioning ratio also increased with increasing pH, suggesting that the partitioning ratio of ^{14}C into the liquid phase depended on the pH of the soil solution.

Soil–soil solution distribution coefficient (K_d) is a commonly used parameter to evaluate behaviors of radionuclides in the environment. In our study, the K_d values were calculated from activities of the ^{14}C in the solid and liquid phases at the end of incubation, and the obtained K_d value was 139 ± 77 ml g^{-1} on average. Negatively charged anions generally have low K_d values because of simple electrostatic interaction. Our value, however, was higher than expected from the chemical form of $^{14}CH_3{}^{14}COO^-$. For example, Kaneko et al. [1] obtained the K_d value of 9.5 ml g^{-1} for the sorption test of acetic acid using cement materials. The reason for our high K_d value is explained next.

26.3 Involvement of Microorganisms in the ^{14}C Behavior

Many microorganisms inhabit rice paddy fields, and they are responsible for nutrient cycling. We studied the involvement of microorganisms in environmental transfer of ^{14}C. Microorganisms in batch cultures were treated with autoclaving (121 °C, 15 min), mixing with glutaraldehyde [final concentration of 2.5 % (vol/vol)], and mixing with cycloheximide (final concentration, 250 μg ml^{-1}). Autoclaving and expose to glutaraldehyde inactivate bacteria and fungi, but exposure to cycloheximide only inhibits fungi. The partitioning ratios of ^{14}C into solid, liquid, and gas phases for each treatment sample are listed in Table 26.1. When microorganisms were treated by autoclaving and exposing to glutaraldehyde, almost all the ^{14}C added remained in the liquid phase; that is, negligible transformation of ^{14}C occurred. On the other hand, the ^{14}C atoms in the control and the cycloheximide-treated sample were partitioned into solid, liquid, and gas phases at certain ratios, and these ratios were similar between the control and the cycloheximide samples. We confirmed fungi made no contribution to partitioning of ^{14}C

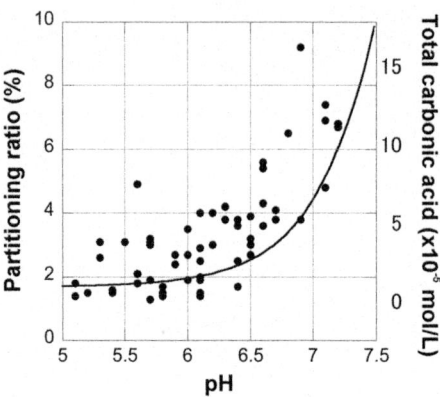

Fig. 26.2 Relationships between pH and partitioning ratios of ^{14}C into the liquid phase (scatter plots). *Solid line* shows the solubility curve of total carbonic acid in water

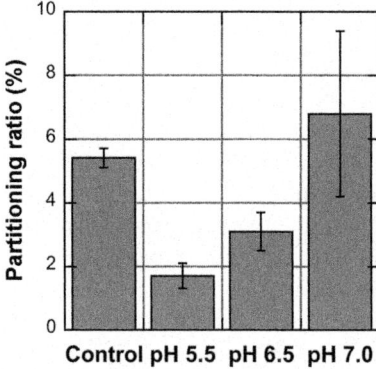

Fig. 26.3 Effect of pH on the partitioning of ^{14}C into the liquid phase

based on these results. We concluded that environmental transfer of ^{14}C in rice paddy fields was driven by bacteria, not by fungi.

To confirm incorporation of ^{14}C into bacteria cells, bacteria that were isolated from a flooding water of a paddy soil sample were cultivated on agar plates containing [1,2-^{14}C] sodium acetate [3]. After cultivation, bacterial colonies were formed, and their autoradiography images showed that all colonies had the ability to take up ^{14}C (Fig. 26.4). In our experimental procedure, bacterial cells were consequently partitioned into the solid phase, and thus the solid phase contains the ^{14}C incorporated by bacteria, which could be one of the reasons for the relatively high K_d values.

26.4 Transfer of ^{14}C from Soil to Rice Plants

Soil-to-rice plant transfer factors (TFs) of ^{14}C, which was defined as ^{14}C concentration in rice grains (Bq/kg-dry) divided by that in soil (Bq/kg-dry), were determined by laboratory and field experiments. In the laboratory experiment using a

Table 26.1 The partitioning ratios of ^{14}C into solid, liquid, and gas phases for each treatment.

Treatment	Partitioning ratio (%)		
	Solid phase	Liquid phase	Gas phase
Control	27.9	4.5	67.5
Autoclaving	0	98.0	2.0
Glutaraldehyde exposure	0	96.8	3.2
Cycloheximide exposure	29.3	4.8	65.9

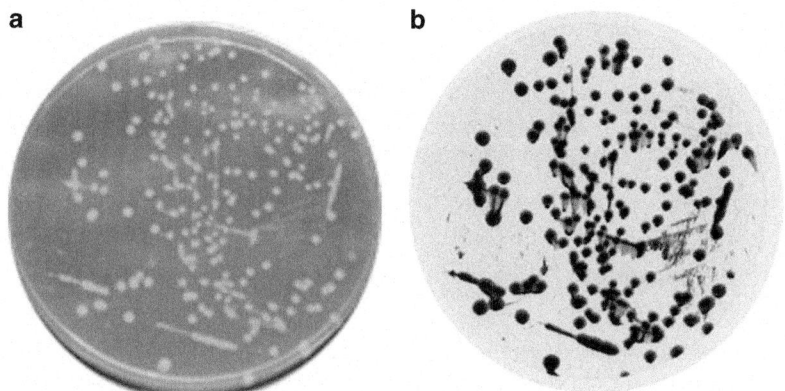

Fig. 26.4 Colonies of bacteria (**a**) and their autoradiography image (**b**). Heterotrophic bacteria have the ability to uptake ^{14}C from an agar medium

growth chamber, we grew rice plants with addition of [1,2-^{14}C] sodium acetate. This ^{14}C compound was supplied once to rice plants in the flooding water just before blooming, and TF of 6.8 ± 2.4 on average was obtained. In these tracer experiments, rice plants were also cultivated without [1,2-^{14}C] sodium acetate as negative controls in the same growth chamber as the ^{14}C-treated rice. Interestingly ^{14}C was detected even from the rice grains of negative control samples. These results suggested that the ^{14}C-bearing gas, which was released from bacterial cells in rice paddy soils, was fixed by the rice plants in the negative controls through photosynthesis.

We also examined the possibility of root uptake of ^{14}C by stable isotope techniques under field conditions [4]. If plant carbon originates from the atmospheric CO_2, the $\delta^{13}C$ values in crops can be calculated using the $\delta^{13}C$ value, $-8‰$ in air [5], and the ^{13}C fractionation ratio in photosynthesis by rice plants of -18 to $-20‰$ [6, 7]. The calculated $\delta^{13}C$ values in our study ranged from $-28‰$ to $-26‰$, and the results implied that no soil carbon contribution occurred for white rice; however, by setting some conditions, for example, ^{13}C fractionation ratio of $19‰$, we obtained the average TF value of 0.11 ± 0.04 for white rice. To compare these TF values obtained in laboratory and field experiments, it is necessary to pay attention to the difference between [1,2-^{14}C] sodium acetate and the actual organic compounds present in the natural soil.

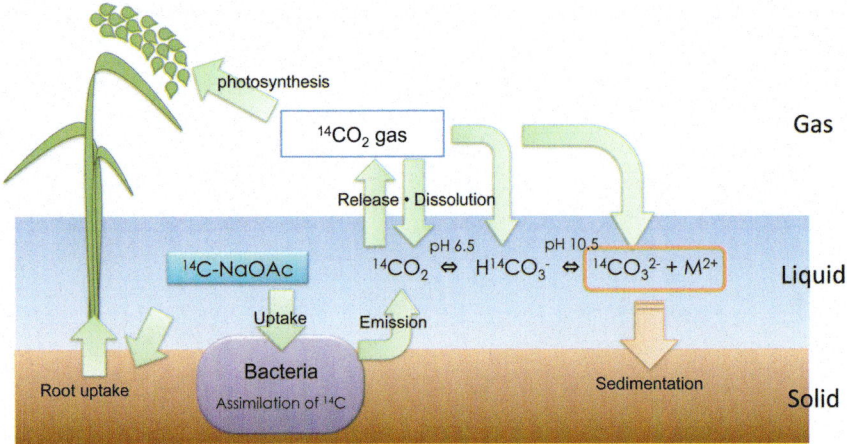

Fig. 26.5 A conceptual diagram for the behavior of ^{14}C in rice paddy fields

26.5 Behavior of ^{14}C in Rice Paddy Fields

From the aforementioned results, the behavior of ^{14}C in rice paddy fields could be considered as follows (a conceptual diagram appears in Fig. 26.5). When irrigation water is contaminated by ^{14}C-bearing sodium acetate, the ^{14}C compound is taken up and metabolized by indigenous bacteria. A part of the ^{14}C is assimilated by the bacterial cells, and the rest of the ^{14}C is released as gaseous compounds from the cells as a result of dissimilation. The dominant chemical species of ^{14}C in gas forms is carbon dioxide, and thus some of the released $^{14}CO_2$ is dissolved in soil solution depending on pH. For example, when the pH of the soil solution is less than 6.5, most of ^{14}C in gas forms is released into the air. The released $^{14}CO_2$ is eventually taken up by rice plants during photosynthesis. When the pH of the soil solution is between 6.5 and 10.5, ^{14}C-bearing bicarbonate ion dominates in the soil solution. In addition, once $^{14}CO_2$ has been released into the air, a part of the $^{14}CO_2$ gas may be redissolved in the soil solution again as bicarbonate ion. When the pH of the soil solution is greater than 10.5, although this is not probable in paddy fields, ^{14}C-bearing carbonate ion dominates in the soil solution. Carbonate ion is thermally unstable and thus precipitates as carbonate minerals such as $CaCO_3$. In these alkaline situations, the ratio of ^{14}C in the solid phase may increase as a result of the precipitation of ^{14}C. Because the root uptake of ^{14}C by rice plants is negligible, gasification of ^{14}C is an important environmental transfer pathway for the safety assessment of TRU wastes, and bacteria are responsible for driving this pathway.

Acknowledgments This work has been partially supported by the Agency for Natural Resources and Energy, the Ministry of Economy, Trade and Industry (METI), Japan.

Open Access This chapter is distributed under the terms of the Creative Commons Attribution Noncommercial License, which permits any noncommercial use, distribution, and reproduction in any medium, provided the original author(s) and source are credited.

References

1. Kaneko S, Tanabe H, Sasoh M, Takahashi R, Shibano T, Tateyama S (2002) A study on the chemical forms and migration behavior of carbon-14 leached from the simulated hull waste in the underground condition. MRS Proc 757:621–626
2. Ishii N, Koiso H, Takeda H, Uchida S (2010) Partitioning of C-14 into solid, liquid, and gas phases in various paddy soils in Japan. J Nucl Sci Technol 47:238–243
3. Ishii N, Uchida S (2011) Bacteria contributing to behavior of radiocarbon in sodium acetate. Radiat Prot Dosimetry 146:151–154
4. Tagami K, Uchida S (2010) Estimation of carbon-14 transfer from agricultural soils to crops using stable carbon isotope ratios. Waste Manag Symp Proc 36(10346):1–6
5. Ciais P, Tans P, Trolier M, White J, Francey R (1995) A large northern hemisphere terrestrial CO_2 sink indicated by the $^{13}C/^{12}C$ ratio of atmospheric CO_2. Science 269:1098–1102
6. O'Leary M (1981) Carbon isotope fractionation in plants. Phytochemistry 20:553–567
7. Lloyd J, Farquhar GD (1994) ^{13}C discrimination during CO_2 assimilation by the terrestrial biosphere. Oecologia (Berl) 99:201–215

Chapter 27
Development of a Rapid Analytical Method for ^{129}I in the Contaminated Water and Tree Samples at the Fukushima Daiichi Nuclear Power Station

Asako Shimada, Mayumi Ozawa, Yutaka Kameo, Takuyo Yasumatsu, Koji Nebashi, Takuya Niiyama, Shuhei Seki, Masatoshi Kajio, and Kuniaki Takahashi

Abstract The separation conditions for iodine species were investigated to analyze ^{129}I in contaminated water and tree samples generated from the Fukushima Daiichi Nuclear Power Station (FDNPS). Inorganic iodine species in the samples from FDNPS were thought to be iodide (I^-) and iodate (IO_3^-); therefore, the behaviors of these species during separation using a solid-phase extraction sorbent, Anion-SR, for water samples and combustion for tree sample were studied. When the amount of I was 1 µg and used within a few hours, I^- was extracted with the Anion-SR in 3 M NaOH and diluted HCl (pH 2) solutions, whereas IO_3^- was only slightly extracted in these solutions. In contrast, 15 ng I^- with a larger amount of IO_3^- (1 µg I) in the diluted HCl (pH 2) and allowed to stand for 1 day was only slightly recovered. It is possibly that I^- was changed to another species in a day in this condition. Iodate was successfully reduced to I^- with $NaHSO_3$ in the diluted HCl solution and extracted with the Anion-SR. Consequently, the solution condition to analyze both I^- and IO_3^- using Anion-SR was observed to be the diluted HCl at pH 2 with a reductant. For the tree samples, a combustion method was applied and

---

A. Shimada (✉)
Nuclear Cycle Backend Directorate, Japan Atomic Energy Agency,
2-4 Tokai-mura, Naka-gun, Ibaraki, Japan

Fukushima Project Team, Japan Atomic Energy Agency,
2-4 Tokai-mura, Naka-gun, Ibaraki, Japan
e-mail: shimada.asako@jaea.go.jp

M. Ozawa • Y. Kameo • T. Niiyama • S. Seki • M. Kajio
Fukushima Project Team, Japan Atomic Energy Agency,
2-4 Tokai-mura, Naka-gun, Ibaraki, Japan

T. Yasumatsu • K. Nebashi
Tokyo Power Technology Ltd., 5-5-13 Toyosu, Eto-ku, Tokyo, Japan

K. Takahashi
Nuclear Cycle Backend Directorate, Japan Atomic Energy Agency,
2-4 Tokai-mura, Naka-gun, Ibaraki, Japan

the rate of temperature increase was optimized to avoid anomalous combustion. Greater than 90 % recovery was obtained for both I^- and IO_3^-, and the chemical species in the trap solutions was observed to contain I^-.

Keywords ^{129}I • Combustion • Fukushima Daiichi Nuclear Power Station • Iodine species • Isotopic exchange • Solid-phase extraction

27.1 Introduction

Because of the accident, a large amount of radioactive waste was generated at the Fukushima Daiichi Nuclear Power Plants (FDNPP). To establish the waste management strategy, the radioactivity inventory has to be evaluated. Iodine-129 is one of the important nuclides of which the radioactivity has to be evaluated. Although I^- is considered a major species of ^{129}I generated in the reactor, IO_3^- and I_2 are possibly generated, depending on the reactor conditions [1]. Furthermore, because seawater was introduced to the reactors for cooling down in the early phase of the accident and seawater contains the natural iodine species, $^{127}IO_3^-$, an isotope exchange reaction between $^{127}IO_3^-$ and $^{129}I^-$ may have occurred. Therefore, analytical conditions to determine total I content, in this case IO_3^- and I^-, in water and tree samples were investigated in the current work.

Presently, contaminated water is accumulating in the basement of the reactor and turbine buildings at FDNPP. The accumulated water-processing equipment was installed to decontaminate and to desalinate. Consequently, secondary waste such as spent zeolite and sludge is generated. To evaluate the radioactivity inventory of the waste indirectly, water samples were collected from the inflow and outflow of the apparatus [2]. The contaminated water contains high levels of radioactivity of ^{137}Cs, ^{90}Sr, and other radionuclides. To limit radiation exposure of the analyst, rapid chemical separation of iodine species from these radionuclides is required. Chemical separation studies using the solid-phase extraction sorbent Anion-SR have been reported to rapidly separate I^- from major fission products such as Cs and Sr in contaminated water samples [3]. However, Anion-SR essentially extracts only I^- and not IO_3^-. Therefore, reduction of IO_3^- to I^- is required to analyze total I. In this study, $NaHSO_3$ was used as the reductant and the solution conditions were studied to reduce IO_3^- to I^-.

Because of the hydrogen explosion of FDNPP, trees on the site were contaminated by the radionuclides. Many of the trees were cut down to provide space to install tanks storing the contaminated water. Consequently, approximately 40,000 m^3 of trees were stored in the site as radioactive waste [4]. A combustion method was used to analyze ^{129}I in cement, ash, and soil samples [5, 6]. To apply a combustion method for the tree samples, there were some subjects: evaporation and deposition of the organic materials and anomalous combustion. Therefore, decomposition of organic material to CO_2 and H_2O using oxidant was examined. In addition, the rate of temperature increase was controlled to avoid anomalous combustion. Furthermore, the influence of the chemical species, IO_3^- or I^-, on recovery was studied.

27.2 Experimental

27.2.1 Reagents

Ultrapure grade NaOH solution (3 M) was purchased from Kanto Kagaku. Ultrapure grade of tetramethyl ammonium hydroxide (TMAH) solution was purchased from Tama Chemicals. The other reagents were all analytical grade or higher and purchased from Wako Chemical. Empore solid extraction disks, Anion-SR, were purchased from 3 M. Standard solution of ^{129}I (35.8 kBq/g ^{129}I; chemical composition, 50 µg/g NaI and 50 µg/g $Na_2S_2O_3$ in H_2O) was purchased from AREVA and diluted with H_2O before use. Potassium iodide (analytical grade, 99.9 %) and KIO_3 (analytical standard material grade, 99.98 %) were purchased from Wako Pure Chemical Industries, and $^{127}I^-$ and $^{127}IO_3^-$ were prepared from these reagents, respectively, because the stable iodine isotope is only ^{127}I.

27.2.2 Separation Using Anion-SR

Figure 27.1 shows the experimental procedure for the solution samples. Sodium iodide-129 (^{129}I: 0.1 Bq = 15 ng) and $K^{127}IO_3^-$ (^{127}I: 1 µg) were added to 50 ml 3 M NaOH solution or 50 ml diluted HCl solution (pH = 2) with and without reductant (0.1 ml 0.1 M $NaHSO_3$) to study iodine species behavior in the analysis using Anion-SR. After addition of the iodine species, the solutions were allowed to stand for 1 day before separation with Anion-SR. The reductant was added approximately 20 min before the separation. The operation of Anion-SR was based on Shimada et al. [3]. Briefly, the Anion SR disk was centered on the base of the filtration funnel and the reservoir was clamped on the top of the disk. The appropriate solution was poured into the reservoir followed by suction filtration The Anion-SR disk was conditioned with acetone, methanol, ultrapure water, 4 w/v% NaOH, and ultrapure water. After conditioning, the sample solution was introduced into the Anion-SR disk and washed with ultrapure water. The extracted I was recovered with 9.5 ml 1 M HNO_3. To oxidize I^- to IO_3^-, 0.1 ml NaClO solution (effective Cl concentration, >5 %) was added to the recovered solution. Additionally, 0.1 ml 2 ppm Rh standard solution was added to the recovered solution as an internal standard. Finally, 1 M HNO_3 was added to the recovered solution to a final volume of 10 ml. The concentration of I was measured by inductively coupled plasma mass spectrometry with dynamic reaction cell (DRC-ICP-MS). In the reaction cell, oxygen gas was collided with ions. Because the order of ionization potential is I > O > Xe, O reacts with Xe to neutralize but I does not react with O. As a result, the count of ^{129}Xe, impurity of Ar gas, was decreased. The experimental conditions of DRC-ICP-MS were consistent with the conditions reported by Kameo et al. [7]. Percent recovery was calculated as in Eq. (27.1).

Fig. 27.1 Schematic diagram of analysis of $^{127}IO_3^-$ and $^{129}I^-$ in the solution samples

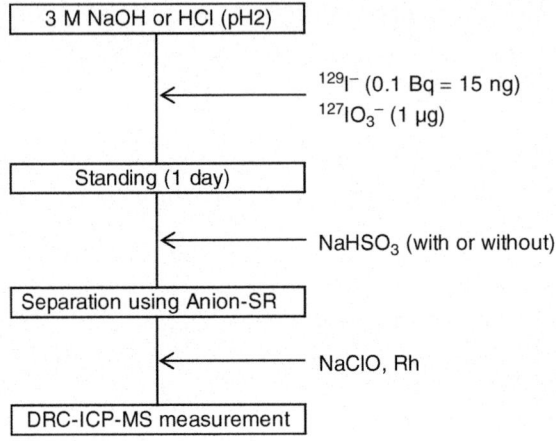

$$\text{Percent recovery} = \frac{\text{Amount of I in the recovered solution}}{\text{Added amount of I}} \times 100 \quad (27.1)$$

The separation experiment to determine percent recovery was carried out twice, and the uncertainty was quantified by the dispersion in these two measurements.

27.3 Combustion Method

A known amount of I^- or IO_3^- was added to 1 g pine bark, representative of tree samples taken at the establishment of the Japan Atomic Energy Agency, and the bark was put in a wet oxygen gas line set in an electric furnace. Because smoking was observed with noncontrolled temperature increase, especially around 240 °C, the rate of temperature increase in the range from 100 °C to 300 °C was controlled by steps and slow. The vaporized I was trapped in three steps of 20 ml 2 % TMAH solution. Because insoluble organic material was deposited in the gas line and the trap, an oxidant (8.2 g hopcalite II) was set between the sample and traps to decompose it (Fig. 27.2). The temperature of the oxidant was set to 500 °C before the temperature increase of the pine bark sample. The temperature of the sample was kept at 500 °C for 1 h and then increased to 900 °C for 1 h to vaporize iodine species in the sample. The extracted I in the traps was measured by DRC-ICP-MS.

Fig. 27.2 Illustration of the combustion apparatus

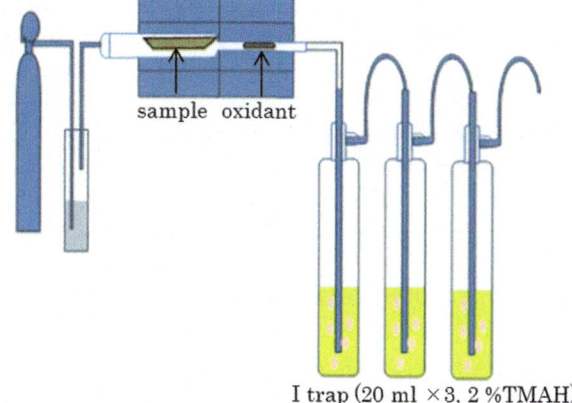

I trap (20 ml ×3, 2 %TMAH)

Table 27.1 Percent recovery of I (initially $^{127}IO_3^-$ and $^{129}I^-$)

Solution condition	Without NaHSO$_3$		With NaHSO$_3$	
	^{127}I	^{129}I	^{127}I	^{129}I
3 M NaOH	2.3 ± 1.9	71.4 ± 2.9	2.1 ± 0.2	86.4 ± 4.3
HCl (pH 2)	8.6 ± 6.3	19.2 ± 4.8	64.0 ± 7.3	59.6 ± 3.1

27.4 Results and Discussion

27.4.1 Separation Using Anion-SR

Good recovery of I^- (as $^{129}I^-$) from the Anion-SR was observed in the 3 M NaOH solution, but only minor amounts of IO_3^- (as $^{127}IO_3^-$) were recovered (Table 27.1). The percent recoveries did not appear to be dependent on the reductant because good recoveries of I^- and poor recoveries of IO_3^- were observed regardless of the presence or absence of NaHSO$_3$. This result supports the expectation that $^{129}I^-$ is extracted and $^{127}IO_3^-$ is not extracted by the Anion-SR, as $^{127}IO_3^-$ was not reduced to $^{127}I^-$ by NaHSO$_3$ in 3 M NaOH solution, and the isotopic exchange reaction between $^{127}IO_3^-$ and $^{129}I^-$ and reduction from $^{127}IO_3^-$ to $^{127}I^-$ were negligible at least for 1 day. It is reported that ^{129}I in seawater offshore of Fukushima is mainly $^{129}I^-$ regardless of the presence of large amounts of natural $^{127}IO_3^-$ [8].

In the diluted HCl solution at pH 2, ^{129}I and ^{127}I were recovered in the presence of the reductant and were not recovered without the reductant (Table 27.1). Both the isotopes behaved similarly in this case despite the difference in the initial chemical species, $^{129}I^-$ and $^{127}IO_3^-$. In contrast, the experiments using 1 μg $^{127}I^-$ in the diluted HCl solution at pH 2 without the reductant and standing, higher recovery of $^{127}I^-$, 72 ± 6 %, was obtained. Although the influence of the I^- amount on recovery cannot be wholly denied, it is possible that a considerable amount of $^{129}I^-$ would be changed to another chemical species in the diluted HCl solution in a day. In addition, the changed chemical species and IO_3^- were reduced to I^- by the reductant in this experimental condition. Consequently, inorganic iodine species were analyzed at pH 2 with the reductant, and only I^- was analyzed in 3 M NaOH. It is possibly to apply speciation methods to analyze I^- and IO_3^-, depending on the solution conditions.

27.4.2 Combustion Method

When 1 g pine bark was combusted by the electric furnace without control, smoking was observed at the temperature range from 200 ° to 300 °C, especially at 240 °C. Therefore, a stepwise and slow rate of temperature increase was introduced. When smoking was observed, the gap of the setting temperature was decreased and holding time at the temperature was prolonged. Finally, smoking was not observed by the controlled rising temperature program shown in Fig. 27.3, in which the increments of 50 °C for 20 min from 100 °C to 200 °C, and in increments of 15 °C for 20 min from 200 °C to 300 °C.

Tetramethyl ammonium hydroxide solutions containing known amounts of I^- or IO_3^- were prepared to make calibration curves for the DRC-ICP-MS measurement. However, the counts of mass number 127 for IO_3^- were essentially at background levels, which indicated that measurement of IO_3^- in 2 % TMAH solution was not available in our instrument. It is considered that I is vaporized as I_2 by the combustion and I_2 is trapped in an alkaline solution as I^-. Therefore, the calibration curve was prepared from I^- solutions.

Combustion experiments of pure pine bark sample and pine bark samples spiked with 50 μg I as I^- or IO_3^- were performed and the amounts of ^{127}I in the traps were determined by DRC-ICP-MS with the calibration curve previously mentioned

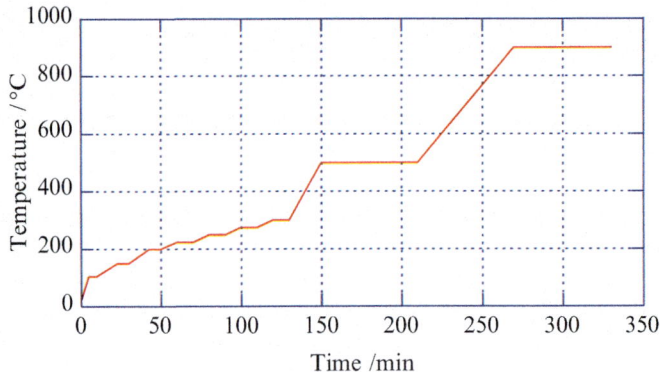

Fig. 27.3 Optimized rising temperature program

Table 27.2 Recovered I (μg) in the traps by the combustion method

Sample	Trap 1	Trap 2	Trap 3
1 g pine bark	0.29	0.052	0.025
1 g pine bark + 50 μg I (I^-)	45	0.030	0.004
1 g pine bark + 50 μg I (IO_3^-)	52	0.070	0.017

(Table 27.2). Greater than 90 % recovery was obtained for both I^- and IO_3^- spiked samples in the first trap. This result suggested that the chemical species of I in the traps was I^- regardless of the initial chemical species.

27.5 Conclusion

To analyze ^{129}I in the contaminated water from FDNPS, separation of I from radionuclides such as ^{137}Cs and ^{90}Sr using Anion-SR and measurement by DRC-ICP-MS are convenient and effective. As a solution condition, a diluted HCl solution of pH 2 with reductant was required to analyze inorganic iodine species using Anion-SR.

To analyze ^{129}I in tree samples from FDNPS, a combustion method is suitable. Both I^- and IO_3^- were vaporized by combustion and trapped in the 2 % TMAH solution as I^- with high recovery (>89 %). Anomalous combustion was avoided by the stepwise and slow temperature increase in the range from 100 ° to 300 °C.

Open Access This chapter is distributed under the terms of the Creative Commons Attribution Noncommercial License, which permits any noncommercial use, distribution, and reproduction in any medium, provided the original author(s) and source are credited.

References

1. Tigeras A, Bachet M, Catalette H, Simoni E (2011) Prog Nucl Energy 53:504–515
2. Tanaka K, Yasuda M, Watanabe K, Hoshi A, Higuchi H, Kameo Y (2013) Proceedings of the 14th workshop on environmental radioactivity, Tsukuba, pp 17–26
3. Shimada A, Sakatani K, Kameo Y, Takahashi K. J Radioanal Nucl Chem. Accepted for publication
4. Tanaka K, Shimada A, Hoshi A, Yasuda M, Ozawa M, Kameo Y (2014) J Nucl Sci Technol 51(7–8):1032–1043
5. Ishimori K, Haraga T, Shimada A, Kameo Y, Takahashi K (2010) Ibaraki (Japan): Japan Atomic Energy Agency; JAEA-technology 2010–016 (Japanese)
6. Toyama C, Muramatsu Y, Uchida Y, Igarashi Y, Aoyama M, Matsuzaki H (2012) J Environ Radioact 113:0116–0122
7. Kameo Y, Ishimori K, Shimada A, Takahashi K (2012) Bunseki Kagaku 61(10):845–849
8. Hou X, Povinec PP, Zhang L, Shi K, Biddulph D, Chang C-C, Fan Y, Golser R, Hou Y, Jeskovsky M, Tim Jull AJ, Liu Q, Luo M, Steier P, Zhou W (2013) Environ Sci Technol 47:3091–3098

Part IX
Treatment of Radioactive Waste: Reduction of the Radioactivity or Volume of Nuclear Wastes

Chapter 28
Consideration of Treatment and Disposal of Secondary Wastes Generated from Treatment of Contaminated Water

Hiromi Tanabe and Kuniyoshi Hoshino

Abstract The earthquake and tsunami on March 11, 2011, caused severe accidents at the several Fukushima Daiichi Nuclear Power Units, and a significant volume of highly contaminated water was generated from the accident. Several methods have been applied to decontaminate the water, including systems from AREVA S.A. and Kurion, Inc., in addition to the SARRY (Simplified Active Water Retrieval and Recovery System) and ALPS [Advanced Liquid Processing System; incorporated in the MRRS (Multi Radionuclide Removal System)] systems from Toshiba Corporation. After the decontamination treatments using these systems, various kinds of sludge and spent adsorbents were generated as secondary wastes. These wastes are now tentatively stored at the site, but further treatment shall be applied to produce appropriate waste forms for interim storage and final disposal in a repository.

Waste management—the treatment, storage, transportation, and disposal of these wastes—is believed to require several decades. The authors examined how to manage these wastes in consideration of the large volume of waste, the variety of waste types, and the long period required to carry out their treatment and disposal in a safe and efficient manner. The requirements for an inventory list and online waste management system; a development strategy for waste treatment, storage, transport, and disposal; formation of an R&D implementation and evaluation team; and long-term knowledge management are discussed in this chapter.

Keywords Contaminated water • Disposal • Fukushima Daiichi Nuclear Power Units • Inventory • Secondary waste • Treatment

---

H. Tanabe (✉) • K. Hoshino
Radioactive Waste Management Funding and Research Center, 1-15-7, Tsukishima, Chuo-ku, Tokyo, Japan
e-mail: tanabe.hiromi@rwmc.or.jp

© The Author(s) 2015
K. Nakajima (ed.), *Nuclear Back-end and Transmutation Technology for Waste Disposal*, DOI 10.1007/978-4-431-55111-9_28

28.1 Introduction

A significant volume of highly contaminated water was generated from the accidents at the Fukushima Daiichi Nuclear Power Units. Several methods have been applied to decontaminate the water. After decontamination treatments using several systems developed by AREVA, Kurion, and Toshiba, various kinds of sludge and spent adsorbents were generated as secondary wastes. These wastes are now tentatively stored at the site, but further treatment shall be applied to produce appropriate waste forms for interim storage and final disposal in a repository. Management of these wastes is believed to take several decades. The authors examined how to manage these wastes in consideration of the large volume of waste, the variety of waste types, and the long period to carry out the treatment and disposal in a safe and efficient manner. The issues identified are discussed in the following sections.

28.2 Requirements for an Inventory List and Online Waste Management System

A fundamental issue is to establish a raw waste inventory list with information on various characteristics, including chemical and physical form and radionuclide inventory (Table 28.1). This is the first step for pursuing further examination of treatment and disposal of these wastes. However, it is very difficult to establish a

Table 28.1 Important characteristics of radioactive waste that may be used as parameters for waste classification [1]

Origin	Chemical properties:
Criticality	Chemical composition
Radiological properties:	Solubility and chelating agents
Half-lives of radionuclides	Potential chemical hazard
Heat generation	Corrosion resistance/corrosiveness
Intensity of penetrating radiation	Organic content
Activity concentration of radionuclides	Combustibility and flammability
Surface contamination	Chemical reactivity and swelling potential
Dose factors of relevant radionuclides	Gas generation
Decay products	Sorption of radionuclides
Physical properties:	Biological properties:
Physical state (solid, liquid, or gaseous)	Potential biological hazards
Size and weight	Bio-accumulation
Compactibility	Other factors:
Dispersibility	Volume
Volatility	Amount arising per unit of time
Miscibility	Physical distribution
Free liquid content	

complete list from the beginning because the volume of waste and variety of waste types is large and certain radionuclides contained in the wastes are difficult to identify and measure. Therefore, stepwise development and evaluation are important. A management system of the waste inventory should be established, and new waste information obtained by several intensive waste characterization projects should be added to the management system, where the information will be shared among concerned organizations.

The management system should include information about treatment method and source term characteristics, including waste form, volume, surface dose rate, ID number, and location of waste package, which are collected when raw wastes are treated. This system should be maintained and used until final disposal of the wastes. When all wastes are disposed of in a designated disposal facility, the inventory list will be used as the waste records of the disposal facility.

28.3 Development Strategy of Waste Treatment, Storage, Transport, and Disposal Technologies

The authors proposed a work flow for selecting a waste management technology [2] (as shown in Fig. 28.1). Establishment of the development criteria for each technology is a fundamental issue and the first step of the development.

For treatment technology, it is important to have a simple system that can be applied to a variety of wastes, and which has a volume reduction factor and is economical. In addition, the system must minimize secondary waste and address the difficulty of residual research and development (R&D) to commercialize the technology. Regarding the stability of waste forms produced by any treatment technology, a low leaching rate is required, especially for waste containing long-lived radionuclides over a certain amount. Easy identification and measurement of the radionuclides in the waste by the treatment process would also be considered an advantage.

Storage technology has many options that are either in operation or under R&D in different countries. The storage cost is an important index to use when selecting an option. If a waste generates a large quantity of heat as a result of containing a large amount of beta- and gamma-emitting nuclides such as Cs-137, the waste storage period should be considered before final disposal to reduce heat generation. If the waste has greater heat generation in disposal, the waste emplacement area must be increased to reduce the temperature of the surrounding engineered barriers or rock to less than the allowable temperature. As a consequence, disposal cost will increase. The storage period and disposal cost have a strong mutual relationship. If the storage period must be increased by more than a few hundred years to reduce heat generation for disposal, certain radionuclides such as Cs-137, whose half-life is less than several decades, would have undergone significant decay by time of disposal, which also raises the possibility of lowering the waste classification level for disposal.

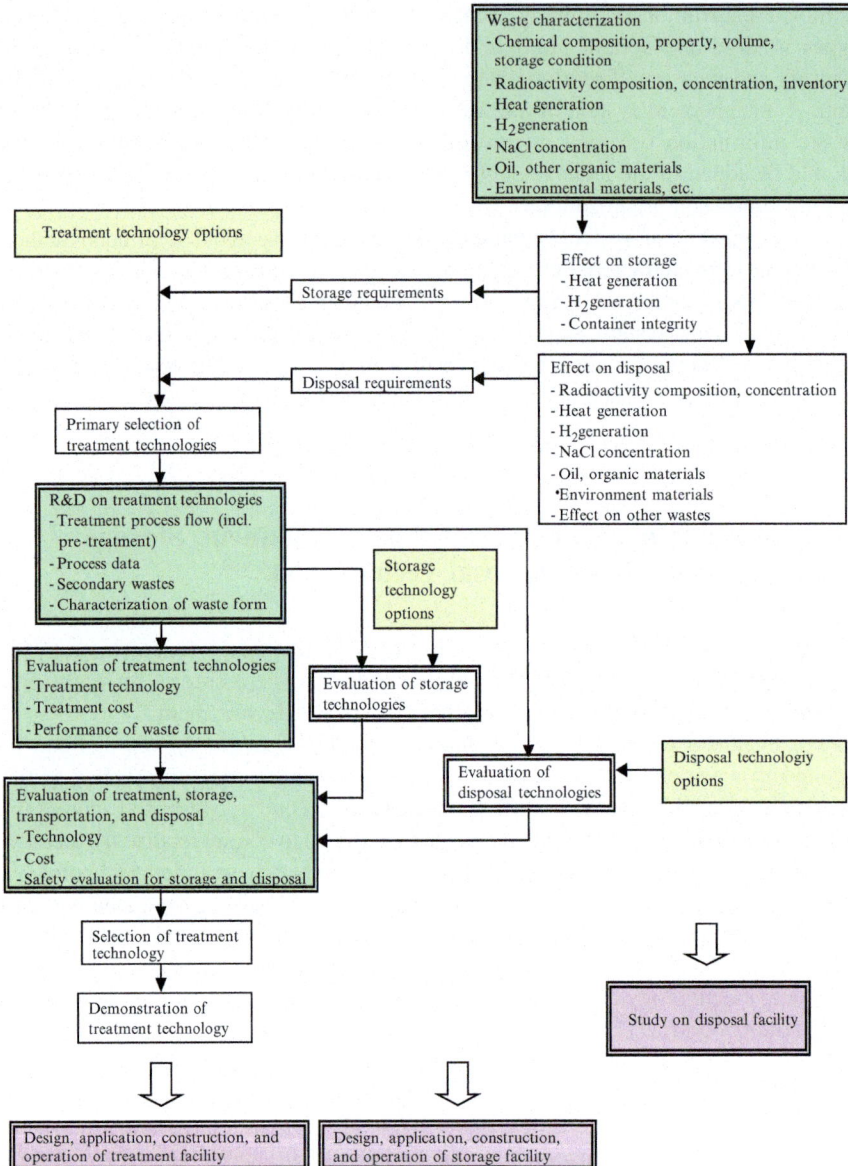

Fig. 28.1 Flow chart for selection of waste management technology [2]

Disposal technology also has many options that are either in operation or under R&D in different countries. The disposal cost depends on the design of the disposal facility, the composition of the engineered barriers, and the depth of the disposal facility, which are strongly dependent on the characteristics of the radionuclides in the waste.

Table 28.2 Items for the management database of treatment technology options

I.D. No.
Name of technology
Description of technology
Applicability to various wastes
Process flow diagram
Description of process flow diagram
Treatment conditions, including temperature, pressure, conditions, material balance, radionuclide balance, and decontamination factor (DF)
Chemical reagents and utilities
Volume reduction ratio of waste
Characteristics of waste form, including configuration, uniaxial compression strength, vacant volume ratio, apparent density, leaching rate of radionuclides, and leaching rate of chemical components
Content and status of NaCl
Status of technology; commercialization, development, fundamental research
Technical issues (R&D for commercialization)
Standards to be applied
References

Technology options for treatment, storage, transportation, and disposal have been proposed by domestic and international organizations. These options should be integrated and managed in a database of technology options. R&D results should also be added in the database as soon as possible. To manage the technology options, a database should be prepared and used to share information among the related organizations. The authors proposed a set of items to be managed in the database of treatment technology options (Table 28.2).

When an appropriate technology is being selected from several options, it is important to evaluate each technology option and compare them. At the end, it is also important to evaluate the combination of technologies from treatment to disposal to make the final selection of an appropriate set of these technologies and to establish a total system for specific wastes. The authors propose a set of indexes for evaluation of the combination of waste management technologies (Table 28.3). Evaluation of the total system should include the long-term safety of disposal because this is the most important issue and goal of waste management.

28.4 Formation of an R&D Implementation and Evaluation Team

The development of a waste inventory database and waste treatment technology and the evaluation of the feasibility of each technology and the total system of waste management requires a team composed of system engineers, researchers,

Table 28.3 Evaluation indexes for combination of waste management technologies

a. Treatment	b. Waste form	c. Storage
–Applicability to various wastes	–Waste classification for transportation and disposal	–Storage method
–Components	–Uniaxial compaction strength	–R&D for commercialization
–Chemical reagent and utilities	–Density	–Storage cost
–Measures to protect against NaCl-induced corrosion	–Vacant volume ratio	
–Material and radionuclide balances	–Leaching rate of radionuclides	
–R&D for commercialization	–Leaching rate of chemical components	
–Treatment cost	–Hydrogen generation rate and measures to manage hydrogen	
	–Temperature resistance	
	–Radiation resistance	
d. Transportation	e. Disposal	f. Evaluation of total system
–Transportation container	–Disposal method (design, engineered barriers, depth)	–R&D for commercialization
–Transportation cost	–R&D for commercialization	–Treatment, storage, transportation, and disposal cost
	–Disposal cost	–Long-term safety of disposal

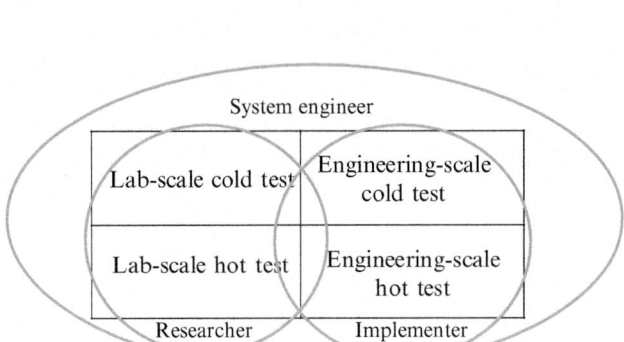

Fig. 28.2 Research and development (R&D) implementation and evaluation team

and implementers from the start of R&D. The authors propose an R&D implementation and evaluation team (Fig. 28.2). The role of each member is as follows: researchers conduct laboratory-scale cold and hot tests for collecting basic information on waste inventory and treatment technology, implementers carry out engineering-scale cold and hot tests for commercializing the technology, and system engineers carry out the system design and cost evaluation of the technology to evaluate the feasibility of the technology. The members of the team should work together on R&D and to check whether the data obtained are sufficient for all members. As a result of this collaboration, it is expected that sufficient data will be obtained to pursue the next phase of testing and to transfer the technical knowledge smoothly to other members. Because the engineering-scale test will be conducted by the implementers, education and training of future operators will be also performed. The team should also evaluate the storage, transportation, and disposal technology. For this purpose, system engineers who perform these evaluations should be involved.

28.5 Requirements for Long-Term Knowledge Management

As already stated, when a nuclear accident occurs there is a much larger volume of wastes and greater variety of waste types than the conventional wastes generated from the usual operation and planned decommissioning of a nuclear power unit. The wastes from a nuclear accident also contain a wider range of concentrations for the various radionuclides. Consequently, it will take a long period to characterize the wastes and to carry out the R&D and evaluation of treatment and disposal technology. It is also expected that the actual treatment and disposal of the wastes will not take place until after more than a few decades. In consideration of this waste management period, long-term knowledge management is needed. The authors propose the formation of an R&D implementation and evaluation team that will manage and retain technical knowledge. The team should involve younger staff members, and education and training of the next generation of staff should be performed by on-the-job training (OJT).

28.6 Conclusion

A significant volume of highly contaminated water was generated from the accidents at the Fukushima Daiichi Nuclear Power Units. Several methods have been applied to decontaminate the radioactivity of the water, and these methods have generated various kinds of sludge and spent adsorbents as secondary wastes. Because long-term waste management is needed, the authors examined a broad range of issues concerning how to manage these wastes in a safe and efficient manner. The requirements for an inventory list and online waste management system; a development strategy for waste treatment, storage, transport, and disposal technology; formation of an R&D implementation and evaluation team; and long-term knowledge management are proposed.

Open Access This chapter is distributed under the terms of the Creative Commons Attribution Noncommercial License, which permits any noncommercial use, distribution, and reproduction in any medium, provided the original author(s) and source are credited.

References

1. International Atomic Energy Agency (2009) Classification of radioactive waste. IAEA Safety Standards Series No. GSG-1. IAEA, Vienna
2. Tanabe H (2013) Consideration on treatment and disposal of secondary wastes generated from treatment of contaminated water, B37. In: Proceedings of the 2013 annual meeting of Atomic Energy Society of Japan, Tokyo

Chapter 29
Volume Reduction of Municipal Solid Wastes Contaminated with Radioactive Cesium by Ferrocyanide Coprecipitation Technique

Yoko Fujikawa, Hiroaki Ozaki, Hiroshi Tsuno, Pengfei Wei, Aiichiro Fujinaga, Ryouhei Takanami, Shogo Taniguchi, Shojiro Kimura, Rabindra Raj Giri, and Paul Lewtas

Abstract Municipal solid wastes (MSW) with elevated concentrations of radioactive cesium (rad-Cs hereafter) have been generated in some areas of Japan in the aftermath of the Fukushima Daiichi Nuclear Power Plant (F1 hereafter) accident. Both recycling and final disposal of the contaminated MSW have become a difficult problem in the affected areas, resulting in accumulation of treated residues in the treatment facilities.

The rad-Cs in MSW, especially fly ash, often showed a high leaching rate. Extraction of contaminated MSW with water or hot oxalic acid followed by selective removal of rad-Cs from the extract using ferrocyanide (Fer hereafter) coprecipitation technique could be an ultimate solution for waste volume reduction. The MSW extracts contain various metal components as well as chelating reagents like oxalic acid, and are often very saline. The composition of the extract varies widely depending on waste sources, applied treatment techniques, and rad-Cs extraction method etc. The applicability of the Fer coprecipitation technique had to be tested and validated before it could be applied for actual treatment.

In this work, we applied the Fer technique and observed removal of cesium (Cs) from water and oxalic acid extracts (all spiked with rad-Cs tracer or stable Cs) of various MSW samples collected from uncontaminated areas. Finally, the Fer technique was applied on site for removal of rad-Cs in the extracts of

---

Y. Fujikawa (✉)
Kyoto University Research Reactor Institute, Asahiro-nishi, Kumatori-cho, Sennan-gun, Osaka 590-0494, Japan
e-mail: fujikawa@rri.kyoto-u.ac.jp

H. Ozaki • H. Tsuno • P. Wei • A. Fujinaga • R. Takanami • S. Taniguchi • R.R. Giri
Osaka Sangyo University, 3-1-1 Nakagaito, Daito-shi, Osaka 594-8530, Japan

S. Kimura
Osaka University of Pharmaceutical Sciences, 4-20-1 Nasahara, Takatsuki, Osaka 569-1094, Japan

P. Lewtas
Edith Cowan University, 270 Joondalup Drive, Joondalup WA6027, Australia

© The Author(s) 2015
K. Nakajima (ed.), *Nuclear Back-end and Transmutation Technology for Waste Disposal*, DOI 10.1007/978-4-431-55111-9_29

contaminated MSW. By modifying coprecipitation conditions according to solution matrix, Cs removal rates of higher than 95 % could be obtained.

Keywords Cesium • Ferrocyanide • Metal • Municipal solid waste • Oxalic acid • pH

29.1 Background and Objectives

The estimated sustainable life period of the existing final disposal sites for municipal solid wastes (MSW) in Japan was only 18 years as of the end of FY2008. Therefore, waste avoidance, waste volume reduction, and recycling of MSW have been a national policy. However, the Fukushima Daiichi Nuclear Power Plant (F1) accident has created an entirely new dimension in environmental pollution problems. Because waste incineration and water treatment are, by their nature, the processes that concentrate pollutants such as radioactive cesium (rad-Cs) in ashes and sludge, MSW containing high concentrations of rad-Cs are produced in some areas where high atmospheric deposition of rad-Cs occurred in the aftermath of the F1 accident. As a result, recycling of MSW as concrete material and compost has become difficult, and their reuse has been often prevented because of public opposition even when rad-Cs concentrations in the wastes are below the clearance level (100 Bq/kg). Most of the citizens in the affected area are in hard opposition to disposal of rad-Cs-containing wastes even if radioactivity of the wastes is below the governmental limit for their disposal in landfills with leachate collection systems (i.e., 8,000 Bq/kg of Cs-134 + Cs-137). As the result, treatment residues are now piling up in many treatment facilities in some area, which may eventually jeopardize the treatment itself and exert serious negative impacts to everyday life. For example, sewage facilities in Fukushima Prefecture stored 74,401 t of dewatered sludge, molten slug, and incinerator ashes as of May, 2014. Therefore, suitable technologies to reduce the volume of such wastes or to decontaminate rad-Cs at low cost are urgently required.

Private companies and agencies have been working on sludge volume reduction through drying combined with granule processing [1] with the purpose of alleviating storage problems at treatment facilities. High-temperature combustion of sludge with an additive for controlling basicity of incineration material also proved effective in condensing rad-Cs in fly ash. The cost of this technique, however, was high and would be justified only when a very strong social need for sludge volume reduction exists [2]. Another tested technique in this regard is extraction of sewage by hot 0.1 M oxalic acid followed by recovery of the extracted rad-Cs by zeolite [3]. The cost of the oxalic acid method is considered acceptable for large-scale sewage treatment facilities, although waste volume reduction is dependent on the amount of zeolite necessary to remove Cs from the extract. The Cs distribution factor value (ml/g) reported for zeolite was a few thousand whereas the values for ferrocyanide (Fer) compounds determined by the in situ Fer coprecipitation method were between 10^4 and 10^6 [4]. Apparently, the use of the Fer coprecipitation technique for rad-Cs removal from waste extract is appropriate to maximize waste volume reduction.

On the other hand, there are concerns on the outcome of using Fer, especially regarding the radiological risk of generating concentrated waste regarding rad-Cs and the chemical hazard from Fer compounds.

The concentration of rad-Cs in insoluble Fer precipitate [Q_{Cs} (Bq/kg)] generated by adding 0.1 mM potassium ferrocyanide (the concentration used in most of our experiments) to the waste extract can be estimated as follows:

$$Q_{cs} = \frac{r}{100} \frac{E}{100} \frac{M}{pV} C_0$$

Here r is percentage of rad-Cs removed from the extract of MSW by Fer technique, E is percentage of rad-Cs extracted from MSW with water or oxalic acid, M is weight (kg) of MSW extracted by V (l) of the solvent, p is weight of Fer precipitate formed per unit volume of the extract (kg/l), and C_0 is rad-Cs concentration (Bq/kg) in original MSW. Assuming that r, E, M, V, and p are 95 %, 90 %, 1 kg, 2.5 l, and 35×10^{-6} kg/l, respectively (the values typically encountered in our on-site tests), Q_{Cs} (Bq/kg) is 9,771 C_0, implying that rad-Cs concentration in the Fer precipitate can be about four orders of magnitude higher than that in the original MSW. The designated wastes with rad-Cs concentration >100,000 Bq/kg are going to be sent to the interim storage facility in Fukushima Prefecture and the waste volume reduction is going to be carried out at the interim storage site before final disposal. The wastes with rad-Cs concentration lower than 100,000 Bq/kg are going to be disposed in a leachate-controlled landfill constructed by the national government or a conventional municipal landfill. The amount of designated wastes stored in 12 prefectures is 140,343 t as of December 31, 2013 [5], but most are less than 100,000 Bq/kg in rad-Cs concentration. The amount of designated waste exceeding 100,000 Bq/kg is predicted to be 9,000 t with rad-Cs concentration varying between 120,000 and 540,000 Bq/kg depending on the origin of the waste [6]. If the extraction of the waste followed by Fer coprecipitation was conducted for 9,000 t of designated waste >100,000 Bq/kg, and r, E, M, V, and p values were the same as discussed early in this paragraph, 790 kg of insoluble Fer waste with rad-Cs concentration 1.2×10^9–5.3×10^9 Bq/kg (total amounts of rad-Cs, 9.2×10^{11} to 4.2×10^{12} Bq) can be generated. By comparison, the content of rad-Cs in a piece of vitrified high-level radioactive waste (weight, 500 kg) can be as high as 4.8×10^{15} Bq [7], that is, three orders of magnitude higher than that from 9,000 t of highly contaminated designated waste. With appropriate instrumentation and management, it is possible to handle the rad-Cs concentrated waste resulting from the volume reduction of designated waste relatively safely.

The chemical risk of using Fer compounds to concentrate rad-Cs also requires attention. Although reagents such as oxalic acid that may be used for the extraction of rad-Cs are biodegradable and the degradation products are nontoxic, Fer compounds contain a cyano group within their structure, and are potentially more hazardous. Chemical toxicity of Fer compounds, especially that of ferric ferrocyanide (Prussian blue, PB hereafter), in mammals has been studied extensively because PB is a decorporation drug to treat internal rad-Cs contamination for both humans and livestock animals [8]. Based on laboratory animal studies, human male volunteer studies, and the experience of actual administration of PB to people contaminated with ^{137}Cs, it was concluded that PB is basically nontoxic. The history of the use of Na-Fer, Ca-Fer, and K-Fer as food additives also indicates that the toxicity of Fer compounds is low. More important is the risk pertinent to the long-term decomposition of Fer and possibility of free cyanide leaching from waste materials. For example, large amounts of Fer compounds have been generated in the coal and petroleum gas purifier used in gas production industries. The used purifier (containing ferric ferrocyanide) was often abandoned around coal pyrolysis plants, etc., and has caused the pollution of soil and groundwater. The problem is widespread: 1,310, 234, and 1,100 to 3,000 sites are known in Germany, Netherlands, and the U.S., respectively [9]. In these sites, groundwater contained cyanide complexes such as Fer rather than free and more toxic CN^- or HCN, probably because Fer was decomposed rapidly only when exposed to daylight and the decomposition of Fer in the dark underground was very slow [10]. Laboratory experiments showed that Fer was eluted from soil at pH > 13 whereas ferricyanide (Fe(III)-CN complex) was easily eluted by freshwater [11]. If Fer is to be used to concentrate rad-Cs, the resulting cyanide complex-containing waste should be managed properly by avoiding exposure to daylight and alkaline reagent. It is also possible to decompose Fer thermally or chemically (e.g., United States Environmental Protection Agency [12]) before the final disposal, depending on the cost allowed for the treatment.

Although the use of Fer coprecipitation technique has to be evaluated from the environmental safety considerations, it is also necessary to know if the technique is applicable to the actual MSW treatment at all. MSW waste extracts contain high concentrations of multiple transition metals (Fe, Mn, Cu, Zn, and Ni), alkali metal ions (Na and K), NH_4^+, alkaline earth ions (Ca and Mg), and anions (F^-, Cl^-, SO_4^{2-}, NO_3^-, and PO_4^{3-}). Mixtures of various kinds of insoluble Fer-metal precipitates can be formed in such solutions, and the substitution of alkali metals in the precipitate should also occur. Solubility of each Fer-metal compound as well as the reaction kinetics between Fer ion and each metal should influence the amount and chemical structure of Fer-metal precipitate thus formed. The types and concentrations of anions in the solution affect the efficiency of coagulation of colloidal Fer solid, and thus the solid–liquid separation. Therefore, the feasibility of the Fer coprecipitation technique has to be tested and validated before its application to the actual treatment.

The objective of this work is to identify the factors that are likely to govern Cs removal from MSW extracts by the Fer coprecipitation technique and to optimize coprecipitation conditions for Cs removal. As detailed information on chemical components in the extracts of rad-Cs contaminated wastes is hard to obtain, we first obtained and analyzed uncontaminated MSW extracts (i.e., do not contain rad-Cs from the F1 accident), applied Fer precipitation techniques to the uncontaminated MSW extracts, and then proceeded to rad-Cs-contaminated waste treatment.

29.2 Principle of Ferrocyanide Coprecipitation for Cs Removal

The reaction of soluble Fer salts (K, Na, or H compounds) with metal (Fe, Cu, Zn, Ni, Cd, Mn, etc.) ions in solution produces insoluble metal-Fer complexes. Fer ion and metal can precipitate as such compounds as $A_2M_3[Fe(CN)_6]_2 \cdot nH_2O$, $A_2MFe(CN)_6 \cdot nH_2O$, $M_2Fe(CN)_6 \cdot nH_2O$, or as mixtures of these, depending on concentrations of alkali metal ions (designated as A^+) and divalent transition metal ions (M^{2+}) in the solution. The elemental composition and crystal structure of precipitates also vary with the combination of soluble Fer salt (e.g., lithium Fer, sodium Fer, or potassium Fer) and transition metal salts (e.g., chloride, nitrate, or sulfate salts) used [13]. Some trivalent metals (e.g., Fe^{3+}) also precipitate with Fer.

The insoluble Fer compounds preferentially incorporate Cs into their structure by multiple mechanisms such as ion exchange, isomorphic substitution, and adsorption. Distribution of Cs to metal-Fer precipitates is known to vary depending on solution pH and chemical characteristics of Fer solids. The distribution coefficient values (ml/g) are in the range of 10^4 to 10^6 for K-Co-Fer, 10^5 to 10^6 for Na-Ni-Fer, 10^5 for Na-Cu-Fer, 10^4 for K-Cu-Co-Fer, and 10^3 for K-Zn-Fer and Zn-Fer [5]. The coefficient value for Cs with Zn-Fer is small compared to those with Fe-Fer, Cu-Fer, and Ni-Fer complexes, but the aforementioned distribution coefficient values could have been underestimated owing to insufficient removal of colloidal Fer solids from solution. Alkali metal (i.e., principal constituents such as Na and K in solution) substitution in metal-Fer complexes also results in changes in Cs distribution [14]. The pH ranges for Cs distribution to preformed Fer complexes of Ni(II), Zn(II), Cu(II), and Fe(III) are reported to be 0 to 10, 1 to 8, 0 to 8, and 0 to 6, respectively [14]. Variations in Cs distribution to solid Fer within these pH ranges are also reported [15]. It is known as well that Fer should not be used in highly caustic and acidic solutions, because it is chemically decomposed in these reaction conditions.

The size of Fer precipitates is important as it determines settling velocity, which is a crucial factor for separation of Cs-containing solids in solution. Iron-Fer precipitates are often found as colloidal particles, and their separation by gravity settling method is difficult whereas Ni-Fer precipitate settles more easily. The physical properties of precipitates depend on preparation procedures. For example, in strongly oversaturated solutions, very fine crystalline particles with a disordered lattice and higher solubility are formed incipiently, whereas an inactive solid phase is formed in slightly oversaturated solutions [16].

Solubility of metal-Fer precipitate is also a governing factor for Cs removal. The reported solubility product values for pure $Fe_4[Fe(CN)_6]_3$, $Zn_2[Fe(CN)_6]$, $Cu_2[Fe$

(CN)$_6$], and Ni$_2$[Fe(CN)$_6$] are 3.3×10^{-41}, 4×10^{-16}, 1.3×10^{-16}, and 1.3×10^{-15}, respectively [17]. In the case of Ni$_2$[Fe(CN)$_6$], Fer concentration in solution should be higher than 10^{-5} M for insoluble Fer precipitate to be formed. Because the solubility of fresh metal-Fer compounds can be higher than that of pure compounds, the minimum Fer concentration used in this study was 10^{-4} M.

There are three different ways to use Fer to remove Cs: (1) addition of soluble Fer salts and metal elements to waste solution (in situ formation of Fer solid), (2) addition of freshly prepared insoluble Fer-metal complex slurry to waste solution, and (3) use of Fer-metal adsorbents in solution. The distribution of Cs to insoluble Fer compounds is highest when the in situ Fer formation method is applied. If appropriately used, the in situ method is the best for both decontamination and waste volume reduction.

29.3 Experimental

The waste materials, extraction methods, and the corresponding measured metal concentrations are summarized in Table 29.1. The distilled water extraction technique was used for fly ash samples because it was reported that rad-Cs in fly ash samples could be extracted with water [18]. Application of a hot oxalic acid (0.1 M) extraction technique to recover rad-Cs from contaminated sewage sludge in a pilot-scale test was reported after the F1 accident [3]. We also tested the method for some samples in this study.

The distilled water extraction technique (for fly ash samples) and hot oxalic acid extraction technique (for dewatered sludge and slug samples) were used for waste samples collected from western Japan unaffected by the F1 accident. Stable (nonradioactive) Cs or Cs-137 tracer (both contained Cs as Cs$^+$ ion) was added to the extracts to estimate Cs removal.

The contaminated fly ash samples from two municipal waste incineration plants and a sewage treatment plant in the affected area were extracted and analyzed for rad-Cs in the facilities where they were produced. The sum of Cs-134 and Cs-137 radioactive materials in the fly ash samples K, N, CM and CI (listed in Table 29.1) were 8×10^4, 2×10^4, 9.9×10^4, and 1.2×10^4 Bq/kg (dry weight basis), respectively. Addition of Cs tracer to the samples was not necessary because it already contained rad-Cs from the F1 accident, removal of which could be estimated by radionuclide analysis.

All reagents used in our study were of analytical grade. The sample pH was adjusted using NaOH and HCl. Potassium ferrocyanide (K$_4$[Fe(CN)$_6$], K-Fer hereafter) was used as a soluble Fer salt source. One metal salt of Fe(III)Cl$_3$, Fe(II)SO$_4$, NiSO$_4$, and ZnSO$_4$ was added to extract when indigenous metals in it did not produce insoluble metal-Fer complex. NaCl was used to vary electrolyte concentration of the solutions.

The procedure for extract treatment by Fer precipitation technique was as follows. Twenty-five milliliters of water extract or 0.5–5 ml of 1 M oxalic acid

Table 29.1 Waste materials used and the extraction procedure

Sample code	Waste material[a]	Extraction procedure	Tracer added	Heavy meal concentration of the extract (mmol/l)					
				Al	Fe	Mn	Cu	Zn	Ni
1	Fly ash H (sewage sludge, scrubber wastewater of melting furnace)	Centrifugation or membrane filtration	Stable cesium 100 µg/l	0.1	0.1	0.1	0.1	1.0	0.0
2	Fly ash M (sewage sludge, melting furnace)	Distilled water extraction for 1 h (3 g dry weight/30 ml)	Stable cesium 100 µg/l	0.5	0.1	0.4	0.3	11	0.0
3	Sludge R (drinking water treatment sludge)	1 M oxalic acid extraction at 90 °C (10 g dry weight waste/25 ml)	Stable cesium 100 µg/l	216.7	32.6	3.8	0.1	0.2	0.1
4	Fly ash O (incinerated sewage sludge)		Stable cesium 100 µg/l	350.4	52.6	3.8	0.8	9.9	0.0
5	Fly ash KO (incinerated sewage sludge)		Stable cesium 100 µg/l	469.5	336.7	19.6	3.9	11.4	0.1
6	Fly ash NA (incinerated sewage sludge)		Stable cesium 100 µg/l	422.9	52.6	3.4	1.7	9.6	0.1
7	Fly ash CL (industrial waste, scrubber wastewater of melting furnace)	Centrifugation or membrane filtration	Stable cesium 100 µg/l	129.4	2.4	0.1	0.0	956.2	0.0
8	Fly ash K[b,c] (municipal waste, melting furnace)	Distilled water extraction for 4 h (3 g dry weight/30 ml) and filtration with membrane filter	None[e]	0.3	$<10^{-3}$	$<10^{-3}$	$<10^{-3}$	0.002	NA[f]
9	Fly ash N[b,d] (municipal waste, incinerator)		None[e]	0.0	$<10^{-3}$	0.016	0.004	1.3	NA
10	Fly ash CM	Distilled water extraction for 1 h (10 g dry weight waste/25 ml)	None[e]	NA[f]	NA	NA[f]	0.2	4.3	NA[f]

(continued)

Table 29.1 (continued)

Sample code	Waste material[a]	Extraction procedure	Tracer added	Heavy meal concentration of the extract (mmol/l)					
				Al	Fe	Mn	Cu	Zn	Ni
11	Fly ash CM	0.1 M oxalic acid extraction at 90 °C for 1 h (10 g dry weight/25 ml)	None[e]	NA[f]	7.7	NA[f]	0.2	6.5	NA[f]
12	Fly ash CI	0.1 M or 0.5 M oxalic acid extraction at 90 °C for 1 h (10 g dry weight/25 ml)	None[e]	NA[f]	0.6	NA[f]	0.0	0.0	NA[f]

[a]The large capital after the waste material name is the abbreviated name of the samples
[b]The waste materials used had been treated with a stabilizing reagent to reduce the leaching of heavy metals. The other waste materials were collected before the stabilization treatment
[c]Na and K, 160 mM each
[d]Na and K 30 mM each, Ca 130 mM each
[e]Sample contained radioactive cesium from TEPCO accident
[f]Not analyzed

Table 29.2 Removal of Cs from the solution under different pH and potassium ferrocyanide concentration

Sample code	0.1 mM potassium ferrocyanide						pH 5	
	pH 3	pH 5	pH 6	pH 7	pH 8	pH 10	0.5 mM[b]	1 mM[b]
1	98	96	96	98	97	98	98	99
2[a]	–	89	–	–	–	–	94	99
3	99	95	97	96	94	92	99	99
4	100	97	98	96	93	92	100	99
5	100	95	94	96	96	92	98	93
6	100	98	98	98	92	90	96	97
7	92	74	53	70	54	50	91	95

[a]0.2 mM $NiSO_4$ was added before adding potassium ferrocyanide
[b]Potassium ferrocyanide concentration used

extract was poured into a 50-ml glass beaker, Cs tracer and/or soluble metal salt was added to it (when necessary), and then solution pH was adjusted to values shown in Tables 29.2, 29.3, and 29.4. Metal hydroxides (ineffective for Cs removal) rather than metal-Fer complexes could be formed in the solutions when pH is in neutral to alkaline regions. Therefore, slightly acidic pH (3 or 5) was used in most cases for Fer precipitation. After addition of soluble Fer salt ($K_4[Fe(CN)_6]$) to pH-adjusted sample, the sample was transferred to a volumetric cylinder and diluted to 50 ml with distilled water. Then, it was mixed once by turning the cylinder fitted with a glass stopper and letting it stand for approximately 1 h, and the cylinder content was centrifuged at $6,400\,g$ ($2,600\,g$ for samples with rad-Cs) at 4 °C for 20 min. Sometimes, Ni-Fer or Cu-Fer precipitates were prepared in a separate bottle and added to the extracts in place of soluble Fer salts for treating extracts of rad-Cs-contaminated wastes.

Concentrations of metals and stable Cs in the extracts were determined using ICP-MS (Agilent) and/or ICP-AES (Shimadzu) before and after Fer precipitation. Metal contents in sample nos. 10 to 12 were analyzed on site using a portable voltammetry instrument (Modernwater) with autosampler as the samples could not be transported out of the site for analyses. A portable Ge semiconductor detector (NAIG) was used to analyze Cs-137 and Cs-134 in samples containing rad-Cs from the F1 accident.

29.4 Results and Discussion

Metal concentrations in waste samples extracted with oxalic acid (samples 3–6, Table 29.1) were in general high. The scrubber wastewater from an industrial waste incinerator (sample 7, Table 29.1) showed very high Zn concentration. Not well metals in samples 8–12 (Table 29.1) could be measured because of the constraint on elemental analyses of rad-Cs-contaminated samples. Nevertheless,

Table 29.3 Removal of cesium (Cs) from the 1 M oxalic acid extract with different dilution factor applied before ferrocyanide (Fer) coprecipitation (the solution pH was 5 and Fer was 0.1 mM for all samples)

Sample code	100 times dilution (~0.01 M oxalic acid)	20 times dilution (~0.05 M oxalic acid)	10 times dilution (~0.1 M oxalic acid)
3	97	99	93
4	97	95	93
5	95	97	96
6	98	95	92

the data suggest that metal concentrations in rad-Cs-contaminated sewage sludge (samples 10–12, Table 29.1) were similar to those in the extracts of uncontaminated sewage sludge (samples 1, 2, 4, 5, and 6, Table 29.1). Samples 8 and 9 (Table 29.1) showed low metal concentrations as they were stabilized waste materials that were treated to reduce heavy metal leaching.

The removals of Cs in different samples and test conditions are summarized in Tables 29.2, 29.3, and 29.4. Table 29.2 shows Cs removal efficiencies (%) for samples 1–7 under different pH and K-Fer concentrations without an addition of metals. The tests were conducted for pH 3–10 and soluble Fer salt concentrations 0.1–1.0 mM. The results indicated that insoluble Fer complexes were formed with metals present in the waste extracts (Table 29.1) upon addition of soluble K-Fer salts, resulting in high Cs removal efficiencies. Fer complexes could be formed with any of the metals such as Fe, Mn, Cu, and Zn present in sufficient concentrations (Table 29.1) to precipitate 0.1 mM Fer ions. Cs removal in sample 7 was lower (e.g., 74 % at pH 5) than those in other samples, although transition metals (particularly Zn) in the sample were abundant for the formation of insoluble Fer complexes. In a control experiment discussed in our previous work [4], we investigated on the effect of Zn concentration on Cs removal, and found that Cs removal by Fer solids tends to be low when Zn is present at pH 5. Formation of the Zn-Fer complex, which is known to have a comparatively low Cs distribution factor, probably reduced Cs removal in the sample. The Cs removal in sample 7 at pH 3 increased to 92 %, possibly because of the formation of iron-Fer complex, which has a high Cs distribution factor [4]. Moreover, Cs removal increased with increasing K-Fer concentration. However, this leads to increased amount of precipitate in the solution, which is not preferable from the aspect of waste volume reduction.

Table 29.3 shows Cs removal from oxalic acid extracts. In a separate experiment, we examined the effect of oxalic acid concentration on Fer coprecipitation method and concluded that for 0.1 mM Fer concentration, oxalic acid concentration should be 0.01 M or less for precipitation of insoluble Fer compounds. In actual waste extracts, oxalic acid is consumed by calcium present in the wastes, and hence actual oxalic acid concentrations are lower than those in the original reagents. The data in Table 29.3 show that Cs removal is possible with 20 times (nominal concentration = 0.05 M) as well as 10 times (nominal concentration = 0.1 M) dilutions of 1.0 M oxalic acid extract.

Table 29.4 Removal of radioactive cesium (rad-Cs) from samples 8–12

Sample code	Fer concentration (mM)	pH	Metal added	Concentration of added metal (mM)	Cs-137 removal (%)	Nominal oxalic acid concentration (M)[a]
		Fly ash K				
8	0.1	3	Fe(III)	1.8	100	0
		3	Fe(III)	0.4	100	
		3	Fe(II)	1.8	93	
		5	Ni	0.2	100	
		3, 5, 7, 9	Zn	0.1–1	0	
		Fly ash N				
9	0.1	5	None	0	23	0
		3	Fe(III)	1.8	52	
		3	Fe(II)[b]	1.8	100	
		5	Fe(II)	1.8	91	
		5	Ni	0.1	58	
		5	Ni	0.2	62	
		5	Zn	0.1	0	
		5	Zn	0.4	26	
		5	Zn	1	3	
		Fly ash CM (water extract)				
10	0.1	0.1	2.2	None	92	0
		5	Nickel ferrocyanide		96	
		Fly ash CM (0.1 M oxalic acid extract)				
11	0.1	3	None	0	36	0.05
	0.3	3	None	0	96	
	0.1	5	Cupper ferrocyanide		82	
	0.1	5	Nickel ferrocyanide		93	
	0.2	5	Nickel ferrocyanide		100	
		Fly ash CI (0.1 M or 0.5 M oxalic acid extract)				
12	0.3	3	Ni	0.6	69	0.05
	0.3	5	Ni	0.6	80	0.05
	0.1	3	None	0	69	0.25
	0.1	5	None	0	62	0.25

[a]In actual waste extracts, oxalic acid is consumed by calcium present in the wastes, and hence actual oxalic acid concentrations are lower. The concentration listed in this table is nominal concentration, not considering the consumption of oxalic acid
[b]Removal of metals: Fe 99 %, Cu 60 %, Zn 0 %

Table 29.4 shows results of coprecipitation tests conducted with fly ash extracts contaminated with rad-Cs as the result of the F1 accident. On-site analysis of samples 8 and 9 using a portable voltammetry instrument revealed that metal concentrations in the samples were relatively low. Addition of K-Fer alone in sample 8 resulted in 23 % rad-Cs removal. In fact, metal salts had to be added before the addition of soluble Fer salts for the formation of insoluble of Fer complexes in the samples. We, therefore, compared removals of rad-Cs with different metals (e.g., Ni(II), Fe(II), Fe(III) or Zn(II)). In sample 9, light green-colored Ni-Fer precipitate was formed when Ni and soluble Fer salt were added, but rad-Cs removals were only 58–62 %. The removal increased to almost 100 % only when 1.8 mM Fe(II) (in excess to 0.1 mM Fer) was used. Apparently, the removal of rad-Cs changed significantly for sample no. 9 depending on the type of iron salt (ferric or ferrous iron) used with K-Fer, but the reason remains unknown at this point. In contrast, the extract of molten fly ash sample 8 showed almost complete removal of rad-Cs with Fer-Fe(II), Fer-Fe(III), and Fer-Ni coprecipitation. The results may be explained by the existence of colloidal, nonionic rad-Cs (e.g., sorbed on suspended particles, but passed through 0.45-μm-pore-size filter) in the incinerator fly ash extract sample 9, because Cs in such form does not precipitate with Fer-Ni, but it precipitates with Fe through coagulation-precipitation mechanisms.

For rad-Cs-contaminated sewage wastes (samples 10–12, Table 29.4), on-site analysis for metal contents showed rather high Zn concentrations for samples 11 and 12 whereas Fe was prevalent in sample 13. We, therefore, conducted co-precipitation tests at pH 3 to produce Fe-Fer rather than Zn-Fer, that has a low Cs distribution factor, or added Ni-Fer in place of K-Fer to prevent formation of Zn-Fer in the samples.

Overall, very high rad-Cs removals (>95 %) were observed for contaminated waste extracts (samples 8–11), although we did not have enough time to optimize coprecipitation conditions for sample 12.

29.5 Conclusion

Selective removal of Cs using Fer precipitation was conducted with extracts of sludge and fly ashes generated from municipal water treatment plants and waste incineration plants in the areas affected by the F1 accident. More than 95 % rad-Cs removals were achieved for an optimized combination of pH, Fer concentration, and type of added metal salts. The chemical form (ionic or particulate) of Cs in waste extracts, heavy metal leaching from the wastes (i.e., whether the waste had undergone stabilization treatment), and Zn concentration influenced Cs removal. The results undoubtedly suggest that knowledge of principal metal content is very important for successful application of the Fer coprecipitation technique to remove rad-Cs from contaminated wastes.

Open Access This chapter is distributed under the terms of the Creative Commons Attribution Noncommercial License, which permits any noncommercial use, distribution, and reproduction in any medium, provided the original author(s) and source are credited.

References

1. Japan Sewage Works Agency (2013) Completion of a temporary plant to dry radioactive sewage sludge. http://www.jswa.go.jp/kisya/h25pdf/0410kisya.pdf. Accessed 15 June 2013
2. Nomura M (2013) The Japan Sewage Works Agency and Great East Japan disaster- measures against radioactive sewage sludge. In: Seminar on the protection of environment under nuclear disaster combined with natural disaster, Osaka, 26 April 2013
3. Ministry of Land, Infrastructure, Transport and Tourism (2011) Measures against radioactivity in the sewage system, Outline of the study and the outcome. http://www.mlit.go.jp/common/000213235.pdf. Accessed 15 June 2013
4. Fujikawa Y, Wei O, Fujinaga A, Tsuno H, Ozaki H, Kimura S (2013) Removal of cesium from the extract of municipal water treatment sludges by precipitation with ferrocyanide solids. In: Proceedings of the 15th international conference on environmental remediation and radioactive waste management, Belgium, ICEM2013-96320, September 8–12 2013
5. Ministry of the Environment (2013) Amount of designated waste stored as of December 31 2013. http://shiteihaiki.env.go.jp/02/02.html. Accessed 11 April 2014
6. Ministry of the Environment (2013) Summary report of the safety of interim storage and environmental impact assessment. http://josen.env.go.jp/soil/pdf/compiled_1310.pdf. Accessed 11 April 2014
7. Aomori Prefecture (2007) Assessment of the safety of the radioactive waste returned after treatment in overseas factories. p 59. https://www.pref.aomori.lg.jp/soshiki/energy/g-richi/files/0701-houkokusyo.pdf. Accessed 11 April 2014
8. Pearce J (1994) Studies of any toxicological effects of Prussian blue compounds in mammals: a review. Food Chem Toxicol 32(6):577–582
9. Mansfeldt T, Rennert T (2003) Iron–cyanide complexes in soil and groundwater. In: Schulz HD, Hadeler A (eds) Geochemical processes in soil and groundwater. Wiley, Weinheim, pp 65–77
10. Meeussen JCL, Kelzer MG, de Haan FAM (1992) Chemical stability and decomposition rate of iron cyanide complexes in soil solutions. Environ Sci Technol 26(3):511–516
11. Matsumura M, Kojima T (2003) Elution and decomposition of cyanide in soil contaminated with various cyanocompounds. J Hazard Mater B97:99–110
12. United States Environmental Protection Agency (1992) EPA handbook vitrification technologies for treatment of hazardous and radioactive waste. United States Environmental Protection Agency, Washington, DC, p 100
13. Haas PA (1993) A review of information on ferrocyanide solids for removal of cesium from solutions. Separ Sci Technol 28(17&18):2479–2506
14. Barton GB, Hepworth JL, McClannaham ED Jr, Moore RL, Van Tuyl HH (1958) Chemical processing wastes: recovering fission products. Ind Eng Chem 50(2):212–216
15. Tutsui T, Kimura S (1980) Removal of ^{60}Co and ^{137}Cs with nickel ferrocyanide complex salts involving alkali metal ion coprecipitation. Jpn J Health Phys 15(1):55–59
16. Stumm W, Morgan JJ (1981) Aquatic chemistry. Wiley Interscience, New York, Chap 5
17. Patnaik P (2004) Dean's analytical chemistry handbook, 2nd edn. McGraw-Hill, New York, Chap 4
18. Nishizaki Y, Miyamae H, Takano T, Izumiya K, Kumagai N (2012) Removal of radioactive cesium from molten fly ash. J Environ Conserv Eng 41(9):569–574

The manufacturer's authorised representative in the EU is Springer Nature Customer Service Centre GmbH, Europaplatz 3, 69115 Heidelberg, Germany. If you have any concerns regarding our products, please contact ProductSafety@springernature.com

Printed and bound by CPI Group (UK) Ltd, Croydon, CR0 4YY
25/03/2026
02078219-0001